NEW CENTURY MATHS

MATHEMATICS STANDARD (PATHWAY 1)

YEAR 11

Sue Thomson

Judy Binns

Series editor: Robert Yen

2ND EDITION

New Century Maths Year 11: Mathematics Standard
Pathway 1
2nd Edition
Sue Thomson
Judy Binns

Publishing editor: Robert Yen
Project editors: Alan Stewart and Anna Pang
Editor: Lisa Schmidt
Cover design: Chris Starr (MakeWork)
Text design: Nicole Melbourne
Project designer: Aisling Gallagher
Cover image: iStock.com/Trifonov_Evgeniy
Permissions researcher: Wendy Duncan
Production controller: Christine Fotis
Typeset by: Cenveo Publisher Services

Any URLs contained in this publication were checked for
currency during the production process. Note, however, that
the publisher cannot vouch for the ongoing currency of URLs.

© 2017 Cengage Learning Australia Pty Limited

For product information and technology assistance,
in Australia call **1300 790 853**;
in New Zealand call **0800 449 725**

For permission to use material from this text or product, please email
aust.permissions@cengage.com

National Library of Australia Cataloguing-in-Publication Data
Thomson, Sue, 1958- author.

New century maths year 11 : mathematics standard pathway /
Sue Thomson and Judy Binns.
9780170413503 (paperback)
For secondary school age.
Mathematics--Australia--Textbooks.
Mathematics--Study and teaching (Secondary)--Australia.
Mathematics--Problems, exercises, etc.
Mathematics--Textbooks.

Binns, Judy, author.

Cengage Learning Australia
Level 7, 80 Dorcas Street
South Melbourne, Victoria Australia 3205

Cengage Learning New Zealand
Unit 4B Rosedale Office Park
331 Rosedale Road, Albany, North Shore 0632, NZ

For learning solutions, visit **cengage.com.au**

Printed in China by China Translation & Printing Services.
1 2 3 4 5 6 7 21 20 19 18 17

C ONTENTS

1

PRESENTING DATA   2

2

USING ALGEBRA   40

3

HEALTHY FIGURES   64

4

EARNING MONEY 90

5

PAYING TAX 122

6

TAKING CHANCES 146

7

MEASUREMENT 176

PREFACE

New Century Maths 11 Mathematics Standard (Pathway 1) has been rewritten for the new Mathematics Standard syllabus (2017). In this 2nd edition, teachers will find those familiar features that have made *New Century Maths* a leading mathematics series, such as clear worked examples, graded exercises, syllabus references, investigations, language activities, chapter summary mind maps, practice exercises and a glossary/index.

The syllabus describes two pathways for Mathematics Standard that begin in *Year 11:*

- a vocational pathway that we will call **Pathway 1**, for students heading towards an optional HSC exam and ATAR, entry into the workforce or further training after school, providing practical mathematical skills for life

- a more traditional and academic pathway that we will call **Pathway 2**, for students heading towards a HSC exam, ATAR and university studies.

This book caters for Pathway 1 in Year 11. Both pathways share a common Year 11 course that splits into the Mathematics Standard 1 and 2 courses in Year 12, but because students taking each pathway have specific learning needs, we have published two levels of texts for *both* Years 11 and 12.

Mathematics Standard 1 is designed for students who require a broad knowledge and understanding of mathematics to apply in employment and everyday life. We have endeavoured to produce a practical text that captures the spirit of the course, providing relevant and meaningful examples of mathematics being used in society and industry.

The *NelsonNet* teacher website contains additional resources such as worksheets, video tutorials and spreadsheets. We wish all teachers and students using this book every success in embracing the new senior mathematics course.

ABOUT THE AUTHORS

Sue Thomson was head of mathematics at De La Salle Senior College, Cronulla, Director of Teaching and Learning at Hunter Valley Grammar School, an examination writer and assessor for the NSW Board of Studies, and a senior HSC marker. An active presenter for MANSW and beyond, Sue's interests are in language development, financial literacy and making mathematics accessible to all.

Sue dedicates this book to the memory of her husband and co-author, **Ian Forster**.

Judy Binns was head teacher of mathematics at Mulwaree High School in Goulburn and has taught at Homebush Boys High School. She has an interest in motivating students with learning difficulties and has wide experience teaching and writing for practical senior mathematics courses. Judy has been co-writing the *New Century Maths* 7–8 series for over 20 years.

Series editor **Robert Yen** has been writing for *New Century Maths* since 1995, as well as writing and presenting for MANSW and co-editing its journal, *Reflections*. Robert now works for Nelson Cengage as the mathematics publisher.

CONTRIBUTING AUTHORS

Megan Boltze (Ashcroft HS) and **Kuldip Khehra** (Quakers Hill HS) wrote and edited many of the *NelsonNet* worksheets.

John Drake, **Katie Jackson** and **Joanne Magner** created the video tutorials.

ISBN 9780170413503

SYLLABUS REFERENCE GRID

Topic and subtopic	New Century Maths 11 Mathematics Standard (Pathway 1) chapter
Algebra	
MS-A1 Formulae and equations	2 Using algebra
	3 Healthy figures
MS-A2 Linear relationships	11 Graphing lines
Measurement	
MS-M1 Applications of measurement	3 Healthy figures
M1.1 Practicalities of measurement	7 Measurement
M1.2 Perimeter, area and volume	9 Measuring area and volume
M1.3 Units of energy and mass	10 Managing a home
MS-M2 Working with time	15 It's about time
Financial mathematics	
MS-F1 Money matters	2 Using algebra
F1.1 Interest and depreciation	4 Earning money
F1.2 Earning and managing money	5 Paying tax
F1.3 Budgeting and household expenses	10 Managing a home
	12 Interesting figures
	13 Buying a car
Statistical analysis	
MS-S1 Data analysis	1 Presenting data
S1.1 Classifying and representing data	8 It's better than average
S1.2 Exploring and describing data	14 Comparing data
MS-S2 Relative frequency and probability	6 Taking chances

MAPPING TO NEW CENTURY MATHS 11 MATHEMATICS STANDARD (PATHWAY 2) CHAPTERS

Useful for using both pathway books in a Year 11 mixed-ability class

New Century Maths 11 Mathematics Standard (Pathway 2) chapter	New Century Maths 11 Mathematics Standard (Pathway 1) chapter
1 Collecting and presenting data	**1** Presenting data
2 Formulas and equations	**2** Using algebra
3 Earning money and taxation	**4** Earning money
	5 Paying tax
4 Probability	**6** Taking chances
5 Measurement	**7** Measurement
	9 Measuring area and volume
6 Managing a home	**10** Managing a home
	3 Healthy figures: 3.01, 3.02
	4 Earning money: 4.07
7 Linear functions	**11** Graphing lines
8 Interest and depreciation	**12** Interesting figures
9 Owning a car	**13** Buying a car
10 Analysing data	**8** It's better than average
	14 Comparing data
11 World locations and times	**15** It's about time
12 Driving safely	**3** Healthy figures: 3.03, 3.04, 3.05
	8 It's better than average: 8.08

ABOUT THIS B●●K

AT THE BEGINNING OF EACH CHAPTER

- Each chapter begins on a double-page spread introducing the **Chapter problem**, outlining the **Chapter contents** with syllabus codes, a list of chapter outcomes titled **What will we do in this chapter?**, and a list of applications titled **How are we ever going to use this?**

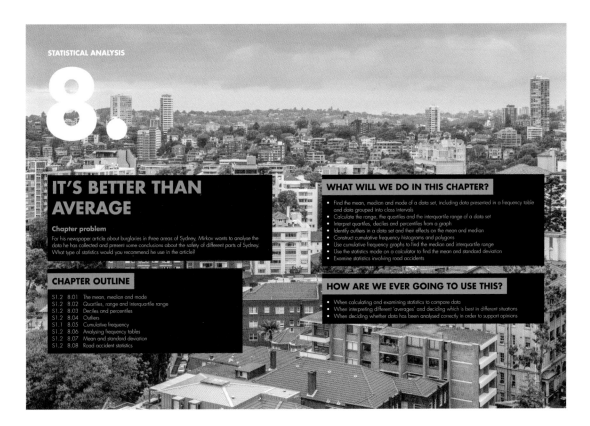

STATISTICAL ANALYSIS

8.

IT'S BETTER THAN AVERAGE

Chapter problem

For his newspaper article about burglaries in three areas of Sydney, Mirkov wants to analyse the data he has collected and present some conclusions about the safety of different parts of Sydney. What type of statistics would you recommend he use in the article?

CHAPTER OUTLINE

S1.2	8.01	The mean, median and mode
S1.2	8.02	Quartiles, range and interquartile range
S1.2	8.03	Deciles and percentiles
S1.2	8.04	Outliers
S1.1	8.05	Cumulative frequency
S1.2	8.06	Analysing frequency tables
S1.2	8.07	Mean and standard deviation
S1.2	8.08	Road accident statistics

WHAT WILL WE DO IN THIS CHAPTER?

- Find the mean, median and mode of a data set, including data presented in a frequency table and data grouped into class intervals
- Calculate the range, the quartiles and the interquartile range of a data set
- Interpret quartiles, deciles and percentiles from a graph
- Identify outliers in a data set and their effects on the mean and median
- Construct cumulative frequency histograms and polygons
- Use cumulative frequency graphs to find the median and interquartile range
- Use the statistics mode on a calculator to find the mean and standard deviation
- Examine statistics involving road accidents

HOW ARE WE EVER GOING TO USE THIS?

- When calculating and examining statistics to compare data
- When interpreting different 'averages' and deciding which is best in different situations
- When deciding whether data has been analysed correctly in order to support opinions

IN EACH CHAPTER

- Important facts and formulas are highlighted in a shaded box.

- Important words and phrases are printed in red and listed in the glossary at the back of the book.

- Graded exercises are linked to worked examples and include exam-style problems and realistic applications.

- **Investigations** and **Practical activities** explore the syllabus in more detail, through group work, discovery and modelling activities.

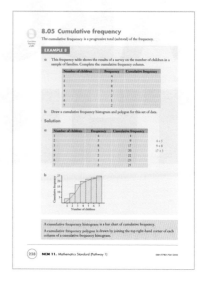

INVESTIGATION

STATISTICS ON A SPREADSHEET

1 a Enter into a spreadsheet the following data about the daily maximum temperatures in Alice Springs for 1 week.

	A	B	C	D	E
1	Day	Temperature (°C)			
2	Sunday	29		Mean	
3	Monday	31		Mode	
4	Tuesday	30		Median	
5	Wednesday	33		Standard deviation	
6	Thursday	29		Maximum	
7	Friday	28		Minimum	
8	Saturday	35		Range	
9					
10					

b Enter each formula into the given cell and save the spreadsheet.

Cell E2: =average(B2:B8)   Cell E3: =mode(B2:B8)

Cell E4: =median(B2:B8)   Cell E5: =stdev.p(B2:B8)

Cell E6: =max(B2:B8)   Cell E7: =min(B2:B8)

Cell E8: =E6−E7   *Sometimes, if you type the first couple of letters, the spreadsheet will suggest the correct word.*

2 a Repeat Question **1** for data for the town or city where you live. Go to the Bureau of Meteorology website to find the data.

b Find data for 1 month instead of for 1 week. Repeat Question **1** for this new data. You will need to adjust the formulas you enter for the statistics.

PRACTICAL ACTIVITY

REACTION TIMES AND DISTANCES

This activity involves three stages: measuring your reaction time, constructing a conversion graph between km/h and m/s, and making a table of values to show the distances a car will travel during your reaction time at different speeds. To complete this activity, you will need graph paper, a ruler and a stopwatch.

1 Measuring your reaction time

You need two people for this activity.

a Position the ruler with the 0 cm measure at the bottom. One person holds the top of the ruler and the other person places their hand at the 0 cm mark, ready to catch it.

b Without warning, the person holding the ruler lets it go and the other person has to catch it. Record the measurement on the ruler where it is caught.

c Change places and repeat.

d Use the table below to read your reaction times.

Measurement in cm	6	8	10	12	14	16	18	20	22	24	26
Reaction time in seconds	0.11	0.13	0.14	0.16	0.17	0.18	0.19	0.20	0.21	0.22	0.23

2 Constructing a conversion graph

This table shows the relationship between km/h and m/s. Create a conversion line graph by graphing the table of values on graph paper.

km/h	0	50	100
m/s	0	13.9	27.8

3 Determining your reaction distance

a To find the distance you will travel during your reaction time at different driving speeds, multiply your reaction time by the speed in m/s. For example, suppose that your reaction time is 0.24 s and you are traveling at 100 km/h. From the graph or table, 100 km/h = 27.8 m/s. You will travel 0.24 × 27.8 = 6.67 m during your reaction time.

b Copy and complete this table of values.

Speed in km/h	40	60	75	80	100
Reaction distance in m					

AT THE END OF EACH CHAPTER

- **Keyword activity** focuses on the mathematical language and terminology learned in the chapter.

- **Solution to the Chapter problem** revisits the problem introduced at the start of the chapter and presents a solution to the problem.

- **Chapter summary** concludes the chapter and includes a mind map exercise.

- **Test yourself** contains chapter revision linked to the relevant exercise set.

- **Practice sets** revise the skills and knowledge of previous chapters.

KEYWORD ACTIVITY

WORD MATCH

average	central tendency	cumulative frequency	histogram
interquartile range	mean	mode	number
often	order	outlier	percentiles
polygon	quartiles	range	spread ten

Copy and complete each sentence using a word or phrase from the list.

1 The _____ is the most frequent score.
2 The mean is the sum of scores divided by the _____ of scores.
3 The difference between the highest score and the lowest score is called the _____.
4 _____ divide the data into four equal parts.
5 Another word for the mean is _____.
6 The mode is the score that occurs most _____.
7 A score that is very different to the other scores is called an _____.
8 Mean, median and mode are measures of _____ _____.
9 The difference between the upper quartile and the lower quartile is called the _____ _____.
10 To find the median, you must first put the scores in _____.
11 The standard deviation is a measure of _____.
12 Deciles divide the data into _____ equal parts.
13 The _____ _____ is a progressive total of the frequencies.
14 When a large amount of data is divided into 100 equal parts, they are called _____.
15 When we have calculated the cumulative frequency, we can draw two graphs: a _____ and a _____.
16 A measure of central tendency the value of which depends on all the scores is the _____.

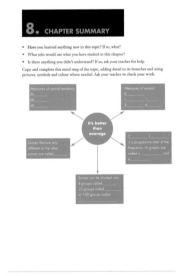

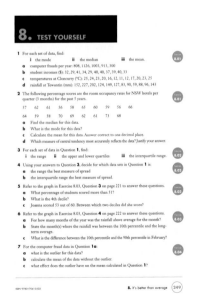

Year 11
Mathematics
Standard
Reference
Sheet

AT THE END OF THE BOOK

- **Glossary and index** includes a comprehensive dictionary of course terminology
- **Answers**

NELSONNET TEACHER WEBSITE

Margin icons link to print (PDF) and multimedia resources found on the *NelsonNet* teacher website, **www.nelsonnet.com.au**. These include:

Worksheet   Puzzle sheet   Skillsheet   Video tutorial   Spreadsheet   Weblink

- **Worksheets** and **puzzle sheets** that are write-in enabled PDFs
- **Skillsheets** of examples and exercises of prerequisite skills and knowledge
- **Video tutorials**: worked examples explained online
- **Spreadsheets**: *Excel* files
- **Weblinks**
- A **teaching program**, in Microsoft Word and PDF formats
- **Chapter PDFs** of the textbook
- **Resource Finder**: search engine for *NelsonNet* resources.

NEW CENTURY MATHS AND
MATHS IN FOCUS 11–12 SERIES

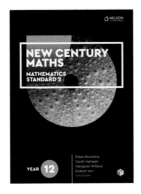

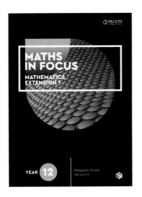

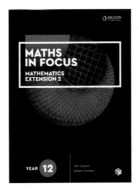

1.

PRESENTING DATA

Chapter problem

During a drought, the water authority produced a graph to show the decreasing supply of water in a dam due to the water usage of local residents. Is the decrease in water supply as bad as the graph seems to indicate? If the graph gives a false impression, identify what has been done to create this impression.

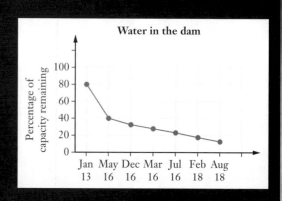

CHAPTER OUTLINE

WHAT WILL WE DO IN THIS CHAPTER?

- Interpret a variety of graphs and identify graphs that are misleading
- Understand the difference between a census and a sample
- Examine the various types of samples (random, stratified, systematic and self-selected) and be able to detect bias in sampling
- Classify data as numerical (either discrete or continuous) or categorical (either nominal or ordinal)
- Organise data into frequency tables, including data grouped into class intervals
- Display data on frequency histograms and polygons, dot plots, stem-and-leaf plots and Pareto charts

HOW ARE WE EVER GOING TO USE THIS?

- When we read information in the media involving statistics or graphs
- When we want to present statistical information in an interesting way
- When we participate in a survey or conduct a survey
- When we interpret the results of surveys
- When we notice a misleading graph in a report or advertisement

Every picture
tells a story

1.01 Interpreting graphs

Exercise 1.01 Interpreting graphs

1 **Bar charts** (also called **column graphs**) are mostly used for data that are categories, not numbers. This bar chart shows the populations of the eight Australian state capitals in 2016.

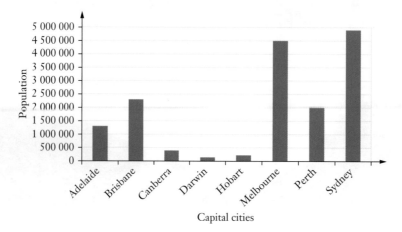

a What does one interval on the vertical axis represent?

b Which city has the largest population? Estimate its population from the graph.

c Which city has the smallest population? Estimate its population from the graph.

d Explain why we can only *estimate* the population from the graph.

e Which city has an approximate population of 1.3 million?

f Use the Internet to find the current populations of each city.

2 This **horizontal bar chart** shows the annual population increase for each Australian state and territory in 2017.

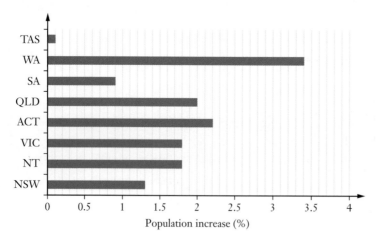

a By what percentage did NSW's population increase in 2017?

b In which region was the percentage population increase 2.2%?

c Which states had the same percentage increase?

d Does this graph tell you anything about the actual population of each of the states? Explain your answer.

e The population of Tasmania in 2016 was approximately 517 400. Calculate the actual increase in Tasmania's population.

f Use the Internet to find the annual percentage population increase in the most recent year.

3 We use a **clustered bar chart** for categories we want to compare. This graph shows the nutritional information for different foods.

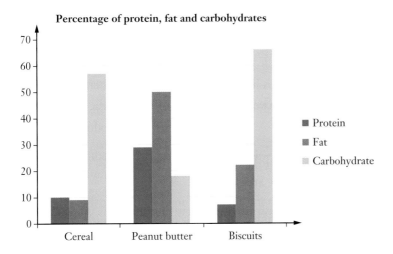

Percentage of protein, fat and carbohydrates

a Which food has the highest percentage of protein?

b Which food has the highest percentage of carbohydrate?

c Which food has the lowest percentage of fat?

d What is the difference in the percentage of fat in peanut butter and biscuits?

e If you were on a low carbohydrate diet, which food could you include in your diet?

f If you were on a low protein diet, which foods could be included in your diet?

g James had a 140 g serving of cereal for his breakfast. Approximately how many grams of protein did the serving of cereal contain?

h Zishan has to limit her carbohydrate intake. What foods should she avoid?

4 This clustered bar chart shows the number of motor vehicle thefts per year for three different areas.

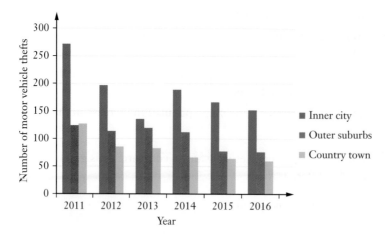

a Describe the trend in motor vehicle thefts in the country town.

b In which year do the outer suburbs have the lowest number of thefts?

c In which year is the difference between the number of thefts in the inner city and the country town the smallest?

d Estimate this difference.

e Write a brief paragraph describing the differences and similarities in the number of motor vehicle thefts in these three areas.

5 A **conversion graph** is used to convert from one unit to another. This graph shows conversions between degrees Celsius and degrees Fahrenheit.

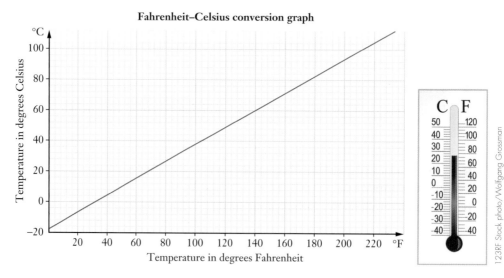

a Convert a temperature of 68°F to °C.

b Water boils at 100°C. What Fahrenheit temperature is this?

c Water freezes at 0°C. What Fahrenheit temperature is this?

NCM 11. Mathematics Standard (Pathway 1)   ISBN 9780170413503

d When Australian temperatures were measured in Fahrenheit, a day when the temperature reached 100°F, or a century, was considered a very hot day. What is this temperature in degrees Celsius?

6 A **step graph** is a graph of 'broken' flat line segments (intervals) that look like steps.

Prem repairs air conditioners. This step graph shows his charges according to the hours of work.

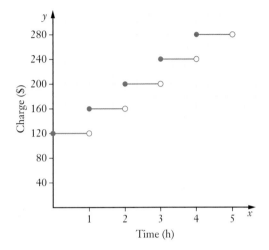

a How much does Prem charge for any time under an hour?

b How much does he charge for exactly 2 hours?

> Remember: read the amount with the filled circle, not the open circle.

c How much would Prem charge for working at your house for 4.5 hours?

d How much does he charge for each additional hour he stays to make repairs?

e Why does he charge a lot more for the first hour than for additional hours?

1.02 Misleading graphs

Sometimes, graphs are deliberately drawn to give a false impression. This can be done by:

- not having a scale on the vertical axis
- showing only part of the scale or an irregular scale on either axis
- not showing zero on the vertical scale or showing only a small part of the vertical axis
- using pictures or three-dimensional figures on the graph to exaggerate the differences.

This graph shows the cost of a 30 second call on three different mobile phone plans. If you view the graph quickly, it appears that the calls are much cheaper on the Green plan than on the other two plans, but a closer look will show that there really isn't a big difference in cost between the three plans. Although the information on the graph is correct, it gives a *misleading* impression.

Showing only a small part of the vertical axis on a graph is the most common way to create a misleading impression. As you work through the questions in the next exercise, you will learn about other ways to create misleading impressions.

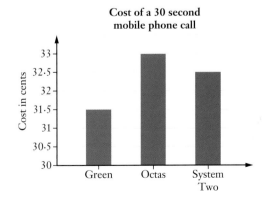

Cost of a 30 second mobile phone call

Exercise 1.02 Misleading graphs

1 Use the misleading graph above to answer the following questions.

   a How much does a 30 second mobile phone call cost on each of the three plans?

   b How much cheaper is a call on the Green plan than on the System Two plan?

   c This graph shows the same information as the above graph.

In which of the two graphs does the cost of a call on the Green plan look much cheaper than the cost on the other plans?

   d Write a sentence to describe how the scales on the vertical axes are different in the two graphs.

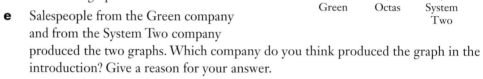

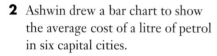

   e Salespeople from the Green company and from the System Two company produced the two graphs. Which company do you think produced the graph in the introduction? Give a reason for your answer.

2 Ashwin drew a bar chart to show the average cost of a litre of petrol in six capital cities.

   a In which city is the price of petrol the cheapest?

   b Which city has the highest petrol price?

   c Use the graph to estimate the price of petrol in Brisbane.

   d Approximately how many cents cheaper is the price of petrol in Brisbane than in Darwin?

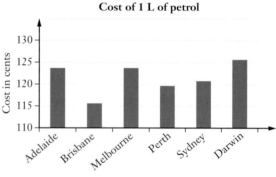

   e Ashwin included this graph in an advertisement for motoring holidays in Brisbane. He claimed that the graph shows that petrol prices are much lower in Brisbane than they are in other capital cities. Explain what Ashwin did to the graph to make the price of petrol look a lot cheaper in Brisbane.

3 This table shows the local shop prices of the same brand of coffee since 2012, and two related line graphs.

Year	2012	2013	2014	2015	2016	2017
Price of coffee	$6.15	$6.60	$7.30	$8.05	$8.90	$9.65
Increase since 2012	$0	$0.45	$1.15	$1.90	$2.75	$3.50

NCM 11. Mathematics Standard (Pathway 1)

ISBN 9780170413503

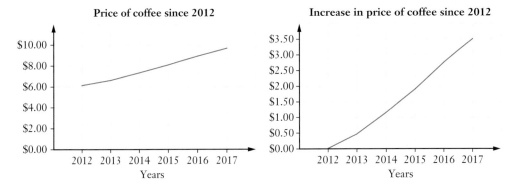

Price of coffee since 2012

Increase in price of coffee since 2012

The two graphs illustrate how graphs can make *increases* look small or big.

a Identify how the graphs are different.

b Give a reason for producing each of the graphs and who might use them.

c Imagine that you want to draw a graph to show how the population in your area has increased. Describe how you could make the increase look small and how you could make it look big.

4 Scott thinks that oil companies are taking advantage of consumers. He decided to investigate the price of petrol and the cost of crude oil and displayed his information on a line graph.

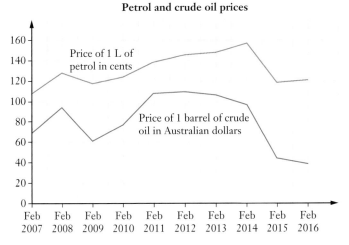

Petrol and crude oil prices

a What was the price of 1 L of petrol in February 2007?

b How much did a barrel of crude oil cost in February 2007?

c Profit = selling price − cost price. The graph shows the oil cost at 109 and the petrol cost at 146 in February 2012, giving the impression that the profit is 146 − 109 = 35. Explain why this impression is completely false.

d What relationship does the graph show between the price of petrol and the price of crude oil?

5 This graph compares the average weekly wages in Malvolia ($790) and Australia ($1580).

Average weekly wages

Malvolia   Australia

  a What misleading impression does this graph give?

  b Explain two things that are wrong about this graph.

  c What should be drawn on the graph instead of pictures?

  d Redraw this graph correctly.

Student
survey form

1.03 Census *vs* sample

Statistics involves the collecting of information, which is then analysed and used to make decisions.

To collect information, we usually survey a representative group. This process is called taking a **sample**.

To collect information about a whole population, *all* people or items must be surveyed. This is called taking a **census**. People who are members of small groups can be missed in a sample. We always use a census when we want to make sure that the views of small groups are included.

The Australian Bureau of Statistics (ABS) conducts a national census every five years, in a year ending in a 1 or a 6.

Sample	Census
• Surveys a selected group of people or items	• Surveys all people or items in the population
• Gives approximate information about the population	• Gives exact information about the population
• Simple and inexpensive	• Complex and expensive
• Can be done quickly	• Takes a lot of time to collect and process the information

EXAMPLE 1

The United Club asked Andrew to evaluate the facilities and services it provides to its 14 100 members. Should he use a census or a sample to gather information?

Solution

A census would be expensive and it would take a long time to process the information.

A sample should be used in this situation.

ISBN 9780170413503

From what population should Andrew take a sample to investigate the reputation of the United Club in the local community?

Solution

Population means who should Andrew be asking.	The population would be the residents of the community where the United Club is located.

Exercise 1.03 Census *vs* sample

1 For each investigation, should a census or a sample be used? Give a reason for your answer.

  a The most popular car colour in Australia

  b The number of retired people living in Canberra

  c The number of people who watch the NRL Grand Final

  d The use of soap versus body wash

  e Testing coffee for taste

  f The population of Broken Hill

  g The number of people using the emergency department at Westmead Hospital on Saturday nights

  h Length of time a certain type of car battery lasts

  i The political party supported by people in NSW

  j The favourite TV show of your 10 best friends

  k Steve Smith's batting average

  l Internet usage by Year 11 students at your school

2 What is the population for each of the following investigations?

  a Are girls better than boys at Maths?

  b Voting intentions for the next State election

  c The best song of the last decade

  d Student attitudes to school uniform at your school

  e Favourite make of car in NSW

  f The amount of money people are prepared to spend on going to the gym each week

  g The venue Year 12 should use for their formal

  h Donating to charities by wealthy individuals

  i The factors influencing the choice of supermarket

  j The batting performance of the Australian cricket team

INVESTIGATION

THE AUSTRALIAN CENSUS

The Australian census is conducted by the Australian Bureau of Statistics. Visit their website and search for 'How Australia takes a Census' to answer these questions.

1 When was the first national census conducted?

2 When was the last census?

3 List five questions that were asked in the last census.

4 Find three questions that have been asked in the past, but were not asked in the most recent census.

5 Were there any questions in the most recent census that have not been asked before? If so, what were they?

6 Are all questions in the Census compulsory?

7 Give three examples of how census information is used.

8 What was different about the Census in 2016? What challenges did this present?

9 Who is NOT required to complete the Census?

1.04 Types of samples

There are four ways to take a sample.

In a **random sample**, every member of the population has an equal chance of being included in the sample. For example, when a computer selects a customer at random to be surveyed about a mobile phone company, every customer of that company has an equal chance of being selected.

In a **stratified sample**, different categories in a population are represented according to their proportion of each category and then members of each category are selected randomly.

For example, a school's population may be made up of 72% junior students and 28% senior students. A stratified sample of school students must also be made up of 72% junior students and 28% senior students.

In a **systematic sample**, subjects are selected on a regular basis. For example, testing every 500th battery produced in a factory to check that the machines producing the batteries are working properly.

In a **self-selected sample**, each person in the population chooses whether or not to participate. For example, a new TV program is asking for people's opinions on public transport and invites its viewers to vote online.

EXAMPLE 3

The United Club has 14 100 members. This table gives a breakdown of members by age group.

Andrew is considering three different ways of choosing a sample of the members for his market research. Which type of sample is each of the following?

Age group	Number of members
Under 30 years	2100
30 to 39 years	2700
40 to 49 years	5100
50 years and over	4200

a   Every 100th member on the list of members

b   Names selected randomly by the computer

c   Placing a pile of surveys at the club's reception desk with a sign asking members to complete the survey if interested

d   21 members under 30 years old, 27 members 30 to 39 years old, 51 members 40 to 49 years old and 42 members aged 50 years or more

Solution

a   Selections are made on a regular basis, in this case, every 100th member.
    Systematic sample

b   Every member has an equal chance of being included and the computer chooses randomly.
    Random sample

c   Members decide whether or not they want to participate, so the sample has not been selected by Andrew.
    Self-selected sample

d   The percentage of members in each age group in the sample is the same as the percentage of members in each age group in the population.
    Stratified sample

EXAMPLE 4

Selina decided to use a stratified sample to survey members at her local gym. There are 970 gym members, 590 are female and 380 are male. She plans to survey 10% of the members.

a   How many members will Selina survey?

b   How many female members should she survey?

c   How many male members should she survey?

Solution

a Calculate 10% of the membership.

$$10\% \text{ of } 970 = \frac{10}{100} \times 970$$

$$= 97$$

Selina should survey 97 members.

b Female members make up 590 out of 970 members.

$$\frac{590}{970} \times 97 = 59$$

Write this as a fraction and find this fraction of the 97 members to be surveyed.

Selina should survey 59 female members.

c Male members make up 380 out of 970 members.

$$\frac{380}{970} \times 97 = 38$$

OR from part **b**

97 members − 59 females = 38 males

Selina should survey 38 male members.

Exercise 1.04 Types of samples

Example 3

1 Which type of sample is described in each case?

  a Selecting every 105th name from the phone book

  b The names of all Year 11 students are placed in a box and 2 are drawn to represent the school at a council function

  c Selecting an appropriate number of students from each year at your local high school

  d Subscribers are sent an email asking for their opinion of the play they have just seen

  e Names of the employees at a bank are drawn out of a barrel

  f The audience at a concert finds prize tickets under every 40th seat in each row

  g Employees are sorted from tallest to shortest and every 5th employee completes a questionnaire

  h A morning TV show asks viewers to vote on whether Australia should become a republic

  i 20 females and 28 males were surveyed out of a group of athletes with 100 females and 140 males in the group

  j An airport customs officer searches every 10th airline passenger

  k Without looking, 5 cards are selected from a pack of cards

  l A medical centre asks for volunteers to complete a survey

  m An import/export business employs 125 women and 250 men. 17 women and 34 men are surveyed about their work hours

n A computer selects names from the electoral (voters') roll

o A bank surveys 5% of its customers in each age group

2 Children from 400 families attend the local primary school. The P&C has raised money for new play equipment for the playground and wants to interview parents about their ideas. The committee would like to survey 40 families. Suggest how they might select the 40 families using a:

    a random sample   **b** stratified sample

    c systematic sample   **d** self-selected sample.

3 For the survey above, which method of sampling is the best to use? Explain your answer.

4 What are the possible disadvantages of using:

    a a self-selected sample   **b** a random sample

    c a systematic sample   **d** a stratified sample?

5 Amanda is going to use a stratified sample to survey the parents of her local netball club. There are 540 children playing netball. There are 380 primary students and 160 high school students. Amanda is going to survey 15% of parents.

    a How many parents should complete the survey?

    b How many parents of primary students should complete the survey?

    c How many parents of high school students should complete the survey?

6 Kieran wishes to survey Year 8 students about their opinions of the school. There are 96 boys and 69 girls in Year 8. He aims to survey 20% of students.

    a How many students should complete the survey?

    b How many boys should complete the survey?

7 Jamiela is going to ask a sample of her Facebook friends about their favourite sport. She has 376 single friends and 416 friends in a relationship. She is going to survey 12.5% of her friends.

    a How many friends will she survey?

    b How many single friends will be in the survey?

    c How many friends in a relationship will be in the survey?

8 Global Communications employs 750 people: 479 males and 271 females. The company intends to survey 75 employees about their working conditions.

    a How many males should be surveyed?

    b How many females should be surveyed?

9 Darren is researching political views in NSW. He stands on George St in central Sydney and interviews every 10th person who walks past.

    a Is this a representative sample of the population of NSW? Why or why not?

    b If not, describe how Darren could achieve a representative sample.

1.05 Bias in sampling

Bias is an unwanted influence in sampling that favours a particular section of the population unfairly. Bias produces unreliable results because the sample results are not truly representative of the population.

Andrew is considering how to find a sample of club members to complete his survey by:

a asking 100 people as they come through the door on a Friday night

b asking 100 people playing Bingo on a Wednesday morning.

In what way are these samples biased? How could you ensure that an accurate sample is found?

Solution

a Young people are more likely to be in a club on a Friday night.　All age groups would not be represented fairly.

b People in the club on a Wednesday morning are probably not working.

Working members would be under-represented in the survey.

In both cases, you could obtain an alphabetical list of the members and choose every 100th member (other answers are possible).

Exercise 1.05 Bias in sampling

For each sample given below, state how:

a the sample is biased

b you could ensure that an accurate sample is found.

When everyone has completed this exercise, share your answers and reasoning in a class discussion.

1 Asking people in the crowd at an AFL match their favourite sport

2 Calling every 6th phone number on a page in the phone book to ask about mobile phone usage

3 Surveying people about their preferred supermarket in a shopping centre with only one supermarket

4 Asking people to phone and vote on a new Australian flag

5 Finding out what people think of a new design for a football club jersey by asking people who walk by the office door

6 Asking teachers what music should be played at the Year 12 formal

7 Surveying the Maths faculty on the best subject at school

8 Asking people in Sydney their view on drought in a survey supported by the Government

9 Assessing voting intentions by asking people attending a Young Liberals conference who they will vote for in the next election

10 Emailing a survey to club members to complete

11 Asking the people in a club on a Thursday morning how many hours they spend on the Internet each week

12 Standing outside one of the coffee shops at a shopping centre and asking people to name their favourite coffee shop

INVESTIGATION

STATISTICS IN THE MEDIA

We live in a world of 24-hour news, through TV, news websites, Facebook, Twitter and blogs. To detect bias, we need to consider where the news comes from and what samples were used. Is the information supplied by a reporter, the police, a company executive, a government official or an opinionated blogger? Are the statistics based on a small sample, a large sample, an unrepresentative sample, or a phone / Internet poll? Each may have a particular bias that will influence how the story is reported.

Find examples of news items or surveys reported in a newspaper or online and investigate the following questions.

1 Where did it come from? What could influence how the story was reported?

2 Who wrote the story?

3 Does it show any bias?

4 Can it be supported by other news providers?

5 What type of sampling was used? Was it representative?

6 How many people were questioned?

7 When was the survey conducted?

1.06 Classifying data

There are two main types of data: **categorical** and **numerical**.

Categorical data

Categorical data are represented by categories, usually in words. There are two types of categorical data:

- **nominal data**: where the categories cannot be ordered; for example, colours of cars – blue, green, black and so on

- **ordinal data**: where the categories can be ordered; for example, your view on changing the Australian flag – strongly disagree, disagree, neutral, agree, strongly agree.

Numerical data

Numerical data are represented by numbers. There are two types of numerical data:

- **continuous data**: where the data is *measured* on a scale with no gaps; for example, height

- **discrete data**: where the data is *counted* in separate units; for example, the number of children in a family.

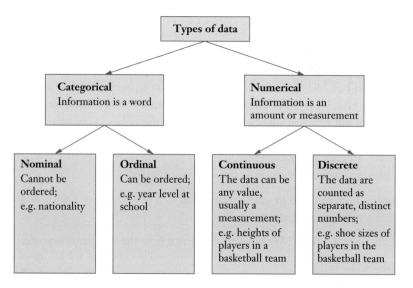

EXAMPLE 6

Classify each type of data as categorical or numerical.

a   Makes of cars

b   Number of pets owned by each person in your class

c   Report grades – A, B, C, D, E

NCM 11. Mathematics Standard (Pathway 1)   ISBN 9780170413503

Solution

a Makes of cars   Categorical

     e.g. Holden, Ford, Mitsubishi, Subaru, Toyota

b Number of pets owned by each person in your class –   Numerical
answers are numbers

c Report grades – A, B, C, D, E   Categorical

EXAMPLE 7

Classify each type of categorical data as nominal or ordinal.

a Colour of hair   **b** Size of popcorn – small, large, jumbo

Solution

a Cannot be ordered   Nominal

b The sizes have an order from largest to smallest   Ordinal

EXAMPLE 8

Classify each type of numerical data as continuous or discrete.

a Number of children in family   **b** Speed of vehicle on a highway

Solution

a The values are whole numbers   Discrete

b Speed is measured on a smooth scale with no gaps   Continuous

Exercise 1.06 Classifying data

1 Classify each type of data as categorical (C) or numerical (N).

a	Brands of computers	**b**	Number of books bought online
c	Prices of pizzas	**d**	Marital status
e	Rainfall in the Sydney CBD	**f**	Daily temperature in Dubbo
g	Number of TVs in the household	**h**	Time spent on the Internet each week
i	Favourite type of music	**j**	Brands of underwear

Example 6

2 Classify each type of categorical data as nominal (N) or ordinal (O).

a Flavours of ice-creams

b The quality of food at your local cafe

c Makes of cars driven past your school

d Preferred prime minister

e Rating the service at your local supermarket

f Year 10 grades

g Gender of Facebook friends

h How much people agree with the Government's education policy

i Employment status of the customers of a local coffee shop

j Rating of a new TV program

3 Classify each type of numerical data as continuous (C) or discrete (D).

a The distances jumped by a team of long jumpers

b The reaction time of car drivers

c The number of points scored by South Sydney in the first 7 games of the season

d The number of cups of coffee people drink each day

e The number of pages in your favourite book

f The weight of each person's luggage on a flight to England

g The amount of water used by a household each month

h Shirt sizes of Year 11 students

i Class sizes at your school

j Time taken to run 200 metres

Australian Bureau of Statistics

INVESTIGATION

DATA IN THE CENSUS

In this investigation, you will find out what sort of data is collected in the Census.

What you have to do

- Visit the website of the Australian Bureau of Statistics.

- Search for the latest Census. A Census is held every five years, such as 2016.

- Find 5 examples of questions that will give categorical answers.

- Find 5 examples of questions that will give numerical answers.

1.07 Frequency tables

Frequency
tables

When information is first collected, it needs to be organised so that people can understand it easily. One simple method is to put information into a frequency table.

EXAMPLE 9

Twenty-five students were asked 'How many children are there in your family?' and the following answers were given:

1	6	7	4	3	3	2	2	5	3	3	3	3
1	7	3	2	3	2	5	4	4	1	1	2	

Arrange this information in a frequency table.

Solution

For each value in the data set, draw a tally mark in the tally column, then count the tally marks and write the answer in the frequency column.

Number of children	Tally	Frequency
1	IIII	4
2	IIII	5
3	IIII III	8
4	III	3
5	II	2
6	I	1
7	II	2
Total		25

Grouped data and class intervals

For a large number of scores, it is not practical to use a frequency table that includes every score. Instead, we can group the data into **class intervals**, then use a grouped frequency table to count the **frequency** of each class interval.

EXAMPLE 10

The daily maximum temperature (in °C) for January and February for Goulburn is given below. Arrange this data into a grouped frequency table, using class intervals of 10–14.9, 15–19.9, 20–24.9, 25–29.9, 30–34.9, 35–39.9 and 40–44.9.

January

28.4	26.4	24.9	22.5	27.6	28.5	31.2	33.1	35.0	32.1	35.9
32.5	39.1	29.5	29.1	34.6	38.5	38.3	23.1	22.0	30.0	32.1
36.1	31.0	25.7	29.0	30.2	35.1	36.2	38.3	36.6		

February

21.6	27.6	31.3	35.4	35.9	34.0	20.2	23.5	34.7	39.0	41.2
32.0	26.0	25.1	29.5	34.0	35.5	26.5	23.4	21.0	26.4	32.5
35.0	30.4	20.1	21.8	25.6	26.1					

Solution

Temperature (°C)	Tally	Frequency
10–14.9		0
15–19.9		0
20–24.9	‖‖‖‖I	11
25–29.9	‖‖‖‖‖‖I	16
30–34.9	‖‖‖‖‖‖I	16
35–39.9	‖‖‖‖‖‖	15
40–44.9	I	1
Total		59

Exercise 1.07 Frequency tables

Keep your answers to this exercise as they will be required for Exercise 1.08.

1 Jasmine asked 30 students in Year 11 'What is your preferred brand of car?' Her results are shown below and coded as follows:

Holden = H, Ford = F, Toyota = T, Mitsubishi = M, Subaru = S, Other = O.

H	H	T	F	F	F	M	T	O	S
T	H	H	H	F	S	M	T	T	T
F	F	H	H	O	S	T	M	H	F

a Arrange Jasmine's data in a frequency table.

b What type of data is this?

2 The number of hamburgers sold each day by a fast food shop between 12 noon and 2 p.m. in August is listed below.

17	27	28	18	18	17	19	19	25	27	17
19	20	19	21	26	28	18	19	20	17	19
23	24	20	18	17	20	19	27	28		

a Arrange this information in a frequency table.

b What type of data is this?

3 This frequency table records the number of letters in each word in a paragraph of text.

a How many words had more than 7 letters?

b How many words had fewer than 5 letters?

c Which number of letters had the highest frequency?

d How many words were in the paragraph?

e What type of data is this?

Number of letters	Frequency
3	4
4	7
5	8
6	14
7	12
More than 7	10

4 This frequency table shows the number of break-and-enter offences each month in a coastal region over two years.

a In how many months were 4 offences recorded?

b In how many months were fewer than 5 offences recorded?

c In how many months were 5 or more offences recorded?

d How many months does this frequency table cover?

e What type of data is this?

Number of offences	Frequency
1	2
2	3
3	2
4	7
5	1
6	4
7	0
8	3
9	4
10	3

5 The PE department measured the heights of 25 Year 11 students in centimetres:

151	167	181	172	179	155	159	162	169	174
178	180	158	166	171	168	157	160	175	172
150	169	163	170	176					

a Arrange this data in a grouped frequency table, using class intervals of 150–154, 155–159, 160–164, 165–169, 170–174, 175–179, 180–184.

b What type of data is this?

Example 10

6 The number of incidents of malicious damage to property per month in the inner city was recorded over three years:

54	41	55	49	37	38	37	48	51	44	52	44
58	70	60	46	63	54	45	43	46	55	55	67
49	66	90	45	66	62	51	51	53	53	38	52

a Arrange this data into a frequency table using class intervals of 36–40, 41–45, 46–50, 51–55, up to 86–90.

b Are there any values that seem to be unusually high or unusually low? If so, what are they?

c What type of data is this?

7 This table shows the speeds of cars passing a school during one hour.

Speed (km/h)	Frequency
51–60	21
61–70	13
71–80	4
81–90	2

a How many cars passed the school in this hour?

b How many cars obeyed the speed limit of 60 km/h?

c Police decided to book anyone travelling at more than 10 km/h over the speed limit. How many drivers did they book?

d What type of data is this?

8 Customers of a coffee shop were asked how many cups of coffee they drank each week. The table shows the results of the survey.

Number of cups of coffee	Frequency
1–5	3
6–10	13
11–15	15
16–20	7
21–25	12

a How many people were surveyed?

b How many people drank fewer than 1 cup of coffee *per day*?

c How many people drink on average at least 1 cup of coffee per day?

d How many people drink an average of 3 or more cups of coffee per day?

e What type of data is this?

ISBN 9780170413503

1.08 Frequency histograms and polygons

Frequency histograms and **frequency polygons** graph **numerical data**, including grouped data. We use these graphs to see the shape of the data.

Frequency histogram	Frequency polygon
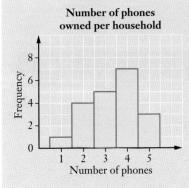	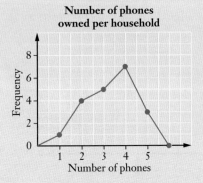

- A bar chart with no gaps between columns

- Shows frequency on the vertical axis and the data values on the horizontal axis

- Has a half-column space at the start

- A line graph that looks like a mountain

- Shows frequency on the vertical axis and the data values on the horizontal axis

- Can be drawn by joining the centres of the columns of the frequency histogram

- Begins and ends on the horizontal axis

EXAMPLE 11

Draw a frequency histogram and polygon for this frequency table from Example 10 on page 22 about the daily maximum temperatures (in °C) for January and February in Goulburn.

Temperature (°C)	Frequency
10–14.9	0
15–19.9	0
20–24.9	11
25–29.9	16
30–34.9	16
35–39.9	15
40–44.9	1

Solution

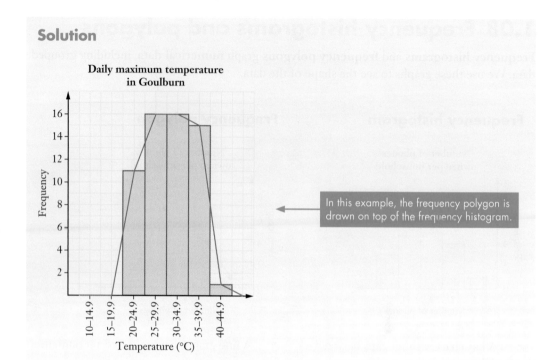

Daily maximum temperature
in Goulburn

In this example, the frequency polygon is drawn on top of the frequency histogram.

Exercise 1.08 Frequency histograms and polygons

In this exercise, we will draw graphs for the frequency tables produced in Exercise 1.07. Time to find those answers!

1 In Exercise 1.07, Question **2**, we looked at the sale of hamburgers in August. For this data, draw a frequency histogram and a frequency polygon.

2 In Question **4**, we looked at the number of break-and-enter offences each month in a coastal region. For this data, draw a frequency histogram and polygon.

3 In Question **5**, we looked at the heights of Year 11 students. For this data, draw a frequency histogram and polygon.

4 In Question **7**, we looked at speeds of cars. For this data, draw a frequency histogram and polygon.

NCM 11. Mathematics Standard (Pathway 1)

ISBN 9780170413503

1.09 Dot plots, stem-and-leaf plots and Pareto charts

Dot plots are used for small sets of **numerical data** that are close together.

EXAMPLE 12

The daily maximum temperature (in °C) in Cairns in July was recorded for 15 days.

32   30   31   32   31   30   31   31

31   31   29   25   28   27   29

Construct a dot plot for this data.

Solution

The temperatures range from 25° to 32°, so the line for the dot plot should go from 24 to 33.

Add a dot above the line for each data value.

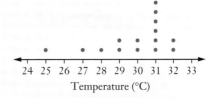

Stem-and-leaf plots are used for larger sets of **numerical data** and list the actual data values.

EXAMPLE 13

Sarah works for law enforcement and recorded the number of malicious property damage reports in the inner city each month over three years.

54	41	55	49	37	38	37	48	51	44	52	44	58
70	60	46	63	54	45	43	46	55	55	67	49	66
90	45	66	62	51	51	53	53	38	52			

Draw an ordered stem-and-leaf plot for this data.

Solution

The stem will be the tens digits. The smallest number is 37 and the largest number is 90, so the stem will go from 3 to 9.

The leaf will be the units digit for each number in the data. This is called an unordered stem-and-leaf plot because the 'leaves' are not in order.

Stem	Leaf
3	7 8 7 8
4	1 9 8 4 4 6 5 3 6 9 5
5	4 5 1 2 8 4 5 5 1 1 3 3 2
6	0 3 7 6 6 2
7	0
8	
9	0

key: 3 | 7 = 37

Then we write the 'leaves' in order. This is called an ordered stem-and-leaf plot.

Fairfax Syndication/Jeffrey Chan

Stem	Leaf
3	7 7 8 8
4	1 3 4 4 5 5 6 6 8 9 9
5	1 1 1 2 2 3 3 4 4 5 5 5 8
6	0 2 3 6 6 7
7	0
8	
9	0

key: 3 | 7 = 37

Pareto charts combine a bar chart and a line graph to represent the same set of data. They are used to display issues or reasons identified in a survey, usually for business, so they are graphs of **categorical data**. The columns in the bar chart show the frequency of each reason, from highest to lowest, while the line graph shows the cumulative frequency (a running total of frequencies).

This Pareto chart illustrates the complaints of customers giving online reviews of a local restaurant, The House of Steak.

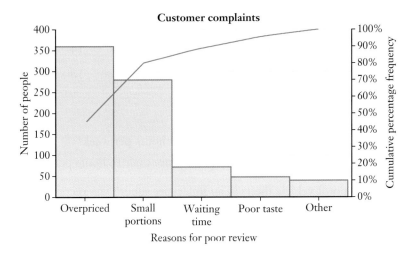

Customer complaints

- The complaints are listed from highest to lowest on the bar graph, with the frequencies listed on the left vertical axis (with a scale of 0 to 400).

- The line graph shows the cumulative frequency (increasing totals of frequencies), with the cumulative percentage frequency listed on the right vertical axis (with a scale of 0% to 100%).

EXAMPLE 14

a   According to the Pareto chart about The House of Steak, what were the two most common complaints about the restaurant?

b   'Customers were more concerned about waiting time for the food than they were about the taste of the food.' True or false?

c   Where should the owners of the restaurant direct their efforts to improve the reviews? Suggest at least two actions they could take.

Solution

a   The two tallest columns on the Pareto chart show the most common complaints.

The two most common complaints about the restaurant are that it is overpriced and the portions are small.

b   The column for waiting time is higher than the column for poor taste.

True, customers are more concerned about waiting time.

c

The owners should look at pricing of the meals and portion size. They could either decrease the prices or increase portion size or both.

Exercise 1.09 Dot plots, stem-and-leaf plots and Pareto charts

1 Ahmed surveyed his class to find how many hours each student spent on the Internet each week:

14	15	17	18	13	16	19	19	18	14
17	15	13	13	14	19	13	14	17	18

    a   Draw a dot plot for Ahmed's data.

    b   How many students were in the class?

    c   How many students used the Internet for 14 hours per week?

    d   How many students used the Internet for more than 16 hours per week?

2 The number of motor vehicle thefts in an inner-city suburb was recorded each month for one year.

15	9	11	16	13	11	13	12	10	15	13	9

    a   Draw a dot plot for this data.

    b   In how many months were 13 motor vehicles stolen?

    c   What percentage of months had 13 motor vehicles stolen?

    d   In how many months were fewer than 12 motor vehicles stolen?

3 Lisa surveyed her friends about the amount of money they spent on fuel last week, rounded to the nearest dollar.

$20	$28	$25	$26	$22	$26	$28	$28	$24	$22	$29	$28

    a   Draw a dot plot for this data.

    b   How many friends did Lisa survey?

    c   What was the most common amount of money spent on fuel?

    d   How many of Lisa's friends spent less than $26 on fuel?

4 Mrs White, the school canteen manager, records the daily number of students visiting the canteen over three weeks.

105	76	97	88	114	86	124	101
112	98	95	105	117	81	112	

    a   Show this information on a stem-and-leaf plot.

    b   On what percentage of days were more than 100 students served? Answer correct to the nearest whole percentage.

NCM 11. Mathematics Standard (Pathway 1)

ISBN 9780170413503

5 A security firm recorded the monthly number of shoplifting incidents over three years.

20	20	23	11	12	33	22	30	16	17	35	48
25	27	25	34	20	23	25	17	12	14	13	13
48	42	55	33	24	39	26	41	33	31	19	55

   a Show this information on a stem-and-leaf plot.

   b How many months had the number of incidents in the thirties?

   c What percentage of the total months was this? Answer correct to the nearest whole percentage.

6 A class of Year 11 students was surveyed on the number of hours of part-time work they did in one month.

| 42 | 16 | 35 | 27 | 9 | 0 | 33 | 21 | 14 | 11 | 26 |
| 29 | 31 | 22 | 8 | 24 | 5 | 0 | 15 | 25 | 17 |

   a Complete a stem-and-leaf plot to show this information.

   b What was the most hours worked by anyone in this sample?

   c How many students worked between 10 and 20 hours in the month?

   d How many students worked more than 20 hours in the month?

   e How many students were in this class?

   f What percentage of students worked more than 20 hours in the month? Answer correct to the nearest whole percentage.

7 A large company compiled the reasons given by their employees for being late for work, and graphed the data on this Pareto chart.

Example **14**

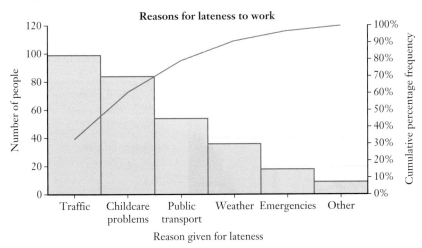

   a What is the most common reason given for lateness?

   b Which reasons explain 79% of latenesses?

   c Considering the reasons given, is the company in a position to do anything about the lateness of its employees? Are there any actions it could take to reduce lateness?

8 Newsentry Direct, an online store, collected data on the problems its customers experienced and showed it on this Pareto chart.

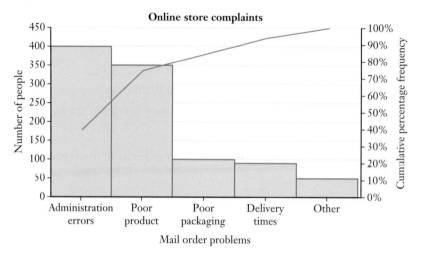

Online store complaints

Number of people (left axis) — *Cumulative percentage frequency* (right axis)

Categories: Administration errors, Poor product, Poor packaging, Delivery times, Other

Mail order problems

a What are the major problems for the company to address if it wants to improve its customer service?

b Using your answer to part **a**, what percentage of the mail order problems would this address?

c Suggest possible solutions for the problems you have identified.

d True or false? 'Concern about quality makes up almost 50% of the problems experienced by mail order customers.'

9 Each year, the causes of road fatalities are analysed. This enables organisations and governments to try to find ways to reduce the road toll. This is the data for one year.

Pareto chart template

a Ask your teacher to download the Pareto chart template from NelsonNet to draw a Pareto chart for the accident data shown below.

b In which two areas should government direct its efforts to reduce the road toll?

Cause of fatal accident	Number of occasions
Speed	152
Driver fatigue	72
Alcohol	70
Not using seat belts	57
Other	11

Fairfax Syndication/Rohan Thomson

STATISTICAL GRAPHS AND DISPLAYS

1 Find at least three examples of statistical graphs and displays. Choose from bar charts, dot plots, pie charts, frequency histograms, frequency polygons, line graphs, stem-and-leaf plots, picture graphs, Pareto charts or any other type you can find.

2 Make a copy of the graphs you find. Write five questions for each graph – use the questions in the exercises in this chapter as a guide.

3 Swap your graphs and sets of questions with friends and answer their questions.

4 Ask your teacher to check your work.

KEYWORD ACTIVITY

DEFINITIONS

Write a definition in your own words for 10 of the words or phrases listed below.

bias	categorical data	census
class interval	continuous data	dot plot
discrete data	frequency histogram	frequency polygon
frequency table	grouped frequency table	nominal data
numerical data	ordinal data	Pareto chart
random sample	sample	self-selected sample
stem-and-leaf plot	stratified sample	systematic sample

GRAPHS AND SPREADSHEETS

In this activity, you will use a spreadsheet to produce some of the more common graphs. Follow the instructions to create each graph, and show it to your teacher.

Ask your teacher to access NelsonNet to download the spreadsheet with the data for this investigation. You should save this spreadsheet to your own computer.

Statistical
graphs

1 BAR CHART

To draw a bar chart that shows the numbers of people immigrating to Australia by their region of origin in 1988–89:

- Select all of the cells from A3 to B8.

- Select the **Insert** tab at the top.

- Under **Charts**, select **Column** and then the first one of the 2D column types.

- Click on the chart title, highlight the text '1988-89' and change it to 'Immigration by region of origin'.

- Ensure the chart is selected (**Chart Tools** appears at top of sheet), select the Layout tab, then axis titles, go to Primary Vertical Axis Title and select Rotated Title. This will insert vertical script on the vertical axis. Highlight this script and change it to 'Thousands'.

- With the chart selected, choose the **Design** tab and select **Move Chart**. In the dialogue box, select the Object in option, and Immigration as the sheet, and then OK. Move the whole chart to the top left of this sheet (click and hold mouse on border of chart, move to desired location and unclick).

> You can also create Pareto charts using a spreadsheet by going to the histogram options.

2 CLUSTERED BAR CHART

Several years of information can be provided in the one graph by using clustered columns. For example, data for both years 1998-99 and 2008-09 can be shown.

- Return to the data sheet.

- Select all cells from B3 to C8.

- Select the **Insert** tab, and under **Charts** choose **Column** and select the first one of the 2D column types.

- Go to the **Layout** tab, click **Chart Title** and select **Above Chart**. Change the title text to the same as for the previous chart.

- Give the vertical axis the label 'Thousands' as for the previous chart.

- Move the chart to the Immigration sheet as you did for the last one and place it to just below the previous column chart.

3 PIE CHART

The same data will be used to draw pie charts of the immigration statistics to show the contributions from the various regions.

- Return to the data sheet and select cells A2 to B8.

- Select the **Insert** tab, and under **Charts** choose **Pie** and select the first 2D pie type.

- Check out the different possible **Chart Layouts** (top of screen). Try them out and choose the chart that you think shows the most information in the way you think looks best.

- Move the chart to the Immigration sheet as you have done previously and place it just to the right of the first simple column chart.

Now do another pie chart for the 2008-09 data. The steps are much the same as before except the initial selections of cells is a bit more complicated.

- Return to the data sheet. Select cells A3 to A8. Then press and hold the Ctrl button while you select C3 to C8 with the mouse.

- Insert the pie chart as before.

- This time, the title needs to be changed. Click the cursor in front of 2008-09 in the title and add 'Immigration to Australia by region of origin' in front of it.

- When you are happy with its format, place it below the first sector graph on the immigration sheet.

4 LINE GRAPHS

Use the vehicle and home theft figures in cells A10 to F13 in the data sheet. Both will be plotted on the same graph.

- Select the cells A11 to F13.

- Choose **Insert**, **Line** and select the first of the 2D line chart types.

- Go to the **Layout** tab, go to **Chart Title** and select **Above Chart**. Then replace the words 'Chart Title' with 'Vehicle and home thefts in the Shoalhaven area'.

- Place the chart in the Lines worksheet.

- Use the data on Bazza's gym from the data sheet (A17 to M20). Draw a line graph for this data. Place the graph in the Lines worksheet.

SOLUTION TO THE CHAPTER PROBLEM

Problem

During a drought, the water authority produced a graph to show the decreasing supply of water in a dam due to the water usage of local residents. Is the decrease in water supply as bad as the graph seems to indicate? If the graph gives a false impression, identify what has been done to create this impression.

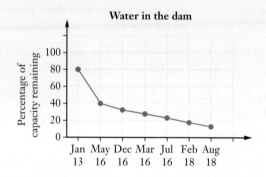

Solution

The graph shows that the supplies of water are running very low. However, the graph is misleading because the scale on the horizontal axis is not consistent. The big drop in dam levels between January 2013 and May 2016, compared with the small drop between February and August in 2018, gives the impression that water was being wasted from 2013 to 2016.

The graph to the right shows the same information in a non-misleading way.

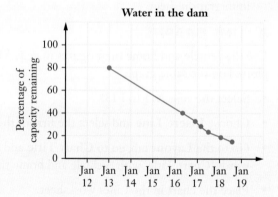

- Give three examples of graphs used in your other subjects. Give the type of graph and what it is used for.

- Explain in your own words when you would use a dot plot, a stem-and-leaf plot and a Pareto chart.

- Is there any part of the topic you didn't understand? If so, ask your teacher for help.

Copy and complete this mind map of the topic, adding detail to its branches and using pictures, symbols and colour where needed. Ask your teacher to check your work.

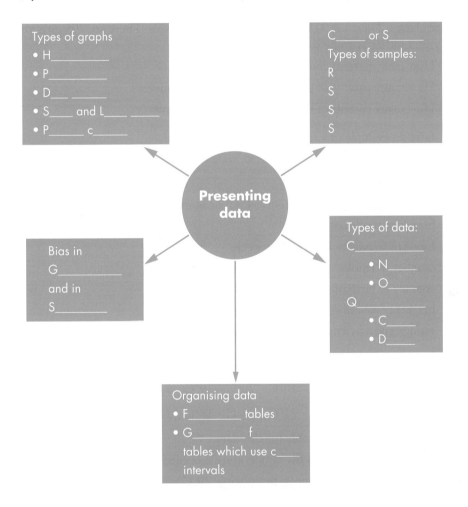

Types of graphs
- H_____
- P_____
- D___ _____
- S____ and L____ _____
- P_____ c_____

C_____ or S_____
Types of samples:
R
S
S
S

Presenting data

Types of data:
C_____
- N_____
- O_____
Q_____
- C_____
- D_____

Bias in
G_____
and in
S_____

Organising data
- F_____ tables
- G_____ f_____
tables which use c____
intervals

1. TEST YOURSELF

Exercise 1.01

1 Use the bar chart in Question **1** on page 4 to estimate:

   a the population of Brisbane

   b the difference between the populations of Brisbane and Melbourne.

Exercise 1.02

2 Redraw the misleading graph in Question **2** on page 8 so that it isn't misleading.

Exercise 1.03

3 Should a census or a sample be used to find the most popular television show in NSW? Give a reason for your answer.

Exercise 1.04

4 Lindsay is going to use a stratified sample to survey the parents of all students in her dance classes. There are 375 children in the dance classes. There are 295 girls and 80 boys. Lindsay is going to survey 20% of parents.

   a How many parents should complete the survey?

   b How many parents of girls should complete the survey?

   c How many parents of boys should complete the survey?

Exercise 1.05

5 a Give an example of a sample that would be biased.

   b For the example you gave in part **a**, state how you could ensure that an accurate sample is found.

Exercise 1.06

6 Classify each type of data as categorical (C) or numerical (N):

   a brands of toothpaste

   b hours of television watched per week

   c number of employees in small businesses

   d country of birth of students at a school

   e heights of school students

   f ratings given by customers for a restaurant.

Exercise 1.07

7 The temperature at 11 a.m. each day is recorded:

15°C 22°C 17°C 21°C 18°C 20°C 19°C 21°C 20°C 19°C 22°C

21°C 21°C 19°C 15°C 21°C 22°C 18°C 21°C 20°C 22°C

Exercise 1.08

   a Complete a frequency table for this data.

   b For how many days was the temperature recorded?

   c How many days had temperatures below 20°C?

   d Draw a frequency histogram for this data.

   e Draw a frequency polygon for this data.

8 These are the results for a quick quiz out of 20 in Mathematics for Jill's Year 11 class:

Exercise
1.09

10   14   13   17   16   11   14   14   12   14   13   16

15   12   16   13   12   15   10   14   13   16   14

a Draw a dot plot for Jill's data.

b How many students are in the class?

c How many students scored more than 14 out of 20 for the test?

9 The Tech-To-Go store records the age of each customer who visits the store on Monday morning:

Exercise
1.09

25   55   36   29   28   50   47   39   52   41   33

50   29   28   56   33   26   35   35   48   32

a Show this information in an ordered stem-and-leaf plot.

b How many customers came into the shop that morning?

c What age was the oldest customer that morning?

d What percentage of customers was under the age of 35? Answer correct to one decimal place.

10 Customers who bought Orange computers were surveyed about any defects they found. This Pareto chart shows the responses.

Exercise
1.09

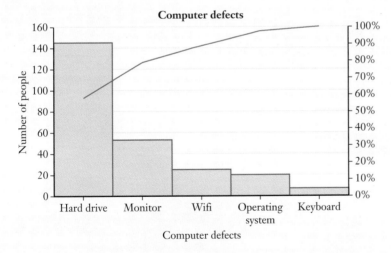

a What are the two major defects for the company to address?

b Using your answer to part **a**, what percentage of the defects would this address?

c What are the two least important defects to address?

d How many customers had wifi defects?

2.

USING ALGEBRA

Chapter problem

Children learn words at an amazing rate. When a child is aged between 1 and 5 years we can predict the typical number of words they will understand using the formula $N = 10(6m + 10)$, where m represents the age of the child in months and N is the number of words.

Abigail, a childcare worker, is worried about Grace, a 20-month-old girl at her daycare centre who doesn't appear to understand many words. She has been tested and understands about 900 words.

Is Abigail's concern justified? Use calculations to support your answer.

CHAPTER OUTLINE

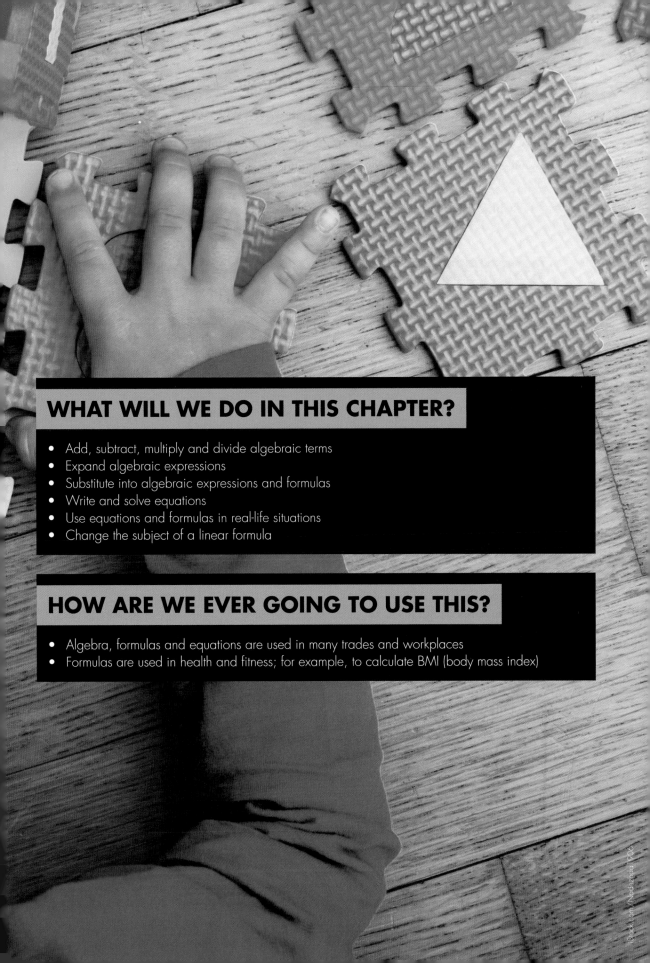

WHAT WILL WE DO IN THIS CHAPTER?

- Add, subtract, multiply and divide algebraic terms
- Expand algebraic expressions
- Substitute into algebraic expressions and formulas
- Write and solve equations
- Use equations and formulas in real-life situations
- Change the subject of a linear formula

HOW ARE WE EVER GOING TO USE THIS?

- Algebra, formulas and equations are used in many trades and workplaces
- Formulas are used in health and fitness; for example, to calculate BMI (body mass index)

2.01 Algebraic expressions

Collecting
like terms

Algebra using
diagrams

Adding and subtracting algebraic terms

Like terms contain exactly the same pronumerals; for example, $4g$ and $\frac{1}{2}g$, $-2pq$ and $5pq$, and x^2 and $-6x^2$. Terms containing different pronumerals are *not* like terms.

Only **like terms** can be added or subtracted.

EXAMPLE 1

Simplify each expression.

a $3k^2 + 5k^2 + 2k^3$

b $3x + 7xy - x + 2xy$

Solution

a The only like terms are the two terms involving k^2. We can add them.

$3k^2 + 5k^2 + 2k^3 = 8k^2 + 2k^3$

b The $3x$ and the $-x$ are like terms and the $7xy$ and $2xy$ are like terms.

Positive and negative signs are owned by the term *following* the sign.

$3x + 7xy - x + 2xy$

$= 3x - x + 7xy + 2xy$

$= 2x + 9xy$

Multiplying and dividing algebraic terms

When **multiplying and dividing terms**, multiply and divide the numbers and pronumerals separately.

EXAMPLE 2

Simplify each expression.

a $6a \times 2b$ **b** $12a \div 2b$ **c** $3a \times 6c \div 12h$

Solution

a Multiply the numbers, then multiply the pronumerals.

$6a \times 2b = (6 \times 2) \times (a \times b)$

$= 12ab$

b Write the division as a fraction, then simplify.

$$12a \div 2b = \frac{12a}{2b}$$

$$= \frac{12}{2} \times \frac{a}{b}$$

$$= 6 \times \frac{a}{b}$$

$$= \frac{6a}{b}$$

c Write the expression as a fraction.

Sometimes it's easier to simplify using the fraction key on a calculator.

$$3a \times 6c \div 12h = \frac{3a \times 6c}{12h}$$

$$= \frac{18ac}{12h}$$

$$= \frac{18}{12} \times \frac{ac}{h}$$

$$= \frac{3}{2} \times \frac{ac}{h}$$

$$= \frac{3ac}{2h}$$

Exercise 2.01 Algebraic expressions

1 Simplify each expression.

a	$5y + y$	**b**	$7h - h$	**c**	$3ab + 11ab - 2ab$
d	$5b^2 + 2b^2 + b^3$	**e**	$3p + 2 + 8p$	**f**	$3 + x + 2$
g	$8w - 2 + w$	**h**	$25 - t - 2t$	**i**	$10k - k$
j	$5f + f^2 + 4f^2$	**k**	$3a - b + a$	**l**	$3z - 6 + z + 11$
m	$15ab + 2a + ab + b$	**n**	$x^2 + x^3 + 4x^2 - 2x^3$	**o**	$8k - 2 + 3k - k$

Example 1

2 Simplify each expression.

a	$5x \times 3y$	**b**	$4h \times 3$	**c**	$5t \times 3s$
d	$15g \div 3b$	**e**	$9t \div 6w$	**f**	$24p \div 12t$
g	$8a \times 3b \div 4c$	**h**	$2a \times 4b \div 5y$	**i**	$4p \times 8q \div 16a$
j	$3m^2 \times 5m$	**k**	$\frac{1}{2}m \times 6m$	**l**	$\frac{1}{5}p \times 20p$
m	$\frac{9l^3}{3l^2}$	**n**	$3a \times 4b \times 5c$	**o**	$\frac{8a}{4}$
p	$\frac{1}{8}h \times 16h$	**q**	$\frac{12ab}{6a}$	**r**	$\frac{4y}{9 \times 2y}$

Example 2

Expanding
algebra

Algebra using
diagrams

2.02 Expanding expressions

When **expanding** an algebraic expression, we multiply out the brackets.

EXAMPLE 3

Expand $5(2a + 3b - 1)$.

Solution

Multiply every term inside the brackets by 5.

$$5(2a + 3b - 1) = 5 \times 2a + 5 \times 3b - 5 \times 1$$
$$= 10a + 15b - 5$$

EXAMPLE 4

Expand and simplify $12b + 8(2 + b)$.

Solution

First multiply the terms inside the brackets by 8 and then add the $12b$.

$$12b + 8(2 + b)$$
$$= 12b + 8 \times 2 + 8 \times b$$
$$= 12b + 16 + 8b$$
$$= 20b + 16$$

EXAMPLE 5

Expand $-(x - 4)$.

Solution

The $-$ sign in the front of the brackets actually means -1.

$$-(x - 4) = -1\,(x - 4)$$
$$= -1 \times x - 1 \times (-4)$$
$$= -x + 4$$

ISBN 9780170413503

Exercise 2.02 Expanding expressions

1 Expand each expression.

a $2(x + 5)$ **b** $3(2a - 5)$ **c** $4(2x + 3y + 2)$

d $a(2a + 3b - 4)$ **e** $4a(2 - 3b)$ **f** $3b(b^2 + b + 7)$

2 Expand and simplify each expression.

a $2b + 3(2 + b)$ **b** $x + 2(4 + x)$ **c** $3d + 4(5 + d)$

d $2(3x + 7) + 10$ **e** $5(4p + q) + 6q$ **f** $8(3w + 4) - 20w$

3 Expand each expression.

a $-(2x + 5)$ **b** $-(3a - 1)$ **c** $-(4 + 5b)$

d $-(1 - x)$ **e** $-(4p - 3q + 10)$ **f** $-(x - y + z)$

4 Expand and simplify each expression.

a $-3(2x + 5)$ **b** $4(-2x + 3)$ **c** $2x(a + 5)$

d $5a + 2(a + 7)$ **e** $-(9 - x)$ **f** $-x(x - 3)$

g $2(3d + 5) + 6$ **h** $y(y - 2) + 4y$ **i** $-(2 - k) + k + 2$

2.03 Substituting into formulas

Andy uses algebraic expressions every day in his electrical business. Regularly, he has to calculate the value of expressions such as $\dfrac{E}{R}$, $\sqrt{PR}$ and I^2R by substituting values for each pronumeral.

Substitution
code puzzle

EXAMPLE 6

Evaluate I^2R when $I = 5$ and $R = 20$.

> **Evaluate** means to work out the value. When we evaluate an expression, the answer is a number.

Solution

Make $I = 5$ and $R = 20$ in the expression I^2R and calculate the answer.

$$I^2R = 5^2 \times 20$$
$$= 500$$

EXAMPLE 7

Andy uses the **formula** $S = V(1 - r)^n$ to calculate the current value of his car. Calculate the value of S if $n = 5, r = 0.3$ and $V = 36\ 000$.

> A **formula** is an algebraic rule that describes a mathematical relationship between pronumerals.

Solution

Replace the pronumerals with their values and then calculate the final amount.

$$S = 36\ 000 \times (1 - 0.3)^5$$
$$= 36\ 000 \times 0.7^5$$
$$= 6050.52$$

The current value of Andy's car is $6050.52.

Exercise 2.03 Substituting into formulas

1 What is the value of I^2R when:

   a $R = 30$ and $I = 10$ **b** $I = 7$ and $R = 28$ **c** $R = 40$ and $I = 12$?

2 Substitute the values $n = 3, r = 0.25$ and $V = 3000$ into the formula $S = V(1 - r)^n$ and evaluate S correct to the nearest whole number.

3 The formula for the perimeter of a rectangle is $P = 2l + 2w$. Calculate P when $l = 12$ and $w = 9.5$.

4 The area of a triangle is given by $A = \dfrac{bh}{2}$. Find A when $b = 12$ and $h = 7$.

5 Use the formula for changing degrees Fahrenheit to degrees Centigrade $C = \dfrac{5}{9}(F - 32)$ to calculate the value of C when $F = 104$.

6 Calculate the value of V in the formula for the volume of a sphere $V = \dfrac{4}{3}\pi r^3$ when $r = 5.2$. Express your answer correct to one decimal place.

7 Evaluate S in the formula $S = \dfrac{D}{T}$ when $D = 240$ and $T = 8$.

8 Use the formula $S = V - Dn$ to evaluate S when $V = 20\ 000, D = 4000$ and $n = 3$.

9 In the formula $A = P(1 + r)^n$, determine the value of A when $P = 3500, \ n = 20$ and $r = 0.025$. Express your answer correct to two decimal places.

10 When $N = 15, H = 4$ and $M = 2$, calculate the value of B in the formula $B = \dfrac{10N - 7.5H}{5.5M}$. Answer correct to one decimal place.

11 In the formula $h^2 = a^2 + b^2$, what is the value of h when $a = 24$ and $b = 7$?

12 Use the formula $v^2 = u^2 + 2as$ to determine the value of v when $a = 6, s = 3.2$ and $u = 8$. Express your answer correct to one decimal place.

2.04 Using formulas

Many people use **formulas** in the workplace. Concrete contractors use formulas to determine the amount of concrete required for paths. Electricians use formulas to calculate voltage changes. Nurses use formulas to calculate solution quantities for injections and medications.

Formulas

EXAMPLE 8

Fried's formula can be used to calculate the infant dose of an adult's medicine.

$D = \dfrac{MA}{150}$ where:

D = the infant dose in mL

A = the adult dose in mL

M = the child's age in months.

The adult dose of a cough medicine is 30 mL. What dose of cough medicine can a parent give to a 12-month-old child?

iStock.com/Sasha_Suzi

Solution

We need to calculate the size of the child's dose which is D.

$D = ?$

$A = 30$

The adult dose is 30 mL.

M is the child's age in months: $M = 12$.

$M = 12$

Substitute the values into the formula.

$D = \dfrac{MA}{150}$

$= \dfrac{12 \times 30}{150}$

$= 2.4$

Write the answer.

The child's dose is 2.4 mL

---

2. Using algebra

EXAMPLE 9

A civil engineer uses the formula
$V = 0.25A(d_1 + d_2 + d_3 + d_4)$ to estimate the
amount of cut and fill required to excavate
a rectangular area. V is the volume of dirt
and rock, A is the area of the site and
d_1, d_2, d_3 and d_4 represent the depth of the
excavation at the four corners.

A rectangular excavation site is 55 m long
by 130 m wide and the depth of the
excavation at each corner is 1.4 m, 0.7 m, 1.5 m and 2.8 m. Use the formula to estimate
the amount of cut and fill required for the excavation.

Solution

Calculate A, the area of the rectangular site. $A = 55 \times 130 = 7150 \text{ m}^2$

Substitute the values of d_1, d_2, d_3 and d_4 into $d_1 = 1.4, d_2 = 0.7, d_3 = 1.5, d_4 = 2.8$
the formula, then calculate the answer.

$$V = 0.25A(d_1 + d_2 + d_3 + d_4)$$
$$= 0.25 \times 7150 \times (1.4 + 0.7 + 1.5 + 2.8)$$
$$= 11\ 440$$

Write the answer. $11\ 440 \text{ m}^3$ is the cut and fill required.

Exercise 2.04 Using formulas

1 The adult dose of a pain relief drug is 24 mL. Use Fried's formula $D = \dfrac{MA}{150}$ from
Example 8 to calculate the dose of the drug that can be given to an 18-month old child.
Answer correct to the nearest mL.

2 The formula $C = \dfrac{5}{9}(F - 32)$ is used to convert temperatures in degrees Fahrenheit (°F) to
temperatures in degrees Celsius (°C). A cake recipe requires an oven set at 350°F. Maria's
oven measures the temperature in °C. At what temperature should Maria set her oven to
cook the cake? Answer correct to the nearest whole degree.

3 The formula $D = S \times T$ relates distance, speed and time. $D =$ the distance travelled,
$S =$ the speed and $T =$ the time. A truck is travelling at an average speed of 95 km/h on
the motorway.

    a Calculate the distance the truck will travel in 3 hours.

    b Express 45 minutes as a fraction of an hour, then calculate how far the truck will
travel in 45 minutes.

4 Young and Clark both have rules for approximating the child's dose of an adult's medicine.

Young's rule: $\text{Dosage} = \dfrac{\text{age of the child in years} \times \text{adult dose}}{\text{age of the child in years} + 12}$

Clark's rule: $\text{Dosage} = \dfrac{\text{weight of the child in kilograms} \times \text{adult dose}}{70}$

Isabella is 5 years old and she weighs 15 kg. The adult dose of the cough medicine Isabella's mother wants to give her is 35 mL.

a Calculate the amount of medicine Isabella should have based on each of Young and Clark's rules.

b Most 5-year-old girls weigh approximately 18 kg. Should Isabella's mother use Young's or Clark's formula to calculate the dose to give her? Justify your answer.

5 The formula $FV = PV(1 + r)^n$ shows the relationship between the future value FV of an investment, the present value PV of an investment, the annual rate of interest r as a decimal and the length of the investment n in years. This morning, Charles invested $5400 at 6% p.a. for 7 years.

a Express 6% p.a. as a decimal.

b Calculate the future value of Charles' investment.

6 The efficiency of a pumping station is given by the formula $E = PM$ where P is the efficiency of the pump, M is the percentage efficiency of the motor and E, P and M are expressed as decimals. Determine the percentage efficiency of the pumping station when the pump is at 85% efficiency and the motor is at 60% efficiency.

> Make sure you express the percentages as decimals.
> Remember: 85% = 0.85

7 Use the excavation formula $V = 0.25A(d_1 + d_2 + d_3 + d_4)$ from Example 9 to determine the volume of soil and rock that must be removed from a square of land 60 m by 60 m. The depth of soil at the corners of the land is 0.3 m, 1.1 m, 1.4 m and 1.8 m.

Example
9

8 When a car's skid marks are in a straight line, the formula $V = 16\sqrt{FS}$ can be used to estimate the car's speed before the brakes were applied. V is the speed of the car in km/h, F is the road surface coefficient of friction and S is the length of the skid marks in metres.

Accident investigators measured the length of a skid mark to be 31 metres. The road's coefficient of friction is 0.85. Calculate the speed of the car that left the skid mark. Answer correct to the nearest km/h.

Shutterstock.com/Kitti Tantibankul

YOUR HEIGHT CAN BE FOUND IN YOUR BONES!

Forensic scientists use science and mathematics to help them determine all kinds of information. For example, the length of one of the four human 'long bones' and the sex of the person is all they require to be able to calculate the height of the person. They use formulas that allow them to determine height from the length of a skeleton's femur, tibia, humerus and radius.

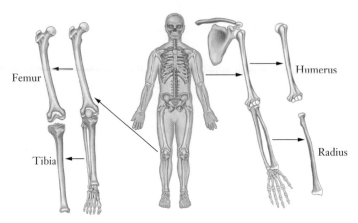

	Females	Males
Humerus	$h = \dfrac{3144H + 64\,977}{1000}$	$h = \dfrac{2970H + 73\,570}{1000}$
Radius	$h = \dfrac{3876R + 73\,502}{1000}$	$h = \dfrac{3650R + 80\,405}{1000}$
Tibia	$h = \dfrac{2533T + 72\,572}{1000}$	$h = \dfrac{2392T + 81\,688}{1000}$
Femur	$h = \dfrac{2317F + 61\,412}{1000}$	$h = \dfrac{2238F + 69\,089}{1000}$

h = the person's height

H = the length of the humerus (upper arm bone)

R = the length of the radius (lower arm bone)

T = the length of the tibia (lower leg bone)

F = the length of the femur (upper leg bone)

All measurements are in centimetres.

1 The bones of two people, a male and a female, were found in a shallow grave. The female's humerus was 33.5 cm long and the male's tibia was 41 cm. What was the height of each person when they were alive?

2 Several bones of an unidentified male skeleton were found in a forest. The humerus was 40 cm, the radius was 31 cm and the tibia was 47.5 cm.

   a Calculate the height of the unidentified male skeleton, based on each bone.

   b What height do you think should be recorded?

   c Can you suggest a reason why the bones don't give exactly the same height?

3 Measure your height and calculate the length of each of your four long bones.

2.05 Solving equations

An **equation** is like a see-saw, balanced about the equal sign. Provided we do the same thing to *both* sides, the equation will stay balanced.

Solve means to find the value of x that makes the equation true.

That value is called the **solution** to the equation.

Equations code puzzle

Working with formulas

EXAMPLE 10

Solve the equation $4x - 11 = 203$.

Solution

When we solve an equation, we want to finish with a statement such as $x = 2$.

$$4x - 11 = 203$$

$$4x - 11 + 11 = 203 + 11$$

Add 11 to both sides of the equation.

$$4x = 214$$

Divide both sides by 4 to find x.

$$\frac{4x}{4} = \frac{214}{4}$$

$$x = 53.5$$

Algebra review

Solving equations by balancing

When the solution is $x = 53.5$, it means that when you substitute 53.5 for x in the original equation, both sides will have the same value.

Solving equations by backtracking

EXAMPLE 11

Solve the equation $15 + 2(4a - 3) = 81$.

Solution

Expand the brackets and simplify.

$$15 + 2(4a - 3) = 81$$

$$15 + 8a - 6 = 81$$

$$8a + 9 = 81$$

Subtract 9 from both sides.

$$8a + 9 - 9 = 81 - 9$$

$$8a = 72$$

Divide both sides by 8.

$$\frac{8a}{8} = \frac{72}{8}$$

$$a = 9$$

Solving equations using diagrams

> If you substitute $a = 9$ back into the equation, both sides have the same value (81).

EXAMPLE 12

Solve the equation $\dfrac{p}{15} - 20 = 165$.

Solution

Add 20 to both sides.

$$\dfrac{p}{15} - 20 = 165$$

$$\dfrac{p}{15} - 20 + 20 = 165 + 20$$

$$\dfrac{p}{15} = 185$$

Multiply both sides by 15.

$$\dfrac{p}{15} \times 15 = 185 \times 15$$

$$p = 2775$$

Exercise 2.05 Solving equations

Solve each equation.

1 $8a - 4 = 420$

2 $3x + 12 = 27$

3 $2p - 18 = 48$

4 $2d - \dfrac{1}{2} = 7\dfrac{1}{2}$

5 $3m - 5 = 22$

6 $5y - 1.6 = 8.4$

7 $16 + 4(2x - 3) = 36$

8 $18 + 2(3a - 1) = 34$

9 $\dfrac{y}{10} - 5 = 3$

10 $\dfrac{g}{3} - 1 = 11$

11 $\dfrac{2t}{5} + 4 = 13$

12 $64 = 16h$

13 $8g + 7 = 27$

14 $\dfrac{4p}{3} = 10$

15 $4t - 23 = 27$

16 $\dfrac{2y}{7} = 5$

17 $\dfrac{10a}{3} = 8$

18 $\dfrac{6p}{11} = 2$

19 $\dfrac{10x}{3} = \dfrac{2}{5}$

20 $\dfrac{8p}{3} = \dfrac{5}{4}$

21 $\dfrac{-3d}{2} = 7$

There is an automatic equation solver at the Wolfram Alpha website. Use it to check your answers.

Wolfram Alpha

ISBN 9780170413503

2.06 Equations and formulas

Equation solving is one of the mathematical tools that allows us to solve problems. If we can convert a problem into a mathematical equation, we can solve the problem.

Formulas and equations

After age 30, a person's height decreases by 0.06 cm per year. This can be modelled by the formula:

$$h = T - 0.06(a - 30)$$

where:

h = height in cm

T = height in cm at age 30

a = age in years.

Grandma was 168 cm tall on her 100th birthday. How tall was she when she was 30 years old?

Solution

Grandma is 100 years old, $a = 100$ and her height is now $h = 168$. We want to find T, her height when she was 30.	$a = 100, h = 168$
Substitute the values into the formula.	$h = T - 0.06(a - 30)$
	$168 = T - 0.06(100 - 30)$
Simplify the equation.	$168 = T - 4.2$
Solve the equation.	$168 + 4.2 = T - 4.2 + 4.2$
	$172.2 = T$
	$T = 172.2$
Answer the question.	Grandma was 172.2 cm tall when she was 30 years old.

Exercise 2.06 Equations and formulas

1 The femur bone is found in the top half of a human leg. The formula that relates an adult female's height, h cm, to the length of her femur bone, F cm, is:

Example 13

$h = 61.41 + 2.32F$. Dagma is 174 cm tall. How long is her femur bone, correct to one decimal place?

2 Use the formula $v^2 = u^2 + 2as$ to determine the value of a when $s = 4$, $u = 5$ and $v = 7$.

3 The cost, C dollars, of a taxi trip over a distance of d kilometres is given by $C = 2 + 1.2d$. What distance did James travel when his taxi trip cost him $32?

4 The perimeter P of a rectangle is given by the formula:

$P = 2L + 2W$, where L is the length and W is the width of the rectangle.

The perimeter of a rectangle is 71 cm and its width is 17 cm. Calculate the length of the rectangle.

5 The formula $P = \dfrac{V}{I}$ relates the power P in watts, the voltage V and the current I in amperes in an electrical circuit. Calculate the voltage in a circuit when the power is 6 watts and the current is 25 amperes.

6 Fried's formula for calculating an infant's dose of an adult's medicine is:

$D = \dfrac{MA}{150}$ where D = the infant dose in mL, A = the adult dose in mL and M = the infant's age in months.

16-month-old Bella had a bad cough. During the night, her mother gave her a 3.2 mL dose of an adult cough medicine to help ease her cough. How much of the medicine can an adult take?

7 The number of times a cricket chirps in a minute is related to the temperature.

The formula $T = \dfrac{n}{8} + 5$ tells us how the number of cricket chirps, n, and the temperature, T in °C, are related.

a On a warm day, a cricket chirped 200 times per minute. What was the temperature?

b How many chirps per minute will a cricket make when the temperature is 20°C?

8 The path a rocket follows on a trip from Earth to Mars is 55 000 000 km long.

a In 1964, the first spacecraft to travel to Mars took 228 days to get there. Use the distance, speed and time formula, $D = S \times T$, to calculate the average speed of the rocket in km/day.

b To minimise health risks to astronauts from cancer-causing cosmic rays, future trips to Mars need to be completed in 150 days. Express 150 days in hours, then calculate the average speed required for flights to Mars in km/h.

ISBN 9780170413503

9 People living on rural properties often have their own sewerage treatment systems. The formula $V = D \times F$ calculates the volume, V litres, of a sewerage tank based on the number of days, D, the sewerage stays in the tank, and the rate at which sewerage enters the tank, F, in litres per day.

Zack's sewerage tank holds 12 000 litres. If the household produces 520 litres of sewerage per day, how many days will the sewerage stay in the treatment tank?

10 In indoor cricket, a batter's score can be calculated using the formula $S = R - 5n$, where R is the number of runs made and n is the number of times the batter was given out.

   a   In one match, Rose made 60 runs and her score was 45. How many times was she given out?

   b   Ibrahim batted well to score 31. He was given out twice. How many runs did he make?

11 The value, $\$V$, of an item when it is n years old is given by the formula $V = P(1 - r)^n$ where P is the original value of the item and r is the annual rate of depreciation as a decimal; for example, $5\% = 0.05$.

   a   Calculate the value of a 6-year-old lounge that was originally valued at $4500. Its annual rate of depreciation is 18%.

   b   After 5 years, Joel's car was worth $12 000. If it has been depreciating at an annual rate of 15%, what was the value of the car when it was new?

12 The speed a car travels during a driver's reaction time is given by the formula:

$D = \dfrac{5vt}{18}$ where:

D = the distance in metres
v = the speed of the car in km/h
t = the driver's reaction time in seconds.
When Pascal was traveling at 90 km/h, his car travelled 45 m during his reaction time. What is Pascal's reaction time?

2.07 Writing equations

Many mathematical problems are expressed in words. We can solve these problems by translating them into algebraic equations.

To solve word problems involving equations:

- Choose a pronumeral such as x.
- Translate the problem into an equation.
- Solve the equation.
- Write a sentence that answers the problem.

EXAMPLE 14

Wayne is thinking of a number. He doubled it and added 5. The answer is 69.
What is Wayne's number?

Solution

Write the problem as an equation.

Let the number be x.

Double x is $x \times 2 = 2x$.

Then, add 5 means + 5. $2x + 5 = 69$

Solve the equation. $2x + 5 - 5 = 69 - 5$

$$2x = 64$$

$$\frac{2x}{2} = \frac{64}{2}$$

$$x = 32$$

Check the solution. $2 \times 32 + 5 = 69$ ✓

Write the answer.   Wayne's number is 32.

EXAMPLE 15

Tom is 8 years older than Susi and the sum of their ages is 22. How old are they both?

Solution

Let Susi's age be n.   Susi's age is n.

Tom is 8 years older. Add 8.   Tom's age is $n + 8$.

Susi's age + Tom's age = 22 $n + n + 8 = 22$

$$2n + 8 = 22$$

Solve the equation. $2n = 14$

$$n = 7$$

Interpret the value of n and check it is   Susi is 7 and Tom is $7 + 8 = 15$.
correct.
$$7 + 15 = 22 ✓$$

Write the answer.   Susi is 7 years old and Tom is 15.

Exercise 2.07 Writing equations

1 Solve each problem by writing an equation and then solving it.

Example 14

   a Five tickets for a film cost $55. How much does each ticket cost? (Let t represent the price of a ticket.)

   b Ten oranges cost $4.80. How much does each orange cost? (Let x represent the cost of one orange.)

   c A number is doubled and the result is 110. What is the number? (Let n represent the unknown number.)

   d A number has 4 subtracted from it and the result is 6. Find the number. (Let y represent the unknown number.)

2 For each word problem, select the correct equation **A**, **B**, **C** or **D**. Then solve the equation to solve the problem.

   a Jarrad has collected 1794 beetles. This is 6 times as many beetles as Lisa has. How many beetles does Lisa have? Let N = the number of beetles Lisa has.

     A $1794 - N = 6$ **B** $6N = 1794$ **C** $N = 6 \times 1794$ **D** $N + 6 = 1794$

   b Kurt mixed 590 mL of white paint with some blue paint. He mixed 1.73 L of paint altogether. How much blue paint did he use? Let N = the amount of blue paint.

     A $N + 590 = 1730$ **B** $N - 590 = 1730$ **C** $\dfrac{N}{590} = 1730$ **D** $590N = 1730$

   c 14 packets of chocolate biscuits are inside every large box. The supermarket sold 546 packets of biscuits. How many large boxes were sold?

     A $14N = 546$ **B** $546 - N = 14$ **C** $N + 14 = 546$ **D** $\dfrac{N}{14} = 546$

   d When N is subtracted from 100, the result is 47. What is N?

     A $N - 100 = 47$ **B** $100 + 47 = N$ **C** $N - 47 = 100$ **D** $100 - N = 47$

3 Translate each of the following problems into an equation, then work out the solution of the equation to solve the problem. Remember! Sometimes drawing a diagram can help you think about the information.

Example 15

   a Student Council is holding a school disco to raise $300. Each ticket is $5 and total costs for the evening are $130. How many tickets must be sold to make the required profit? (Let n stand for the number of tickets sold.)

   b Mr Abdul says, 'I will not tell you my age but I will tell you this: If you add 15 to my age and multiply the answer by 7, you'll get 294.' How old is Mr Abdul? (Let a stand for his age.)

   c The perimeter of a rectangle is 100 cm and its width is 17 cm. What is the length? (Let l represent the length of the rectangle.)

   d I am thinking of a number. If I take away 13, multiply by 6 and add 5, the answer is 95. What is the number? (Let y represent the number.)

e The area of a rhombus is calculated by multiplying the diagonals and dividing by 2. One rhombus has an area of 44 cm² and its longer diagonal is 11 cm. What is the length of the other diagonal? (Let *d* represent the length of the diagonal.)

f Garudi is a salesperson who earns $200 per week plus one-fifth of the value of her sales for that week. Last week, her pay was $750. What was the value of her sales? (Let *x* stand for the value of her sales.)

Changing the
subject of a
formula

2.08 Changing the subject of a formula

The pronumeral on its own on the left side of a formula is called the **subject** of the formula. For example, in the formula for the volume of a rectangular prism, $V = lwh$, the subject of the formula is V. Sometimes, it is useful to rearrange a formula and make a different variable the subject. We do this by using the same methods we use for solving equations.

Exercise 2.08   Changing the subject of a formula

1 The formula relating distance, speed and time is $D = ST$. Make S the subject of this formula.

2 The volume of a rectangular prism is given by $V = lwh$. Make h the subject of this formula.

3 Fried's formula for giving medicine is $D = \dfrac{MA}{150}$. Make A the subject of this formula.

4 $S = R - 5n$ gives the score for a batsman in indoor cricket. Make R the subject.

5 The formula $C = 2\pi r$ gives the circumference of a circle. When we know the circumference we can use the same formula to find the radius. Make r the subject of this formula.

NCM 11. Mathematics Standard (Pathway 1)

ISBN 9780170413503

6 Straight lines on the number plane are given by the equation $y = mx + c$. Make x the subject of this equation.

7 The cost of a taxi trip is given by $C = 2 + 2.1d$. Make d the subject of this formula.

8 $P = 2L + 2W$ is the formula for the perimeter of a rectangle. Make W the subject.

9 Electricians use a formula relating power, voltage and current in an electrical circuit. The formula is $P = \dfrac{V}{I}$. Make I the subject of this formula.

10 The area of a triangle is given by $A = \dfrac{bh}{2}$. We can also use this formula to find the height of a triangle when we know its area and the length of its base. Make h the subject of the formula to help us find the height.

11 The formula $A = 180(n - 2)$ gives the angle sum for any polygon with n sides. Make n the subject of this formula.

KEYWORD ACTIVITY

COMPLETE THE BLANKS

Use this word list to determine which word goes in the blanks below.

divide   expand   expressions   formula   like   negative

pronumeral   quantities   same   subject   term   value

In algebra, we use pronumerals in place of unknown **1**_____ and we call what we write 'algebraic expressions'. We can only add or subtract **2**_____ terms but we can multiply and **3**_____ any algebraic **4**_____.

When a question asks us to **5**_____ brackets we have to multiply every term inside the brackets by the **6**_____ immediately in front of the brackets. We must be very careful with the positive and negative signs when the term in front of the brackets is **7**_____.

A **8**_____ is a mathematical rule written algebraically. Sometimes, we change the **9**_____ of a formula to help us with questions.

When we solve an equation, we are finding the value of a **10**_____. The solution of an equation is the value of the pronumeral that gives both sides of the equation the same **11**_____. When we solve equations we do the **12**_____ thing to both sides of the equation.

SOLUTION TO THE CHAPTER PROBLEM

Problem

Children learn words at an amazing rate. When a child is aged between 1 and 5 years we can predict the typical number of words they will understand using the formula $N = 10(6m + 10)$, where m represents the age of the child in months and N is the number of words.

Abigail, a childcare worker, is worried about Grace, a 20-month-old girl at her daycare centre who doesn't appear to understand many words. She has been tested and understands about 900 words.

Is Abigail's concern justified? Use calculations to support your answer.

Solution

The child is 20 months old. Substituting $m = 20$ into the formula $N = 10(6m + 10)$:

$$N = 10 \times (6 \times 20 + 10)$$

$$= 1300$$

A typical 20-month-old child understands 1300 words.

Grace only understands about 900 words, so Abigail's concern for her is justified.

ISBN 9780170413503

2. CHAPTER SUMMARY

- Which parts of this chapter did you remember from previous years?

- What formulas have you used or seen other people use outside of school?

- Are there any parts of this chapter that you're not sure about? If yes, ask your teacher for extra help.

Copy and complete this mind map of the topic, adding detail to its branches and using pictures, symbols and colour where needed. Ask your teacher to check your work.

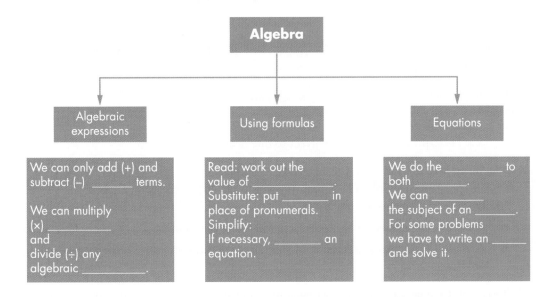

Algebra

Algebraic expressions

We can only add (+) and subtract (–) _____ terms.

We can multiply (×) _____ and divide (÷) any algebraic _____.

Using formulas

Read: work out the value of _____.
Substitute: put _____ in place of pronumerals.
Simplify:
If necessary, _____ an equation.

Equations

We do the _____ to both _____.
We can _____ the subject of an _____.
For some problems we have to write an _____ and solve it.

2. TEST YOURSELF

1 Simplify each expression.

a $5p + 3 + 8p$ **b** $3 + 4x + 12$ **c** $w - 12 + w$

d $t + t^2 + 3t$ **e** $78ab + 2ab + 2$ **f** $6m + 3n - 2m + 5n$

g $20m - m$ **h** $12p \times 3q \div 9$ **i** $12a \times 2b \div 8y$

j $24k \div 2k \times 6$ **k** $8p \times 4q \div 16t$ **l** $6y \div 2y \times 3$

2 Expand and simplify each expression.

a $5a + 3(a + 2)$ **b** $-(8 - y)$

c $-x(x - 5)$ **d** $5(3p + 5) + 2$

3 **a** Calculate the value of V in the formula $V = \dfrac{4}{3}\pi r^3$ when $r = 3.5$. Express your answer correct to one decimal place.

b Evaluate S in the formula $S = \dfrac{D}{T}$ when $D = 450$ and $T = 18$.

c Use the formula $v^2 = u^2 + 2as$ to determine the value of v when $a = 4$, $s = 2.2$ and $u = 11$. Express your answer correct to one decimal place.

4 The formula $D = S \times T$ relates distance, speed and time. $D =$ the distance travelled, $S =$ the speed and $T =$ the time. A motorcycle is travelling at an average speed of 115 km/h on the motorway.

a Calculate the distance the motorcycle will travel in 4 hours.

b Express 15 minutes as a fraction of an hour, then calculate how far the motorcycle will travel in 15 minutes.

5 The total distance it takes a car to stop at different speeds on a dry asphalt road can be calculated by using the formula $D = \dfrac{5Vt}{18}(5t + 2.27)$ where:

$D =$ the total distance travelled in stopping, in metres

$V =$ the car's velocity (speed) in km/h

$t =$ the driver's reaction time in seconds.

Maya's reaction time is 0.9 seconds. Calculate the distance it takes Maya's car to stop when she is driving at 80 km/h.

6 Solve each equation.

a $5x - 11 = 209$ **b** $\dfrac{p}{2} - 1 = 10$ **c** $\dfrac{24}{x} = 5.5$

62   **NCM 11.** Mathematics Standard (Pathway 1)   ISBN 9780170413503

7 The formula $P = N(1 - R)^t$ can be used to calculate population numbers of endangered animals, where P = the current population, N = the previous number of animals, R = the annual rate of decrease in numbers as a decimal, and t = the time in years.

African elephants are being killed by poachers for their ivory tusks and their numbers have been decreasing at an annual rate of 3.5%. In 2017, only 350 000 African elephants remained. Use the formula to calculate the population of African elephants in 1979, 38 years before 2017. Answer to the nearest thousand.

8 I am thinking of a number. If I divide the number by 8, then add 5, I get 9. What is the number? Use an equation to find it.

9 The formula for calculating simple interest is $I = \dfrac{PRT}{100}$. Make T the subject of this formula.

10 The equation relating the cost of catering for a number of guests is given by $C = 340 + 50n$. Make n the subject of this formula.

3.

HEALTHY FIGURES

Chapter problem

Brett is usually a very alert, safe driver. His reaction time is 0.16 seconds. However, when his blood alcohol content (BAC) is 0.05, his reaction time increases to 0.22 seconds.

Brett is driving at 100 km/h when he suddenly needs to brake. How much further does his car travel during his reaction time when his BAC is 0.05, compared to when he isn't drinking?

CHAPTER OUTLINE

WHAT WILL WE DO IN THIS CHAPTER?

- Learn about kilojoules and calories as units of energy
- Solve problems involving energy in food, exercise and activity
- Use rates and formulas in health and safe driving
- Solve problems involving speed, stopping distance, BAC and safe driving

HOW ARE WE EVER GOING TO USE THIS?

- Tables and charts contain information that is essential in maintaining a healthy, fit body
- Formulas and equations are used in many trades and workplaces
- Formulas, tables and charts are used in health and fitness

3.01 Burning energy

Kilojoules (kJ) are metric units of energy. If we use more energy than we eat, we lose body weight. However, if we eat more energy than we use, this extra energy is stored as fat.

This table shows the amount of energy our body needs each day, depending on our age, gender and level of activity.

Age (years)	Lifestyle	Men (kJ/day)	Women (kJ/day)
18–35	Inactive	10 500	8000
	Active	12 500	9000
	Very active	14 800	10 500
36–70	Inactive	10 000	8000
	Active	11 800	8800
	Very active	14 300	10 400
Pregnant women			10 100
Breastfeeding women			11 800

EXAMPLE 1

Chloe is 22 years old, works in an office and has an inactive lifestyle. Every day on her way to work she has a cappuccino at the coffee shop. The cappuccino contains 940 kJ.

a How many kilojoules of energy does Chloe need each day?

b What percentage of her daily energy needs does Chloe have in her morning cappuccino?

Shutterstock.com/ChameleonsEye

Solution

a Chloe is 22, a woman and has an inactive lifestyle. Her energy needs are shown in the top row and right column in the table.

8000

Chloe needs 8000 kJ per day.

b $\dfrac{\text{kJ in the cappuccino}}{\text{kJ per day}} \times 100\%$

$\dfrac{940}{8000} \times 100 = 11.75\%$

Chloe's cappuccino contains 11.75% of her daily energy needs.

Calories (cal) are an older unit of energy that we use sometimes. Calories are bigger than kilojoules.

> Sometimes in food packaging, you will see the word 'kilocalories'. When describing energy in food, kilocalories and calories are the same thing.

> 1 calorie = 4.2 kilojoules

EXAMPLE 2

A 500 mL chai tea latte with full cream milk contains 240 calories. How many kilojoules is this?

Solution

Kilojoules are smaller than calories. To change to a smaller unit, multiply by the conversion factor.

$$240 \text{ cal} = 240 \times 4.2 \text{ kJ}$$
$$= 1008 \text{ kJ}$$

Exercise 3.01 Burning energy

1 Mark has a very active lifestyle as a 20-year-old bricklayer.

  a Use the table on the previous page to find how much energy he needs per day.

  b Mark eats a hearty breakfast containing 3250 kJ to give him energy for the day's work. What percentage (to the nearest whole number) of his daily energy requirements does Mark have for breakfast?

2 Shauna is 36 years old. During the week she is inactive and on the weekend she is very active.

  a How much energy does she need each day during the week?

  b How much more energy does she require per day on the weekend than during the week?

3 Courtney, aged 18, is very thin and refuses to eat more than 6000 kJ per day. Every afternoon she works out at the gym for 2 hours. Courtney's mother is worried and knows that the type of gym workout Courtney does burns 2500 kJ/hour.

  a How much energy does Courtney use at the gym each day?

  b How much energy does she have left from her diet for the remaining 22 hours in the day?

Example 2

4 Use the conversion 1 cal = 4.2 kJ to copy and complete each statement.

    a   500 cal = ___ kJ   **b**   360 kJ = ____ cal   **c**   68 cal = ___ kJ

    d   25 kJ = ____ cal   **e**   2460 kJ = ___ cal   **f**   10 800 kJ = ___ cal

5 a The treadmill at the gym shows that Samantha has used 550 cal. How many kJ is this?

    b Samantha wants to use 2500 kJ on the machine. How many more calories does she have to burn?

6 This graph shows children's daily energy requirements.

    a Who needs more energy per day: boys or girls?

    b At which age do boys and girls need approximately the same amount of energy per day?

    c How many more kJ/day do 16-year-old boys need than 16-year-old girls?

    d At what age do boys require approximately 7500 kJ/day?

    e Calculate the weekly energy requirements for a 12-year-old girl.

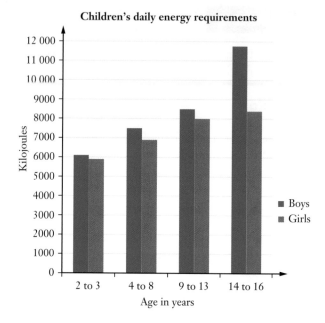

Children's daily energy requirements

    f Jon is an 8-year-old boy who usually eats 8600 kJ of food per day. Is he eating an appropriate amount?

    g If Jon continues to eat 8600 kJ/day, will his weight increase or decrease?

7 This table shows the average length of time it takes an 18-year-old to burn 1000 kJ for different activities.

    a How much energy does Leo use while swimming for 1 hour?

    b How long will it take Divya to use 500 kJ while she is sleeping?

Activity	Time required to use 1000 kJ
Sleeping	4 hours
Eating	3 hours
Working in class	2 hours 30 minutes
Studying	
Watching TV	
Walking	1 hour
Bike riding	50 minutes
Swimming	30 minutes

c Suzie, aged 18, leads a very active life. The pie chart shows the number of hours each day that she spends on different activities.

  i According to the table on page 66, how many kilojoules does Suzie need per day?

  ii If Suzie stays within this recommendation, will she eat a sufficient amount to meet her energy requirements?

  iii What advice can you offer to Suzie concerning her diet?

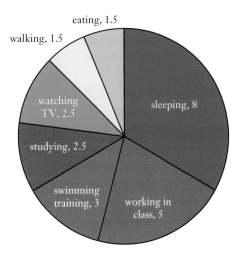

Hours Suzie spends on different activities per day

3.02 Food and energy

The food we eat supplies us with energy. This table shows the amount of energy stored in some normal-sized serves of food.

Investigating fast food

Food	kJ	Food	kJ	Food	kJ
Roast lamb, gravy	1064	Apple pie	1380	Apple	243
Potato bake	1175	One slice of buttered bread	520	Large tomato	120
				Small tomato	30
Mixed grill	2600	Yoghurt	315	Chocolate biscuit	493
Steak	3900	Muesli	1470	Can of soft drink	372
Bacon (2 slices)	640	Eggs (2)	735	Orange juice (glass)	206
Fish	340	Cheese	115	Milk (glass)	628
Grilled chicken breast	1264	Ham (2 slices)	224	Milk for cereal	400
Chips	1425	Mixed nuts (100 g)	2640	Coffee with milk and sugar	295
Ice cream	810	Broccoli	98	Banana	546
Sauce for steak or chicken	246	Sauce for fish	265	Beef sausage	176

EXAMPLE 3

On a plane, Lucas was served a meal of grilled chicken breast with sauce, chips and broccoli. How many kilojoules were in this meal?

Solution

Use the table to look up the energy content of each food item.

Chicken breast: 1264

Sauce: 246

Chips: 1425

Broccoli: 98

Add them.

Total kJ = 1264 + 246 + 1425 + 98

= 3033

Write your answer.

The meal contained 3033 kJ.

EXAMPLE 4

Miriam ate 200 g of mixed nuts. How long will it take her to use the energy from the nuts when she is working out at the gym at the rate of 30 kJ/minute?

Solution

200 g of nuts = 2 serves (from the table)

Total energy = 2640 × 2

= 5280 kJ

Divide the number of kJ by the rate to find the time.

Time = 5280 ÷ 30

= 176 minutes

= 2 hours 56 minutes

Write your answer.

Miriam will have to work out for 2 h 56 min to use the energy from the nuts.

 ISBN 9780170413503

EXAMPLE 5

Athletes running at 15 km/h use 75 kJ/h per kg of their body weight. Claire's body weight is 58 kg. How much energy does she use when she runs for 30 minutes at 15 km/h?

Solution

Claire burns 75 kJ per hour per kg of her body weight. She weighs 58 kg, so each hour she burns 75×58.	Energy burned per hour $= 75 \times 58$ $= 4350$ kJ
30 minutes is half of an hour. Divide by 2.	Energy burned in 30 min $= 4350 \div 2$ $= 2175$ kJ
Write your answer.	Claire burns 2175 kJ of energy.

Exercise 3.02 Food and energy

1   Use the table on page 69 to calculate the energy content of each meal.

a   Danica's breakfast contains a serving of muesli with milk, an apple and a banana.

b   Felix's breakfast comprises two slices of ham, two beef sausages, one egg and one slice of buttered bread.

c   Milan ate a steak with sauce, a large tomato and a serving of chips for his lunch.

d   Emma had some grilled fish with sauce and a large sliced tomato for her lunch.

Example
3

2 George is 16 years old. This is what he eats in a typical day.

Breakfast	Lunch	Dinner
2 slices of bacon	Sandwich with 2 slices of ham	Mixed grill
2 eggs	Apple	Chips
Muesli with milk	2 chocolate biscuits	Broccoli
2 slices of toast with butter	Yoghurt	Apple pie
Glass of orange juice	Can of soft drink	Ice cream
		Coffee, milk, sugar

a Calculate the total number of kilojoules George eats in a typical day.

b An average 16-year-old boy should eat 11 800 kJ per day. What effect will George's typical food intake have on his body weight?

3 Amanda requires a daily diet of 8800 kJ. Use the food items in the table on page 69 to plan a healthy day's menu for Amanda.

4 A large chocolate cake contains 27 600 kJ. Mai cut the cake into 8 equal slices.

a How many kilojoules are in each slice of cake?

b How many minutes of a gym workout at 30 kJ/minute will it take to work off the kilojoules from one slice of the cake?

5 This table shows the energy an average person uses per minute for different activities.

Calculate the amount of energy a person uses when:

a gardening for 30 minutes

b ironing for 45 minutes

c bricklaying for an hour

d circuit training for $1\frac{1}{4}$ hours

e sleeping for 8 hours

f walking for 50 minutes every day for a week.

Activity	kJ/minute
Sleeping	4
Cleaning	15
Ironing	17
Bricklaying	17
Playing tennis	31
Gardening	23
Circuit training	53
Walking	23

6 The more you weigh, the more kilojoules you will burn doing different activities. This table shows the amount of energy per kilogram of body weight that teens and adults use in one hour during various activities.

Example
5

Activity	kJ used per hour per kg of weight
Sleeping	3.7
Driving a car	6.4
Schoolwork	6.8
Walking	18
Skiing (15 km/h)	43
Swimming (45-second laps)	29

a Amy weighs 45 kg. How much energy does she use during a 1-hour maths lesson?

b Martin's body mass is 75 kg. How many more kilojoules does he use during 8 hours sleep than Amy?

c Fahim started walking every day to try to lose weight. He currently weighs 128 kg. How much energy does he use when he walks for 30 minutes?

d Martina is a keen skier who weighs 63 kg. One Saturday she drove her car for 3 hours to the snowfields. She then skied at an average speed of 15 km/h for 5 hours. How much energy did she use in the 8 hours?

e Every morning, as part of his training routine, Scott spends an hour swimming 45-second laps of the pool. During the hour, Scott uses approximately 2200 kJ. What is Scott's approximate weight?

INVESTIGATION

MY DAILY ENERGY REQUIREMENTS
What you have to do

- Keep an activities/exercise diary for one week. Record the number of hours you spend on different activities each day. Classify your activities such as sleeping, eating, doing schoolwork, watching TV, walking, swimming, running or going to the gym into low-level, medium-level and high-level activities.

- Keep a second food diary where you record the number of kilojoules you eat each day. Remember to include 'hidden' kilojoules like those in tea/coffee, biscuits and other snacks.

- Use the results in your activities diary to calculate the number of kilojoules you need, on average, per day. If you don't know the kilojoules involved in any of your activities, use the Internet to find out.

- Do your activities and food intake balance your kilojoules? What changes might you need to make?

3.03 Speed, distance and time

The question 'How long will it take to get there?' can be answered easily with a little mathematics. The values for the distance covered, the speed and the time taken are related by a formula. When we know 2 of the values we can calculate the third value.

Distance covered = speed × time

or $D = S \times T$

The units for speed tell us the units for distance and time.

When the speed is in km/h, the distance is in kilometres and the time is in hours.

Some students find it easier to use the 'distance, speed and time triangle' to solve problems involving speed. Place the letters D for distance, S for speed and T for time in alphabetical order in the triangle.

To calculate the **speed**, cover up S, which leaves $\dfrac{D}{T}$.

This means that $S = \dfrac{D}{T}$, or $S = D \div T$.

To calculate the **time**, cover up T, which leaves $\dfrac{D}{S}$.

To calculate the **distance**, cover up D, which leaves $S \times T$.

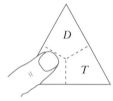

EXAMPLE 6

A racing greyhound runs at a speed of 18 m/s.

a How far will it run in 4 seconds?

b How long will it take the greyhound to complete a 1200 m race? Answer correct to the nearest 0.1 s.

Solution

a Use the triangle. To find the distance, cover the D.

$S = 18$ m/s and $T = 4$ s.

$D = S \times T$

$= 18 \times 4$

$= 72$ m

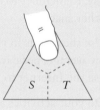

$D = S \times T$

Write your answer.

The greyhound will run 72 m.

b   To find the time, cover the T.

$D = 1200$ m and $S = 18$ m/s.

$$T = \frac{D}{S}$$

$$D = \frac{1200}{18}$$

$$\approx 66.7 \text{ s}$$

$T = \dfrac{D}{S}$ or $T = D \div S$

Write your answer.

The greyhound will complete the race in 66.7 seconds.

EXAMPLE 7

A kangaroo bounds at a speed of 48 km/h. How far will a kangaroo bound in 20 minutes?

Solution

To find the distance, cover the D.

$S = 48$ km/h and $T = 20$ min. The speed is in km/h, so we need the time in hours as well. Divide 20 min by 60 to change it to hours.

$$T = \frac{20}{60} = \frac{1}{3} \text{ h}$$

$$D = S \times T$$

$$= 48 \times \frac{1}{3}$$

$$= 16 \text{ km}$$

Write your answer.

The kangaroo will bound 16 km in 20 min.

Exercise 3.03 Speed, distance and time

1 a   Use the formula $T = \dfrac{D}{S}$ to determine the value of T when $D = 80$ and $S = 16$.

b   In the formula $D = S \times T$, what is the value of D when $S = 80$ and $T = 3$?

2 a   Corrina is driving at a speed of 60 km/h. How far will she drive in 3 hours?

b   How long will it take her to drive 240 km?

Example 6

3   Go-karts race at a speed of 110 km/h. At this speed, how many kilometres can a go-kart travel in a $2\frac{1}{2}$-hour race?

4   Wasim is driving through heavy traffic at a speed of 32 km/h. How far will he travel in 15 minutes?

Example 7

---

5 An ambulance is racing at a speed of 100 km/h to the scene of a serious freeway accident 15 km away.

    a How long will the ambulance take to reach the accident? Express your answer as a decimal of an hour.

    b Multiply your answer to part **a** by 60 to change the time to minutes.

6 A whitewater rafting team completed three sets of rapids and 1.6 km of calm water in 30 minutes. The lengths of the sets of rapids were 150 m, 80 m and 170 m.

    a Calculate the distance that the rafting team covered in 30 minutes. Express your answer in kilometres.

    b Explain why you can't use $T = 30$ in the equation $S = \dfrac{D}{T}$ to calculate the speed of the raft in km/h.

    c Calculate the raft's average speed in km/h.

7 Cheetahs are the fastest animals on land and can run at a speed of 31 m/s in short bursts.

    a How far can a cheetah run in 9 seconds?

    b How long does it take a cheetah to run 140 m? Answer in seconds correct to one decimal place.

8 A peregrine falcon is the fastest animal overall, reaching a top speed of 90 m/s.

    a How far can the falcon fly in 1 minute?

    b How long will it take the falcon to fly 1 km?

9 This table shows the distances in kilometres between several cities.

	Albury	Brisbane	Canberra	Goulburn	Sydney	Tamworth
Albury	–	1610	190	380	600	1040
Brisbane	1610	–	1300	1225	1020	575
Canberra	190	1300	–	95	300	750
Goulburn	380	1225	95	–	205	660
Sydney	600	1020	300	205	–	460
Tamworth	1040	575	750	660	460	–

    a How far is it from Canberra to Tamworth?

    b How long will it take to drive from Canberra to Tamworth at an average speed of 75 km/h?

    c Glen took 5 hours to drive from Goulburn to Albury. What was his average speed?

    d Max and Sanjay left Albury at 6 a.m. on Monday to drive to Brisbane. They shared the driving and completed the trip at an average speed of 70 km/h. At what time did they arrive in Brisbane?

10 The speed limit on the motorway is 100 km/h, but when it is raining the speed limit is 90 km/h. How much longer does it take to travel an 18-km stretch of freeway when it is raining compared to when it is fine? Select the correct answer **A, B, C** or **D**.

    A 1 min 12 s   **B** 1 min 48 s   **C** 10 min 48 s   **D** 12 min 0 s

NCM 11. Mathematics Standard (Pathway 1)   ISBN 9780170413503

REACTION TIMES AND DISTANCES

This activity involves three stages: measuring your reaction time, constructing a conversion graph between km/h and m/s, and making a table of values to show the distances a car will travel during your reaction time at different speeds. To complete this activity, you will need graph paper, a ruler and a stopwatch.

1 Measuring your reaction time

You need two people for this activity.

a Position the ruler with the 0 cm measure at the bottom. One person holds the top of the ruler and the other person places their hand at the 0 cm mark, ready to catch it.

b Without warning, the person holding the ruler lets it go and the other person has to catch it. Record the measurement on the ruler where it is caught.

c Change places and repeat.

d Use the table below to read your reaction times.

Measurement in cm	6	8	10	12	14	16	18	20	22	24	26
Reaction time in seconds	0.11	0.13	0.14	0.16	0.17	0.18	0.19	0.20	0.21	0.22	0.23

2 Constructing a conversion graph

This table shows the relationship between km/h and m/s. Create a conversion line graph by graphing the table of values on graph paper.

km/h	0	50	100
m/s	0	13.9	27.8

3 Determining your reaction distance

a To find the distance you will travel during your reaction time at different driving speeds, multiply your reaction time by the speed in m/s. For example, suppose that your reaction time is 0.24 s and you are traveling at 100 km/h. From the graph or table, 100 km/h = 27.8 m/s. You will travel $0.24 \times 27.8 = 6.67$ m during your reaction time.

b Copy and complete this table of values.

Speed in km/h	40	60	75	80	100
Reaction distance in m					

3.04 Braking fast

Stopping distance is the distance a vehicle travels from the time when a driver decides to stop to when the car comes to a complete stop. It depends on:

- the **reaction time** of the driver

- the current speed of the vehicle

- the texture of the road surface.

A formula can be used to calculate stopping distance, D, measured in metres:

$$D = \frac{V}{1000}(210 + 97R)$$

where D = distance covered in metres, V = velocity (speed) of the car in km/h, R = road surface index (see table).

This formula does not take into account the driver's reaction time.

Road surface index, R	
Dry asphalt	1.3
Wet asphalt	1.6
Gravel	2
Hard snow	6.6
Ice	14.3

EXAMPLE 8

Kelly was driving at a speed of 60 km/h on a gravel road. What distance did she require to stop?

Solution

$V = 60, R = 2$

Substitute into the formula.

$$D = \frac{V}{1000}(210 + 97R)$$

$$= \frac{60}{1000}(210 + 97 \times 2)$$

$$= 24.24 \text{ m}$$

Write the answer.

Kelly will need a bit more than 24 metres to stop. She should allow 25 m.

Exercise 3.04 Braking fast

Round all answers to one decimal place.

1 How far will Kelly take to stop when she is travelling at 80 km/h on a gravel road?

2 Levi was driving on a dry asphalt road at 60 km/h.
   a What distance did his car travel after he decided to stop?
   b How much further would it take him to stop on a wet asphalt road at 60 km/h?
   c What advice would you give Levi if it started raining when he was driving on an asphalt road?

NCM 11. Mathematics Standard (Pathway 1)

ISBN 9780170413503

3 When Kim was driving at 50 km/h on a dry asphalt road, a school student ran onto the road 18 metres ahead.

   a Can Kim stop before hitting the student? Use a calculation to justify your answer.

   b Would the situation have been different if it was raining? Explain your answer.

4 Aaron likes to go to the snow.

   a How far will it take him to stop his car at a speed of 60 km/h on hard snow?

   b Aaron drove slowly on the road to the ski lodge because the road was covered with ice. How far will it take him to stop at 15 km/h on the ice?

5 Taylor's family often travels in the country. In dry weather, on an asphalt road, Taylor usually drives at 100 km/h. What is a safe distance for Taylor to leave between her car and the car in front? Explain your answer.

6 Matthew said 'If I double my speed, I need to allow double the stopping distance'. Explain why you agree or disagree with Matthew's statement.

7 The maximum speed, V km/h, a car can be traveling to stop within a distance D m, on a road with surface index R, can be found by using the formula:

$$V = \frac{1000D}{210 + 97R}$$

When she is driving on country roads, Xanthe likes to allow herself 32 m to stop.

   a What is the maximum speed Xanthe can travel on a gravel road?

   b How much faster can she travel on a dry asphalt road than on a gravel road, with a 32 m stopping distance?

Use your reaction time from the previous Practical activity to answer Questions 8 and 9.

8 The distance (in metres) a car travels during a driver's reaction time (the distance a car travels from when the driver needs to stop to when the brakes are applied) can be calculated with the formula $D = \dfrac{5Vt}{18}$ where V = the car's velocity in km/h and t = the driver's reaction time in seconds.

Calculate your own reaction distance at:

   a 50 km/h   **b** 60 km/h   **c** 100 km/h   **d** 120 km/h.

9 The total distance, D m, it takes to stop a car at different speeds on a dry asphalt road can be found by using the formula $D = \dfrac{5Vt}{18}(5t + 2.27)$.

   a When driving on a dry asphalt road at 60 km/h, what is your car's stopping distance?

   b The speed limit for school zones is 40 km/h. How much shorter is your stopping distance on dry asphalt at 40 km/h than at 60 km/h?

   c Why do you think that this special speed limit is necessary?

3.05 Staying under 0.05

Alcohol affects people's reaction times and their ability to make good decisions. Every country has regulations prohibiting people from driving when they are affected by alcohol. Most countries use **blood alcohol content (BAC)** as a measure to determine whether a person can drive. A BAC of 0.05 means that there is 0.05 g of alcohol in 100 mL of blood. In Australia, it is illegal to drive with a BAC of 0.05 or over.

BAC is affected by the amount of alcohol consumed, the weight of the drinker, their fitness, gender, liver function and age. There are formulas for estimating BAC.

Blood alcohol content (BAC) formulas

$$BAC_{male} = \frac{10N - 7.5H}{6.8M}$$

$$BAC_{female} = \frac{10N - 7.5H}{5.5M}$$

where:

N is the number of standard drinks consumed,

H is the number of hours of drinking and

M is the person's mass in kilograms.

EXAMPLE 9

Nicky weighs 60 kg. In two hours on Saturday night she drank 5 standard drinks of wine. Calculate Nicky's BAC correct to three decimal places.

Solution

Use the formula for BAC_{female}.

In this situation, $N = 5$, $H = 2$ and $M = 60$.

$$BAC_{female} = \frac{10N - 7.5H}{5.5M}$$

$$= \frac{10 \times 5 - 7.5 \times 2}{5.5 \times 60}$$

$$= \frac{35}{330}$$

$$= 0.10606 \ldots$$

$$\approx 0.106$$

Write the answer.   Nicky's BAC is 0.106.

EXAMPLE 10

Dylan's mass is 65 kg. How many standard drinks can he have in 3 hours and keep his BAC under 0.05?

Solution

Use the formula for BAC_{male}.

In this situation, $H = 3$, $M = 65$ and BAC < 0.05.

We need to find the value of N.

$$\text{BAC}_{\text{male}} = \frac{10N - 7.5H}{6.8M}$$

$$\frac{10N - 7.5 \times 3}{6.8 \times 65} < 0.05$$

$$\frac{10N - 22.5}{442} < 0.05 \quad \boxed{< \text{ means 'less than'}}$$

Multiply both sides by 442.

$$\frac{10N - 22.5}{442} \times 442 < 0.05 \times 442$$

$$10N - 22.5 < 22.1$$

Add 22.5 to both sides.

$$10N - 22.5 + 22.5 < 22.1 + 22.5$$

$$10N < 44.6$$

Divide both sides by 10.

$$\frac{10N}{10} < \frac{44.6}{10}$$

$$N < 4.46$$

N needs to be a whole number because it is the number of drinks.

$N = 4$.

Write your answer.

Dylan can have, at most, four standard drinks in 3 hours.

Fairfax Syndication/Rohan Thomson

EXAMPLE 11

The table below shows the BAC for males at different masses and numbers of standard drinks.

Mass	Number of standard drinks									
	1	2	3	4	5	6	7	8	9	10
45 kg	0.038	0.075	0.113	0.150	0.188	0.225	0.263	0.300	0.338	0.375
50 kg	0.036	0.071	0.107	0.142	0.178	0.214	0.249	0.285	0.320	0.356
55 kg	0.039	0.039	0.101	0.135	0.169	0.203	0.237	0.270	0.304	0.338
60 kg	0.033	0.066	0.099	0.132	0.165	0.198	0.231	0.264	0.297	0.330
65 kg	0.031	0.062	0.093	0.124	0.155	0.186	0.217	0.248	0.279	0.310
70 kg	0.028	0.056	0.085	0.113	0.141	0.169	0.197	0.226	0.254	0.282
75 kg	0.026	0.052	0.079	0.105	0.132	0.158	0.184	0.210	0.237	0.263
80 kg	0.025	0.049	0.074	0.098	0.123	0.147	0.172	**0.196**	0.221	0.245
85 kg	0.023	0.045	0.068	0.090	0.113	0.136	0.158	0.181	0.203	0.226
90 kg	0.021	0.041	0.062	0.083	0.104	0.124	0.145	0.166	0.186	0.207
95 kg	0.019	0.038	0.057	0.076	0.095	0.113	0.132	0.151	0.170	0.189
100 kg	0.017	0.034	0.051	0.068	0.085	0.102	0.119	0.136	0.153	0.170

Phillip's mass is 79 kg and he consumed 8 standard drinks in 5 hours. Estimate Phillip's BAC after the 5 hours if his BAC decreases by 0.015 per hour of drinking.

Solution

Phillip's mass of 79 kg is not in the table. Use 80 kg, the closest mass to 79 kg.

Phillip's BAC would be 0.196 if he consumed all the drinks at the same time.

A mass of 80 kg and 8 standard drinks gives 0.196.

Subtract 0.015 for each hour of drinking.

$$BAC = 0.196 - 0.015 \times 5$$
$$= 0.121$$

Write your answer.

Phillip's BAC was 0.121 when he finished drinking.

> Our kidneys remove alcohol from our blood at a constant rate, so BAC should decrease by the same amount every hour.

Exercise 3.05 Staying under 0.05

1 Find, correct to three decimal places, the value of $BAC_{female} = \dfrac{10N - 7.5H}{5.5M}$ with values $N = 4$, $H = 3$ and $M = 55$.

2 Billy's mass is 84 kg. During a 3-hour period, he consumed 6 standard drinks. Calculate his BAC correct to three decimal places.

3 Holly drank too much at a party. She consumed 8 standard drinks in 2 hours. Her mass is 58 kg. Calculate Holly's BAC correct to three decimal places.

4 John and Celine each have a mass of 75 kg. They both consumed 6 standard drinks in 4 hours. How much more was Celine's BAC than John's? Answer correct to four decimal places.

5 Olivia wants to keep her BAC under 0.05. Her mass is 62 kg. How many standard drinks can she consume in 4 hours?

6 Calculate the number of standard drinks each person can consume to keep their BAC under 0.05.

	Person	Gender	Mass	Number of hours drinking
a	Zara	Female	52 kg	3
b	Jake	Male	95 kg	4
c	Raphael	Male	78 kg	2
d	Sophia	Female	65 kg	5

7 The formula, $N = \dfrac{BAC}{0.015}$ gives the number of hours a P-plate driver must wait before driving (until their BAC is zero). Swaetha is a P-plate driver and her BAC is 0.03. How long must she wait before she drives a car?

8 Joungwoo is a P-plate driver. At 9 p.m. his BAC was 0.05. What is the earliest time he can legally drive a car?

9 Karin is a P-plate driver who weighs 62 kg. Between 4 p.m. and 7 p.m. she consumed 6 standard drinks. Calculate her BAC correct to two decimal places. Will Karin be able to drive home at 11.30 p.m? Justify your answer.

10 Ben had a two-hour dinner in a restaurant with friends. He enjoyed four 150-mL glasses of wine from a 750-mL bottle of wine with his meal. The bottle contained 7.5 standard drinks. Ben's mass is 85 kg and he is a P-plate driver.

    a Calculate the number of standard drinks Ben consumed.

    b Calculate Ben's BAC correct to three decimal places.

    c How many hours and minutes would Ben have to wait before he could drive home?

Example
11

11 Use the table on page 82 to calculate each BAC in the table below.

	Body mass	Number of standard drinks	Time spent drinking	BAC
a	51 kg	5	4 hours	
b	62 kg	7	2 h 30 min	
c	87 kg	3	90 min	

12 In each situation, state whether the BAC will increase, decrease or stay the same.

    a Jake had the same number of drinks over a longer time period.

    b Sally had more drinks in the same time.

    c A person with the same body mass had more standard drinks in the same time.

    d Kate had the same number of drinks in the time, but they were bigger than standard drinks.

    e A person with a smaller body mass had the same number of standard drinks in the same time.

IARD

INVESTIGATION

BAC IN DIFFERENT COUNTRIES

In NSW, the legal BAC limit for full licence drivers is 0.05. Is this limit the same in other Australian states and overseas?

1 Visit the website of the International Alliance for Responsible Drinking and search for BAC limits.

2 Search for the legal BAC limit in 12 different countries, such as the USA, United Kingdom, Sweden, New Zealand and Singapore.

3 Which country has the highest legal BAC limit?

4 What countries have a zero legal BAC limit?

FIND-A-WORD PUZZLE

Make a copy of the puzzle, then use the hints to find the keywords from this chapter.

1 The time between when a driver needs to stop the car and when the driver applies the brakes (two words)

2 Blood alcohol content (initials)

3 A measure of how fast you're travelling

4 The technical term for how fast you're travelling, begins with 'v'

5 An old unit for measuring energy

6 The modern unit for measuring energy

7 The number of metres it takes for a car to come to a complete stop (two words)

8 You can calculate this by multiplying speed by time

9 *N* stands for the number of these when using the BAC formula (two words)

10 We use kilojoules to measure this

11 This is a measure of the texture of a road, which tells us how hard it is to stop on it (three words, initials RSI).

Q	S	B	J	E	T	U	Q	R	L	X	I	R	W	G	M
P	T	E	T	Y	X	O	C	E	N	E	R	G	Y	M	H
R	O	A	D	S	U	R	F	A	C	E	I	N	D	E	X
F	P	T	A	T	V	Z	N	C	U	H	N	Q	G	X	V
R	P	S	J	A	R	M	D	T	C	A	S	B	A	C	R
A	I	P	B	N	G	T	T	I	Z	P	Y	L	R	A	J
X	M	E	Z	D	K	I	L	O	J	O	U	L	E	U	T
J	G	R	P	A	I	P	Q	N	W	S	M	V	B	R	W
S	D	M	S	R	E	J	K	T	C	S	Z	H	P	V	M
Y	I	I	L	D	U	S	U	I	Y	P	R	J	C	X	B
Q	S	N	Z	D	O	D	W	M	T	E	N	T	A	Z	R
V	T	U	H	R	Y	N	F	E	D	E	V	M	L	P	A
K	A	T	O	I	Q	N	T	E	W	D	T	N	O	S	F
I	N	E	Y	N	F	S	K	U	B	X	P	E	R	W	K
O	C	S	A	K	G	I	N	V	E	L	O	C	I	T	Y
L	E	A	S	S	Q	D	I	S	T	A	N	C	E	Z	S

SOLUTION TO THE
CHAPTER PROBLEM

Problem

Brett is usually a very alert, safe driver. His reaction time is 0.16 seconds. However, when his blood alcohol content (BAC) is 0.05, his reaction time increases to 0.22 seconds.

Brett is driving at 100 km/h when he suddenly needs to brake. How much further does his car travel during his reaction time when his BAC is 0.05, compared to when he isn't drinking?

Solution

We can use the formula $D = \dfrac{5Vt}{18}$ (from page 79) to calculate the distance a car travels during a driver's reaction time at different speeds and different reaction times where:

D = distance in metres

V = the car's speed in km/h

t = the driver's reaction time in seconds.

Brett is driving his car at a speed of 100 km/h.

When $t = 0.16$, reaction distance, $D = \dfrac{5 \times 100 \times 0.16}{18}$
$= 4.44$ m correct to the nearest cm.

When $t = 0.22$, reaction distance, $D = \dfrac{5 \times 100 \times 0.22}{18}$
$= 6.11$ m correct to the nearest cm.

The car travels an additional $6.11 - 4.44 = 1.67$ m during Brett's reaction time when his BAC is 0.05.

- Which part of this chapter is the most relevant to your life outside school?

- Have you used any of the calculations in this chapter as part of your everyday life?

- Are there any sections of this chapter that you don't understand? If yes, ask your teacher to explain.

Copy and complete this mind map of the topic, adding detail to its branches and using pictures, symbols and colour where needed. Ask your teacher to check your work.

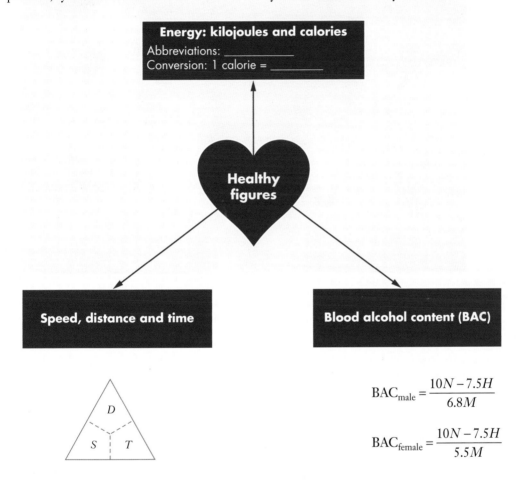

Energy: kilojoules and calories
Abbreviations: _____
Conversion: 1 calorie = _____

Healthy figures

Speed, distance and time

Blood alcohol content (BAC)

$$BAC_{male} = \frac{10N - 7.5H}{6.8M}$$

$$BAC_{female} = \frac{10N - 7.5H}{5.5M}$$

Exercise 3.01

1 a The rowing machine at the gym shows that Nicolai has used 480 calories. Use the conversion 1 Cal = 4.2 kJ to calculate the number of kilojoules Nicolai has used.

b Nicolai wants to use 2800 kJ on the machine. How many more calories does he have to burn?

Shutterstock.com/fakoburito

Exercise 3.02

2 Driving a car uses 6.4 kJ/h of energy for every kg of body mass. Stuart's body mass is 85 kg. How many kilojoules of energy will he use during a 6-hour drive from Newcastle to Ballina?

3 On Boxing Day, the traffic on the motorway was very heavy and slow. It took Paba 90 minutes to travel 53 km from Wahroonga to Gosford. Calculate her average speed in km/h correct to the nearest whole number.

Exercise
3.03

4 The total distance, D metres, it takes to stop a car at different speeds on a dry asphalt road can be found by using the formula $D = \dfrac{5Vt}{18}(5t + 2.27)$ where:

Exercise
3.04

V = the car's velocity (speed) in km/h

t = the driver's reaction time in seconds.

a David was distracted by a text message on his phone while driving on a dry asphalt road at 60 km/h. It took him 2.1 seconds to react. Calculate David's stopping distance to the nearest metre.

b How much longer would his stopping distance have been if he had been driving at 100 km/h than at 60 km/h?

5 Ken's weight is 78 kg and he wants to keep his BAC below 0.05. How many standard drinks can he have in 4 hours?

Exercise
3.05

$$BAC_{male} = \frac{10N - 7.5H}{6.8M}$$

4.

EARNING MONEY

Chapter problem

Hugo earns $17.04 per hour for a 37-hour week in his job in a wholesale plant nursery. He is paid time-and-a-half for the first 5 hours of overtime per week and double time after that. If he is required to work any unscheduled overtime, he receives a $10.68 meal allowance per shift. In addition, he receives a $1.60 allowance per hour when he is required to work in wet areas.

In the week ending 29 May, he worked 43 hours, which included one unscheduled overtime shift and 6 hours working in a wet area.

Hugo thinks his pay for the week ending 29 May is wrong. Is his gross pay correct?

Pay slip: Hugo Mendozia	
Week ending 29 May	
Normal pay	$630.48
Overtime	$153.36
Allowances	$20.28
Gross pay	$804.12

CHAPTER OUTLINE

WHAT WILL WE DO IN THIS CHAPTER?

- Calculate wages and salaries
- Calculate overtime payments, annual leave loading, bonuses and allowances
- Determine payments from commission, piecework and royalties
- Calculate government allowances and pensions
- Prepare a household budget

HOW ARE WE EVER GOING TO USE THIS?

- Checking pay is correct
- Checking holiday loading
- Applying for Centrelink payments and checking payments
- Managing a budget

4.01 Wages and salaries

Wages and salaries

A **wage** is an amount paid for each hour worked.

A **salary** is a fixed amount per year that does not depend upon the number of hours worked. Teachers and other professionals earn a salary.

KRISTINE – CHILDCARE WORKER

When I was at school I earned money by babysitting. It was a job and it was fun. When I left school I decided to go into childcare. There are lots of jobs in childcare because most mums need to go to work. I did a Certificate III in Childcare at TAFE and now I have a job I love. The best part of my job is helping children learn.

EXAMPLE 1

Kristine earns $18.23 per hour working in a long-day childcare centre. She works a 38-hour week.

a How much does Kristine earn per fortnight?

b Calculate the amount Kristine earns per year.

Solution

a Calculate Kristine's earnings for 1 week.

Amount Kristine earns per week $= 38 \times \$18.23$
$= \$692.74$

A fortnight is 2 weeks.

Amount Kristine earns per fortnight $= 2 \times \$692.74$
$= \$1385.48$

Each fortnight, Kristine earns $1385.48.

b There are 52 weeks in a year.

Amount Kristine earns per year $= 52 \times \$692.74$
$= \$36\,022.48$

Each year, Kristine earns $36 022.48.

EXAMPLE 2

Madeleine is a social worker. Her annual salary is $59 000. Calculate:

a her monthly pay

b her fortnightly pay.

Solution

a There are 12 months in a year.

'Annual' means per year. ←

Monthly pay = $59 000 ÷ 12

= $4916.67

b There are 26 fortnights in a year.

Fortnightly pay = $59 000 ÷ 26

= $2269.23

> 1 year = 12 months
>
> 1 year = 52 weeks
>
> 1 year = 26 fortnights

Watch out! 1 month is not 4 weeks and 1 year is not 48 weeks. This is a common mistake and it's WRONG.

PROFILE

PAUL – AN AUSTRALIAN VOLUNTEER ABROAD

I completed a building apprenticeship when I left school, but when I finished my trade I wasn't ready to settle down. I wanted to travel and see the world. I saw a news report about volunteers helping in developing countries and I decided to do my bit to help others. The aid agency paid my travel costs and provides my accommodation and the equipment I need. They also pay me a modest stipend. As a volunteer, I'm making a difference in this community.

A **stipend** (pronounced 'sty-pend') is similar to a salary but it is usually for a relatively small amount. People in religious orders and some volunteers receive a stipend.

Exercise 4.01 Wages and salaries

1 Scott is a qualified ambulance paramedic. He is paid $35 per hour for a 38-hour week.

   a How much does Scott earn per week?

   b How much is Scott paid per fortnight?

   c Calculate Scott's annual pay.

2 Suzanne is a solicitor. Her salary is $82 500 p.a. ←   p.a. = per annum = per year

   a How much does Suzanne earn per month?

   b Calculate Suzanne's fortnightly pay.

   c How much does Suzanne earn per week?

3 Lance is paid a salary for being an office IT manager. Each week he earns $1300.

   a Calculate Lance's annual salary.

   b Explain why Lance's monthly pay is *not* $1300 × 4.

   c Divide Lance's annual salary by 12 to calculate his monthly pay.

   d Lance's salary is based on 7 hours work per day, 5 days per week and 52 weeks per year. Calculate the pay rate per hour that is the basis of Lance's salary.

4 Zheng earns $15.61 per hour at a Chinese take-away store.

   a Last week, Zheng worked 16 hours. How much did he earn?

   b Today, Zheng earned $70.25. How long did he work?

5 Ulla receives a yearly stipend of $22 860 from the university to assist her with her postgraduate study and research.

   a How much does Ulla receive per fortnight from the stipend?

   b The stipend isn't enough to cover all of Ulla's living expenses. She also works as a waitress for 4 hours per night, 2 nights per week. She earns $18.20 per hour as a waitress. Calculate Ulla's total fortnightly income.

6 The minimum wage for a trainee pest inspector is $595.70 for a 38-hour week. What is the minimum pay per hour for a trainee pest inspector?

7 Carlos earns $320 per day as a relief teacher. The table shows the number of days he worked during 5 weeks.

How much did Carlos earn over the 5 weeks?

Dates	Number of days worked
30 April – 4 May	2
7 May – 11 May	1
14 May – 18 May	5
21 May – 25 May	3
28 May – 1 June	2

8 Ashok is a casual office worker. He is paid $178 per day irrespective of the number of hours he works. Usually, he works about 12 days per month.

a How much did Ashok earn for working from 8 a.m. to 1 p.m. on Monday?

b During February, Ashok earned $1958. How many days did he work in February?

c The office offers Ashok a permanent 38-hour a week job at $16 per hour. Do you think he should take the permanent job? Why or why not?

9 Tori is trying to decide which one of three jobs to take.

	Conditions	Pay
Job 1	38-hour week, 5 days per week, possibility of overtime	$19/hour
Job 2	75 hours per fortnight, work 9 days per fortnight	$1450 per fortnight
Job 3	Salary, based on a 35-hour week	$38 800 p.a.

a Ignoring any overtime, which job pays the most per year?

b If you were Tori, which job would you take? Why?

Fair Work

INVESTIGATION

AWARD WAGES

Log onto the **Fair Work** website and search for A-Z list of awards.

- Research the minimum wage for three jobs that interest you.
- Calculate the minimum weekly pay in each job.

WAGES BY SPREADSHEET

Ask your teacher to download the 'Wages' spreadsheet from the NelsonNet website.

1 Jessica doesn't work on Mondays or Tuesdays. Each day from Wednesday to Sunday she works from 7 a.m. to 2.30 p.m. and she has an unpaid, 30-minute lunch break. Jessica's wage is $17.60 per hour. Enter the information about Jessica's job in the spreadsheet and determine her total weekly pay.

2 a One of the formulas used in the spreadsheet is **=SUM(F11:F17)**. What is this formula calculating?

b What spreadsheet formula in cell F19 could be used to determine the total amount Jessica is paid each week?

3 The following table shows the hours worked during the first week in February and the corresponding pay rates for the employees in a small office.

Employee	Pay rate per hour	Number of hours worked	Pay
Imran	$12.51	20	
Sofia	$15.25	35	
Cathy	$20.70	35	
Mike	$16.30	40	
Anita	$16.30	32	
		Total wages bill	

a The hours each employee works per week and their hourly rate of pay could change. Construct a spreadsheet that will allow you to calculate each employee's wage and the total office wage bill when the number of hours worked and the rates of pay could change.

b During the second week in February, each employee received a $4 per hour wage increase. Imran worked 32 hours and the other employees worked for the same number of hours as in the previous week. Use the spreadsheet you constructed to determine the total wages bill for the second week in February.

4.02 Working overtime

Overtime

Overtime is working beyond usual working hours or days, and is paid at a higher rate such as 1.5 times the normal pay (**time-and-a-half**) or twice the normal pay (**double time**). Only people who work for a wage are paid for overtime (it doesn't apply to salary earners).

PROFILE

ALYSSA – AN AGED CARE WORKER

I helped my mum look after my grandfather who has dementia when I couldn't get a job after I left school. When we took Pa for a one-week holiday in a respite centre, the centre manager told me that I had the right attitude and I could consider working in the care service industry. I didn't need any special training because the employer provided on-the-job training and, with Australia's ageing population, there are lots of jobs to choose from. The best thing about my job is the variety. I deal with different clients and do different things every day.

iStock.com/PeopleImages

EXAMPLE 3

Alyssa's normal rate of pay is $16.81 per hour. She is paid time-and-a-half on Saturdays and double time on Sundays.

a How much does Alyssa earn per hour on Saturdays?

b Calculate the amount Alyssa will earn for working 4 hours on a Sunday.

Solution

a When she works on a Saturday, Alyssa is paid $1\frac{1}{2}$ times her normal rate.

$$\text{Pay per hour at time-and-a-half} = 1.5 \times \$16.81$$
$$= \$25.22$$

b When Alyssa works on Sunday, she earns $2 \times \$16.81$, or $33.62 per hour.

$$\text{Pay for 4 hours on Sunday} = 4 \times \$33.62$$
$$= \$134.48.$$

> **Time-and-a-half** is $1\frac{1}{2}$ times normal pay.
>
> **Double time** is 2 times normal pay.

Hasid earns $18 per hour.

a How much will he earn for working a 35-hour week?

b When Hasid works for more than 7 hours per day he is paid overtime. For the first 3 hours he works overtime he is paid time-and-a-half. Any additional overtime hours are paid at double time. How much will Hasid earn for working 12 hours in one day?

Solution

a Multiply hourly rate by 35.

Pay for a 35-hour week = $35 \times \$18$

$= \$630$

b When Hasid works for 12 hours in one day, his time is broken into 7 hours normal + 3 hours at time-and-a-half + 2 hours at double time.

Pay = $7 \times \$18 + 3 \times 1.5 \times \$18 + 2 \times 2 \times \$18$

$= \$279$

Hasid's pay for a 12-hour day is $279.

Exercise 4.02 Working overtime

1 Complete the missing values in the table.

Normal pay per hour	Pay per hour at time-and-a-half	Pay per hour at double time
$17.20	a	b
$14.36	c	d
$24.60	e	f
$31.25	g	h

2 Jenny's normal pay is $16.40 per hour. How much will she earn when she works 5 hours at time-and-a-half?

3 Mike's normal pay is $15.30. How much will he earn when he works for 4 hours on a Sunday at double time?

4 How much will Sancia earn when she works 3 hours at time-and-a-half and 4 hours at double time? Her normal pay is $21.80 per hour.

5 Tuan is a plumber's assistant. He works a 35-hour week at $22.80 per hour. His overtime is paid at time-and-a-half for the first 5 hours overtime in a week and double time for any hours after that. This week, Tuan worked 42 hours.

   a How many hours did Tuan work at double time?

   b Calculate Tuan's pay for the week.

6 Mercia has a holiday job supervising children in a resort. She earns $19 per hour Monday to Friday, time-and-a-half on Saturday and double time on Sunday.

   a The table shows the times Mercia worked during one week in January. What are the missing values in the table?

Day	Hours worked	Pay rate per hour	Pay
Weekdays	21	i	iv
Saturday	4	ii	v
Sunday	6	iii	vi

   b Calculate Mercia's pay for the week.

7 a Casey earned $108 when he worked for 3 hours at double time. What is Casey's normal pay per hour?

   b How much will Casey earn when he works for 3 hours at time-and-a-half?

8 Elise earned $120 when she worked on Sunday for 4 hours at double time. How much does Elise earn for a normal 7-hour day?

9 Callum works for the council. He looks after the grass in parks and at sporting venues. Callum doesn't work any overtime on Monday to Friday. All of the hours he works on Saturday are paid at time-and-a-half and Sunday work is at double time. For Callum's time and pay sheet below, find the values of **a**, **b**, **c**, **d**, **e** and **f**.

Time and pay calculation sheet				
Callum			Normal pay per hour	$16.90
Day	**Start**	**Finish**	**Unpaid breaks**	**Pay**
Monday	7 a.m.	3:30 p.m.	30 minutes	a
Tuesday	b	5:30 p.m.	1 hour	$152.10
Wednesday	–	–		
Thursday	8 a.m.	4 p.m.	c	$118.30
Friday	6:30 a.m.	d	1 hour	$135.20
Saturday	7 a.m.	11 a.m.	Nil	e
Sunday	8 a.m.	f	Nil	$101.40

4.03 Bonuses and allowances

Some jobs include **allowances** for doing unpleasant work, for working under difficult conditions, or to cover expenses such as uniform and travel.

Some jobs pay **bonuses** (extra pay) for doing good work, meeting targets or deadlines.

> **PROFILE**
>
> **KAITLYN** – COOK IN THE AUSTRALIAN NAVY
>
> I joined the navy because I didn't want a 9-to-5 job and I wanted to travel. I get good pay and conditions as well as job security. I'm a fully qualified cook and the navy provided all my training and arranged my TAFE qualifications. I've got good friends in the navy and I've been around the world. I was surprised at the variety of jobs in the navy; jobs I'd never considered like being a waiter or a chaplain. The navy even has permanent jobs for musicians in the navy bands!

EXAMPLE 5

Kaitlyn's basic salary in the navy is $43 434 and she receives an annual $12 128 service allowance as well as an annual $419 uniform maintenance allowance. When she's at sea she receives an additional $11 758 annually.

a Calculate Kaitlyn's weekly pay when she is working on land.

b How much does Kaitlyn earn per fortnight when she's at sea?

Solution

a Kaitlyn's annual salary on land = basic salary + service allowance + uniform allowance.	Salary = $43 434 + $12 128 + $419 $$= \$55\ 981$$
Divide by 52 for weekly pay.	Weekly pay on land = $55 981 ÷ 52 $$\approx \$1076.56$$
b Kaitlyn's annual pay at sea = basic salary + service allowance + uniform allowance + sea allowance.	Salary = $43 434 + $12 128 + $419 + $11 758 $$= \$67\ 739$$
Divide by 26 for fortnightly pay.	Fortnightly pay at sea = $67 739 ÷ 26 $$\approx \$2605.35$$

ISBN 9780170413503

EXAMPLE 6

Sonia is paid $15.48 per hour for her work as a security guard. Each week, she receives an additional $61.05 for her guard dog and $6.75 for her torch. She receives $14.15 per shift travel allowance.

Sonia works a 4-hour shift, 6 nights per week. How much is she paid per week?

Solution

Sonia's total weekly pay = wages + allowances + dog + torch

Wages = $4 \times 6 \times \$15.48$

$= \$371.52$

Travel allowance = $6 \times \$14.15$

$= \$84.90$

Sonia's total weekly pay = $\$371.52 + \$84.90$
$+ \$61.05 + \$6.75$

$= \$524.22$

Exercise 4.03 Bonuses and allowances

1 Tristan's base salary as an air force trainee is $37 485 p.a. In addition, he receives the Australian Defence Force annual allowance of $12 128 and an annual $419 uniform allowance. He also receives $9531 p.a. when he is deployed overseas.

  a Calculate Tristan's weekly pay when he is working in Australia.

  b Determine Tristan's fortnightly pay when he is deployed overseas.

2 Zoran works for a pest control company. He is paid $14.93 per hour and receives an extra $2.81 per day for handling poisons. Zoran works for 8 hours per day, 5 days per week. Calculate his weekly pay.

3 Ryan earns $721 per week as a mobile mechanic. In addition, he receives $29 per week for work-related mobile phone calls and $0.60 per kilometre for work-related travel. Calculate Ryan's pay for a week in which he drove 420 km in his truck for work.

4 Kate is the manager of a fast food chain. She is paid $28 per hour for a 35-hour week plus $8.30 per week laundry allowance. She receives a $30 bonus for every accident-free week at the shop and another $95 bonus per week if the shop makes $100 000 or more in weekly sales.

Last week, the shop was accident-free and sales were $110 000. How much was Kate paid last week?

5 Zack drives a furniture removal truck. He is paid $15.12 per hour Monday to Friday, time-and-a-quarter on Saturday and double time on Sunday. In addition, he receives a flat fee of $12.59 per day for handling heavy furniture. Calculate Zack's pay for a week when he delivered heavy furniture for 33 hours Monday to Friday, 6 hours on Saturday and 3 hours on Sunday.

6 Raina has a job driving disabled children to school. She is paid $16.20 per hour plus $3.65 per day for assisting children. In addition, she receives 65 cents for every work-related kilometre she drives in her car. Calculate Raina's pay for a week when she worked 4 hours each day from Monday to Friday and she used her car for 360 work-related kilometres.

7 Sam is a casual junior baker at a hot bread shop. A casual junior baker earns $12.32 per hour. From midnight Friday to midnight Saturday all bakers receive their normal pay plus 50%. From midnight on Saturday to midnight on Sunday casual bakers receive 98% more than their normal pay per hour.

    a   The table shows the times Sam worked last week. Find the missing values **i** to **xii**.

Shift	Starting time	Finishing time	Unpaid breaks	Number of hours worked	Pay per hour	Pay
1	Thursday 10 p.m.	Friday 6:30 a.m.	30 minutes	i	v	ix
2	Saturday midnight	8 a.m.	1 hour	ii	vi	x
3	Saturday 8 p.m.	Midnight	0	iii	vii	xi
4	Sunday 6:30 p.m.	Midnight	30 minutes	iv	viii	xii

    b   Calculate Sam's total pay.

INVESTIGATION

MY FUTURE CAREER

Earning an income can occupy a lot of your time, so it's important to find a job that you are going to enjoy. In this investigation, you are going to complete some online questionnaires to help you determine the type of occupation that suits your skills and interests.

My Future

1 Visit the **My Future** website.

2 You will need to 'Sign up' as a new user in order to enter the website, and then log in each time you use the site. Remember your password.

3 In the 'My career profile' section of the website there are some questionnaires. Complete a questionnaire, then explore the careers that the website suggests might interest you in the 'Career insight' or 'Occupations' sections of the website. You may be unfamiliar with some of the careers to which you may be suited. Take the time to learn about these careers. It could be the best hour you ever spend!

ISBN 9780170413503

4.04 Annual leave loading

Annual leave loading or **holiday loading** is an extra payment to employees given at the start of their holidays. It is usually calculated as $17\frac{1}{2}\%$ of 4 weeks pay.

Percentage
calculations

Mental percentages

EXAMPLE 7

Briana earns $750 per week as a vet nurse. When she takes her 4 weeks annual holiday she receives an extra $17\frac{1}{2}\%$ of 4 weeks pay as holiday loading in addition to her normal 4 weeks pay.

a Calculate Briana's holiday loading.

b Determine the total value of Briana's holiday pay.

Solution

a Brianna receives $17\frac{1}{2}\%$ of 4 weeks pay.

$17\frac{1}{2}\% = 0.175$

Pay for 4 weeks $= \$750 \times 4$

$= \$3000$

Briana's holiday loading $= 0.175 \times \$3000$

$= \$525$

b Holiday pay = 4 weeks pay + holiday loading

Briana's holiday pay $= \$3000 + \$525$

$= \$3525$

Exercise 4.04 Annual leave loading

1 Calculate $17\frac{1}{2}\%$ of each amount.

   a $350   **b** $1264   **c** $3325   **d** $6895

2 The Edmondson Park Motel pays its employees a $17\frac{1}{2}\%$ holiday loading on their 4 weeks annual leave.

   a   Vicki, the chef, earns $695 per week. Calculate her annual leave loading.

   b   James earns $565 per week as a barman at the hotel. How much will James be paid for his 4-week annual holiday?

3 Angus works part time and he earns $743 per week. For his holidays he receives a loading of $17\frac{1}{2}\%$ of 4 weeks pay. Calculate the total value of his 4-week holiday pay.

4 As a result of a wage claim based on an increase in the cost of living, all workers were granted a 4.2% increase in their pay.

   a   Liam works in data processing and he earns a salary of $58 200 p.a. Calculate his new salary.

   b   How much will Liam be paid for 4 weeks work after the wage rise?

   c   Calculate Liam's new annual leave loading.

5 Phillipa's annual salary is $72 320. She receives a loading of $17\frac{1}{2}\%$ of 4 weeks pay with her holiday pay. Calculate the total value of Phillipa's holiday pay.

6 Linda earns $890 per week. She receives 6 weeks holidays at the end of each year, but her leave loading is only $17\frac{1}{2}\%$ of 4 weeks pay. Calculate Linda's holiday pay.

7 Jon's wage increased from $620 to $700 per week. By how much will his 4-week holiday pay, including $17\frac{1}{2}\%$ loading, increase?

8 P.L. Insurance had a very successful year. In addition to the normal $17\frac{1}{2}\%$ annual leave loading, they decided to pay their employees a 'thank-you' bonus based on the number of years of service. They paid this bonus at the same time as the loading.

Years of service	Bonus as a percentage of annual salary
1–5	0.4%
6–8	0.65%
Over 8	0.9%

Katrina is paid $2152 per fortnight and she has worked for the company for 7 years.

   a   How much is Katrina's bonus?

   b   Calculate the total amount Katrina was paid, before tax, for her 4 weeks holiday, including the bonus.

9 Nate earns $640 per week. At the end of the year, he receives 5 weeks holiday with a $17\frac{1}{2}\%$ loading on 4 weeks. How much more does Nate get paid for taking 5 weeks holidays than for working 5 weeks?

10 Ask your teacher to download the 'Holiday pay' spreadsheet from NelsonNet to answer this question.

Holiday pay

a Yasmin earns $11.25 per hour for a 35-hour week. If the annual leave loading increases from $17\frac{1}{2}\%$ to $22\frac{1}{2}\%$ of 4 weeks pay, by how much will Yasmin's 4-week holiday pay increase?

b What spreadsheet formula could be used to determine the value of the following amounts?

   i normal pay for 1 week

   ii normal pay for 4 weeks

   iii leave loading for 4 weeks

4.05 Commission, piecework and royalties

Salespeople are often paid by **commission**, which is a percentage of the value of the items they've sold.

Piecework is a type of work where a person is paid per item produced or processed.

A **royalty** is a payment to an author, singer or artist for each copy of their work sold. Usually, a royalty is a percentage of the total sales amount.

Percentage calculations

Mental percentages

PROFILE

SARINA – SYDNEY ARTIST

It's great to be able to work doing something you love, but it hasn't always been easy. In the beginning, as an unknown artist, I had to work nights as a waitress because my art didn't pay very much. If I sold the same painting now as I did 10 years ago I'd get a lot more money for it. My current work involves projects where people ask me to paint something specific, and my agent negotiates a fee. I also earn royalties when my artwork is used on commercial items such as placemats, drink coasters, plates, cups, serviettes, calendars and T-shirts.

You can view some of Sarina's work at www.sarina.net.

Photo: Sue Thomson

EXAMPLE 8

Sarina receives a 5% royalty on the wholesale price of serviettes featuring her art. Packets of serviettes wholesale for $3.20 each and sell for $5.80. Calculate Sarina's royalty for the sale of 8000 packets of serviettes.

Solution

Find the value of the serviettes sold.	Wholesale value = $8000 \times \$3.20 = \$25\,600$
Calculate the royalty.	Sarina's royalty = 5% of $25 600
	$- \$1280$

EXAMPLE 9

Jordan is a used car salesman. He is paid a $170 monthly retainer plus 5% commission on his monthly sales over $50 000. Calculate his pay for a month when his sales totaled $80 000.

A **retainer** is the amount of money a salesperson is paid that does not depend on his sales.

Solution

Find the value of the sales for which he is paid commission.	Sales over $50 000 = $80 000 − $50 000
	= $30 000
Calculate the commission.	Commission = 5% of $30 000
	= $1500
Total earnings = retainer + commission.	Total earnings = $170 + $1500
	= $1670

EXAMPLE 10

Danielle earns commission for selling cosmetics at the following rates:

Commission on monthly sales	
First $1000 of sales	5%
On the next $2000	4%
Remainder of sales	3.5%

These different rates are sometimes called a 'sliding scale'.

This month, Danielle's sales totalled $5200. Calculate her commission.

Solution

Commission on the first $1000 = 0.05 × $1000

$$= \$50$$

Commission on the next $2000 = 0.04 × $2000

$$= \$80$$

Danielle's remaining sales = $5200 − $1000 − $2000

$$= \$2200$$

Commission on the remaining $2200 = 0.035 × $2200

$$= \$77$$

Danielle's total commission = $50 + $80 + $77

$$= \$207$$

Exercise 4.05 Commission, piecework and royalties

1 Calculate each percentage amount.
  a 9% of $25 000
  b 5% of $800
  c 2% of $300 000
  d $2\frac{1}{2}$% of $500 000
  e $3\frac{3}{4}$% of $175 200
  f 0.95% of $60 000

2 Marco earns 7% commission on all his sales. Find his commission on a sale of $1675.

3 Sarina receives a 3% royalty on the wholesale price of calendars featuring her art. How much royalty will she receive for 15 500 calendars with a wholesale price of $9.90?

4 Assam sells window shutters and is paid a retainer of $120 per week to cover his expenses, and a commission of 15% of all sales he makes. Assam's sales for the first week in April totalled $2896. Calculate his pay for that week.

5 In her job as a real estate agent, Pauline is paid a retainer of $600 per month plus a commission of 2% of her sales over $800 000. How much did Pauline earn for a month when her sales totalled $1 300 000?

Example
10

6 Tanika sells cosmetics. She earns commission at the following rates.

Calculate Tanika's commission for each of the following monthly sales figures.

Commission on Tanika's monthly sales	
First $500 of sales	5%
On the next $1000	4%
Remainder of sales	3.5%

a $360   **b** $1400   **c** $4200

7 Emily earns monthly commissions when she sells perfumes according to these rates.

Calculate Emily's commission in a month when her total sales were valued at:

Monthly sales	Commission
$800	5% of sales
$801 to $1200	$40 plus 4.5% of sales over $800
$1201 and over	$58 plus 4% of sales over $1200

a $ 360   **b** $998   **c** $5100.

8 Monique has hired Trevor to re-tile her home. Trevor told Monique she required 6 m² of tiles for the kitchen walls, 48 m² of slate for the lounge-room floor and 7 m² of slate for the stairs. How much will Trevor charge to lay the tiles and slate?

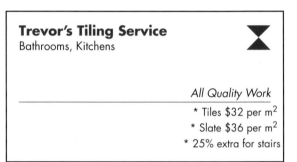

Trevor's Tiling Service
Bathrooms, Kitchens

All Quality Work
* Tiles $32 per m²
* Slate $36 per m²
* 25% extra for stairs

9 Holly enjoys cooking scones which she sells at the local Devonshire Tea shop. She buys her ingredients in bulk and it costs her $12 to make 5 dozen scones in 2 hours. She values her labour at $16 per hour. How much should Holly charge the Devonshire Tea shop for 10 dozen scones?

1 dozen = 12

10 Basam is selling his house for $420 000. The real estate agent's commission is 2% on the first $200 000 and 1.5% on the balance of the sale price. How much will Basam receive from the sale of his house?

11 Magda writes books that sell for $56 each. Each year, she receives 10% royalty on the first 4000 copies and 12.5% royalty on the remaining copies of sales of her books. Calculate the royalty Magda receives in a year when:

a 3650 copies of her books are sold

b 7000 copies of her books are sold.

 ISBN 9780170413503

12 Milan puts letters in envelopes and attaches postage stamps for a marketing company. He is paid 24 cents per letter. Milan can process 70 letters per hour.

   a How much does Milan earn per hour?

   b How much will he earn for processing 260 letters?

   c How many letters does Milan need to process in order to earn over $100?

13 Renuta is a self-employed antique furniture auctioneer. On every item she sells she charges 15% commission on the first $2000 of the sale price and 12.5% of the amount over $2000. How much will Renuta charge for selling an antique dining room suite that sold for $22 600?

14 The table shows the rates of royalty Sarina receives when her art is used on plates. The wholesale price of the plates is $3.40.

Number of plates sold	Royalty rate
First 2000	4% of the wholesale price
From 2001 to 10 000	$272 plus 3% of the wholesale price for the number of plates sold in excess of 2000
10 001 or more	$1088 plus 2.5% of the wholesale price for the number of plates sold in excess of 10 000

Calculate Sarina's royalty for the sale of the following numbers of plates.

   a 1500   **b** 8600   **c** 25 000

4.06 Government allowances and pensions

Earning money

PROFILE

BEN – STUDENT

I have to look after my mum because there's just the two of us and she's got bipolar 1 disorder. When she takes her medication she's OK, but I still need to do the shopping, cook our food and do the housework. It's really hard when I've got assignments and assessments, especially if Mum's unwell. I can't get a job because I have no spare time. Centrelink gives me a carer's pension and a carer's allowance. Mum receives sickness benefits and rent assistance.

If you or someone you know is caring for someone, visit the Disability and Carers page on the Australian Government Department of Social Services website to access help for carers. The site provides information about support and assistance in your local area.

Department of Social Services

EXAMPLE 11

To receive a pension, you need to pass an income and assets test. Joan is a single, aged pensioner who owns her home. She is allowed to have assets up to $250 000 and retain a full pension of $808.30 per fortnight. For every $1000 over $250 000 in assets, her fortnightly pension reduces by $3. At present, Joan's assets are valued at $198 000 and she is about to inherit $75 000. What effect will the inheritance have on her pension?

Solution

Find Joan's assets after her inheritance.

Joan's assets = $198 000 + $75 000

= $273 000

Find the value of Joan's assets over $250 000.

Amount = $273 000 − $250 000

= $23 000

For every $1000 over $250 000, Joan's pension reduces by $3.

$23 000 ÷ $1000 = 23

Joan's pension = $808.30 − 23 × $3.00

= $739.30

Inheriting $75 000 will decrease Joan's pension to $739.30 per fortnight.

Exercise 4.06 Government allowances and pensions

1 Cameron receives $244 per fortnight in youth allowance. He pays his mother $60 per week for board and his fortnightly public transport costs to travel to TAFE are $42. How much of his Youth Allowance is left each fortnight after Cameron pays his board and transport expenses?

2 The maximum fortnightly age pension payable to eligible people over the age of 65 is $808.30 for singles and $609.30 per person for couples. How much less does a couple receive per fortnight than two single people sharing accommodation?

3 Senior health care cardholders and pensioners are entitled to a government telephone payment. Every 3 months they receive a $28.20 phone allowance. Calculate the annual value of the phone allowance.

NCM 11. Mathematics Standard (Pathway 1)   ISBN 9780170413503

4 Jim receives a disability pension and he lives in a public housing, rent-subsidised unit. Jim has to pay 15% of his pension in rent. His pension is $609.30 per fortnight. How much rent does Jim pay per year?

5 Gail receives a disability pension because she is too sick to work. Her fortnightly pension is $797.90.

   a Calculate Gail's annual pension.

   b Gail's pension includes $6.20 per fortnight for medications and she pays $6.30 per prescription medicine. Gail takes a lot of medication, but after she has paid for 58 prescriptions per year, all further prescriptions are provided free. How much more than her fortnightly medication allowance does Gail have to pay for her medications each year?

6 The age pension payment for a single person is $808.30 but for every dollar of income they receive over $164 per fortnight the pension reduces by 50 cents.

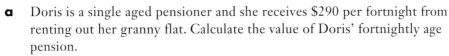

   a Doris is a single aged pensioner and she receives $290 per fortnight from renting out her granny flat. Calculate the value of Doris' fortnightly age pension.

   b Nanna thinks Centrelink is overpaying her for the age pension. For the last 8 weeks she has been receiving $520 per fortnight pension and she also receives a fortnightly income of $1070 from her investments. She knows that Centrelink will make her repay any overpayments she receives. How much will Nanna have to repay Centrelink?

7 All of the residents in a block of units are aged pensioners: 17 of them are single and 26 are couples. Use the information in Questions **2** and **3** to calculate the total maximum annual age pension they receive and the total annual phone allowance.

8 Bill and Rose are a couple both aged 20 years. Bill's fortnightly taxable income is $721.70 but Rose is unable to work and she receives a disability support pension of $201.35 per fortnight. Recently, Rose received a letter from Centrelink informing her that for the last year she has been overpaid. The details included in the letter are:

   - Pension received: $5235.10

   - Entitled: $3614.25

   - Overpaid amount: $1620.85.

The maximum, fortnightly disability pension for a member of a couple aged under 21 years is $562.20 and this amount reduces by 50% of the couple's fortnightly taxable income.

Are the details included in the letter correct? Justify your answer.

Centrelink

FINANCIAL SUPPORT THROUGH CENTRELINK

The Australian Government provides financial support and assistance to Australian citizens through social security. The staff at Centrelink are trained to assist Australians when they are in need.

In this activity, you are going to find out about your social security entitlements.

1 Go to the Centrelink website.

2 Click on the appropriate option to research the answers to the following questions.

3 a What is the maximum fortnightly Youth allowance Centrelink will pay a person who is under 18 years of age and living at home?

b What are 'approved activities'?

c Do you have to be involved in approved activities to receive a Youth allowance?

d What other assistance can someone receiving a Youth allowance receive?

e Is there an income or an asset test for the Youth allowance?

4 a If you are single with no dependent children, what is the maximum amount of rent assistance you can receive from Centrelink per week?

b If you are single and share a residence but have no dependent children, can you get rent assistance if your fortnightly rent is $190? If so, how much?

5 a Who can receive an ABSTUDY payment?

b Jake is an 18-year-old student who is going to live away from home to study at university. He is qualified to receive an ABSTUDY payment. How much will his fortnightly ABSTUDY payment be?

c Can Jake get any money from Centrelink in advance of his ABSTUDY payments? Explain your answer.

6 Melinda's grandmother died in her sleep while Melinda was living with her. Use the Centrelink website to find out what Melinda should do first.

4.07 Budgeting

Have you ever wondered what happens to your money? It's a good idea to have a plan so that you don't waste it. A **budget** lists your expected income and expenses, and can help you to manage your money.

Income covers all the money you might earn.

Expenses covers all the ways you might spend that money, and there are two types:

- **Fixed expenses** are costs that are essential and must be paid. Some are the same amount each time, such as rent. Others aren't always the same, such as food.

- **Discretionary expenses** are amounts that you often spend but which aren't essential, such as entertainment or magazines.

A budget needs to balance your income and expenses so that you have enough money for everything you require and some left over to save for special items, such as a car or holiday.

EXAMPLE 12

Ashleigh works part-time while studying. She receives an allowance from her parents of $100 per week and she earns $120 from her job. She pays $80 per week in rent and spends $30 per week on food. She averages $10 per week for her mobile phone and $20 per week on clothes. She divides the remainder equally between entertainment and savings.

a Create a budget for Ashleigh for a week.

b Ashleigh's rent is increased by $25 per week. How would she need to adjust her budget for this increased expense?

Solution

a List income and expenses and make sure total expenses equals total income.

Income		Expenses	
Allowance	$100	Rent	$80
Earnings	$120	Food	$30
		Mobile phone	$10
		Clothes	$20
		Entertainment	$40
		Savings	$40
Total income	$220	**Total expenses**	$220

Total of fixed expenses = $80 + $30 + $10 + $20 = $140

Remainder available for entertainment and savings = $220 − $140 = $80

Divided equally = $80 ÷ 2 = $40

b The $25 increase in rent means Ashleigh now has $25 less to spend on entertainment and savings. She has $80 − $25 = $55 to divide between entertainment and savings. She could still divide this amount equally between the two ($27.50 each) or she could spend less on entertainment to continue to save $40 per week.

Exercise 4.07 Budgeting

1 Lily owns a car and has the following expenses each year:
- registration $349
- Compulsory Third Party (CTP) insurance $795
- comprehensive insurance $1110
- maintenance bills of $790.

She spends $53 per week on petrol.

a How much does Lily spend on her car each year?

b How much should she set aside in her weekly budget to cover her car expenses?

2 Mitchell works in an office during the week and in the bar at the local club on weekends. He earns $620 per week from the office job and $215 from the club. He pays $280 per week in rent and spends an average of $60 per week on food. His smartphone costs him $20 per week and travel expenses are $60 per week. The remainder of his income has to be divided between entertainment, clothes and savings.

a Create a budget for Mitchell for one week.

b Mitchell is considering buying a car. He would no longer have public transport expenses but he would need to allow $100 per week to pay off a loan and $75 for car expenses. Create a new budget for Mitchell.

3 Marko is an apprentice mechanic. His take-home pay is $790 per week. This table shows his weekly expenses.

a How much does Marko save each week?

b Calculate his net annual income.

c How much is Marko able to save each year?

Item	Amount
Board	$120
Mobile phone	$21
Clothes	$65
Car	$112
Entertainment	$72
Other expenses	$88
Savings	
Total	$790

4 This table shows Shania's monthly budget.

Income		Expenses	
Part-time job	$290	Clothes	$140
Babysitting	$130	School needs	$32
		Entertainment	$50
		Mobile phone	$55
		Fares	$23

a Calculate Shania's monthly income and expenses.

b Calculate the amount she is able to save each year.

c Shania would like to increase her savings so that she can go on an end-of-year holiday. Suggest three ways she could do this.

5 Sanjeev has taken a second job to save for a new car. His budget for a week is shown below.

Income		Expenses	
Main job	$750	Rent	$225
Second job		Travel	$56
		Food	$117
		Clothes	$55
		Entertainment	$75
		Bills	$157
		Savings	
Total	$908	Total	$908

a Calculate how much Sanjeev earns from his second job.

b Calculate how much he can save each week.

c If the car costs $25 000, how long would it take Sanjeev to save this money?

d Suggest ways Sanjeev could save more per week so that he can buy his car sooner. Create a new budget for Sanjeev.

MY BUDGET

In this investigation, you are going to prepare two budgets: one for a typical school leaver and one for yourself based on your own choice of job when you leave school.

Part A: Typical school leaver

INCOME:

- You have finished school.

- After school, you enrolled in a TAFE course for a year.

- You are now working and earn $27 040 p.a.

- You don't have a partner or children, but you need to move out of home.

- Your new home will be 4 km from where you work (fast 50-minute walk).

1 Download the 'My budget' worksheet from NelsonNet which contains the **Weekly tax tables**, **Budget guidelines** and the **Lifestyle costs** provided for this activity.

2 Copy and complete this budget form based on the information, or use the worksheet.

Weekly income		$
Gross weekly income	Equals yearly wage ÷ 52	
Deduct tax (weekly)	From table (tax-free threshold, no leave loading)	
Net weekly income		
Regular weekly expenses		**$**
Housing	Mortgage, rent or share	
Transport	Car running costs, registration, CTP insurance etc. or train, bike or walking	
Personal spending	Clothing	
	Hair, grooming, cosmetics etc.	
Food	Groceries, including pet food	
Utilities	Phone connection + landline calls, mobile phone expenses	
	Electricity, gas	

My budget

Total regular expenses		$
Discretionary weekly expenses		$
Insurance	Home and/or contents	
Insurance	Car – comprehensive	
Insurance	Health	
Entertainment	Pay TV, books, magazines, music, movies etc.	
Recreation	Sport, holidays	
Technology	Internet, laptop	
Total discretionary expenses		$
Net weekly income		
Total weekly expenses (regular + discretionary)		
BALANCE		$

Part B: You in your preferred occupation

Now you will repeat the activity in Part A but to create a budget based on your chosen occupation.

- The aim of a good budget is not simply to maximise savings.

- It is more to do with providing cost amounts that reflect the sort of lifestyle you want and ensuring these are within the income you have at your disposal.

- Your budget costing needs to cover the costs of the essentials of living (e.g. regular expenses such as food, shelter).

- For discretionary expenses, weekly amounts should allow you to afford the extra activities you wish to pursue, whether it is entertainment, holidays, fashion, or a more expensive car.

- You also need to consider carefully, within your budget, the pros and cons of taking out insurance of various types and explain the decisions you make.

- Consider carefully the advantages and disadvantages of taking out different types of insurance, whether you can afford them and explain the decisions you make.

- Assume you are single and have bought or are renting your new home, located 4 km from where you work (fast 50-minute walk).

1 Write your preferred occupation and determine the gross wage for that occupation as follows.

- Go to the My Future website.

- Click on **Occupations**, then find your occupation.

- Select **Prospects**, then **Full time weekly earnings**.

- Write the **Weekly income** into your budget.

My Future

2 Find the tax to be deducted from gross pay, using ATO *weekly* tax tables. These can be found on the 'My budget' worksheet or downloaded from the Australian Tax Office website and searching for the table of 'Weekly withholding amounts'.

3 Regular income and Weekly expenses (regular) *must* have an amount in the last column.

4 Weekly expenses (discretionary) can be left blank as a means of saving money, but you must explain and justify your decision.

5 For this exercise, you are *not* to make up your own figures; you *must* use the prices listed in the **Budget guidelines** and **Lifestyle costs**.

6 Some of the prices/costs are per week, others are per fortnight, per month, or per year. You must convert *all* amounts to a weekly cost.

For *every cost category* (both regular and discretionary), you will need to show:

- a short explanation of what choice you have made and why;

- the calculations/working you have done to arrive at the amount you have used.

KEYWORD ACTIVITY

WORD MATCH

Match each word in the first column to its correct meaning in the second column.

Word		**Meaning**
1 Allowance	**A**	Yearly
2 Annual leave loading	**B**	Pay based on the number of hours worked
3 Bonus	**C**	Pay based on the number of items made or processed
4 Double time		
5 Income	**D**	1.5 times the normal rate of pay
6 Overtime	**E**	A payment to authors, artists or others who create items
7 Per annum (p.a.)		
8 Piecework	**F**	Extra amount paid for holidays, usually 17.5% of 4 weeks pay
9 Royalty	**G**	Extra pay for doing good work
10 Salary	**H**	A fixed amount paid per year
11 Time-and-a-half	**I**	Twice the normal rate of pay
12 Wage	**J**	Additional payment for work under difficult conditions or for doing unpleasant tasks
	K	Money that is received or gained, usually regularly
	L	Working more hours per day or week than normally

SOLUTION ^{TO}_{THE} CHAPTER PROBLEM

Problem

Hugo earns $17.04 per hour for a 37-hour week in his job in
a wholesale plant nursery. He is paid time-and-a-half for the
first 5 hours of overtime per week and double time after that.
If he is required to work any unscheduled overtime, he receives
a $10.68 meal allowance per shift. In addition, he receives
a $1.60 allowance per hour when he is required to work
in wet areas.

Pay slip: Hugo Mendozia	
Week ending 29 May	
Normal pay	$630.48
Overtime	$153.36
Allowances	$20.28
Gross pay	$804.12

In the week ending 29 May, he worked 43 hours, which included one unscheduled
overtime shift and 6 hours working in a wet area.

Hugo thinks his pay for the week ending 29 May is wrong. Is his gross pay correct?

Solution

Hugo worked 37 hours normal time, 5 hours overtime at time-and-a-half and 1 hour at
double time.

Normal pay	$37 \times \$17.04$	$630.48
Time-and-a-half	$5 \times 1.5 \times \$17.04$	$127.80
Double time	$1 \times 2 \times \$17.04$	$34.08
Total overtime	$127.80 + $34.08	$161.88
Wet area allowance	$6 \times \$1.60$	$9.60
Meal allowance		$10.68
Total allowances	$9.60 + $10.68	$20.28
Gross pay	$630.48 + $161.88 + $20.28	$812.64

The calculations for Hugo's normal pay and allowances are correct, but the
overtime calculation is wrong. Hugo has been underpaid by $161.88 − $153.36 = $8.52
(or $812.64 − $804.12 = $8.52).

- What things did you learn in this chapter?
- How do members of your family earn or make money?
- Do you know anybody who works for commission or royalties?
- Did you have any difficulties with the content of this chapter? If so, ask your teacher for assistance.

Copy and complete this mind map of the topic, adding detail to its branches and using pictures, symbols and colour where needed. Ask your teacher to check your work.

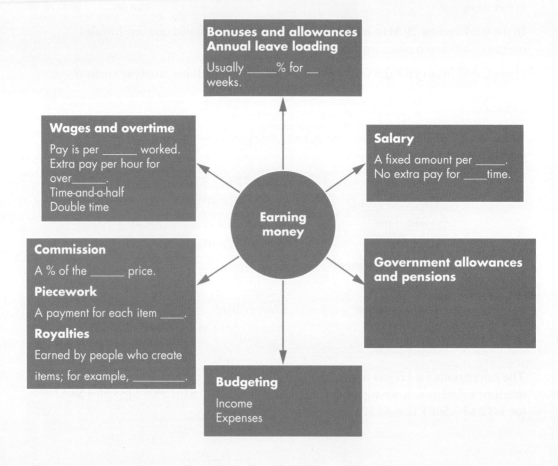

Bonuses and allowances
Annual leave loading
Usually _____% for __ weeks.

Wages and overtime
Pay is per _____ worked.
Extra pay per hour for over_____.
Time-and-a-half
Double time

Salary
A fixed amount per _____.
No extra pay for ____time.

Earning money

Commission
A % of the _____ price.
Piecework
A payment for each item ____.
Royalties
Earned by people who create items; for example, _____.

Government allowances and pensions

Budgeting
Income
Expenses

4. TEST YOURSELF

1 Shelly earns $18.75 per hour for a 35-hour week.

   a How much does Shelly earn per week?

   b Calculate the amount she earns per fortnight.

   c What is Shelly's annual pay?

Exercise 4.01

2 Marcus is an accountant. His annual salary is $96 000.

   a How much does Marcus earn per month?

   b Calculate Marcus' fortnightly pay.

Exercise 4.01

3 Suyin works in an aged-care facility. Her normal pay is $20.24 per hour.

   a How much does she earn per hour when she works at time-and-a-half?

   b When Suyin works the night shift on weekends she is paid double time. Calculate her pay for working 7 hours on the late shift on a Saturday night.

Exercise 4.02

4 George earned $168 for working 4 hours at double time. Calculate George's normal rate of pay.

Exercise 4.02

5 Jamie is a plumber's assistant. His normal pay is $15.65 per hour but when he has to work in wet or muddy conditions he receives a $2.10 allowance per hour. Jamie worked 7 hours today and for three of the hours he was digging in mud. Calculate Jamie's pay for today.

Exercise 4.03

6 Zoe is a plumber. For emergency late night call-outs, she charges a call-out fee of $500 and labour at $125 per hour. Last night, she attended an emergency call at 2 a.m. Calculate the amount she charges for working for 3 hours to fix the problem.

Exercise 4.03

7 Nuala is employed as an early childhood music teacher. Her normal weekly pay is $1325. Calculate her holiday pay for a 4-week holiday including a 17.5% leave loading.

Exercise 4.04

8 Izak is an artist whose work features on fun park admission tickets. He receives 5% commission on the sale of each ticket. How much commission will Izak receive from the sale of 18 000 tickets priced at $45 each?

Exercise 4.05

9 Luke lays tiles for a living. He charges $46/square metre of tiles he lays. How much will Luke charge for laying 17 square metres of tiles?

Exercise 4.05

10 The maximum age pension for a single person is $808.30 per fortnight. Calculate the maximum annual pension for a single person.

Exercise 4.06

11 Every January, Julie has to register and insure her car. Each year, she has trouble paying these expenses because she is short of money after Christmas. This year, her January car registration and insurance costs totaled $2040. What do you recommend she do this year to be ready for these expenses next January?

Exercise 4.07

5.

PAYING TAX

Chapter problem

Hamish is preparing his income tax return. The annual payment summary shown below is for his day job only. Hamish also earned $3640 and paid $1040 tax in his part-time job. He calculated that he has $460 in allowable tax deductions.

Will Hamish receive a tax refund?

PAYG payment summary		
Employee	Hamish Donald	
Employment period	1 July 2017–30 June 2018	
	Total income	$48 230
	Total tax withheld	$8461

CHAPTER OUTLINE

WHAT WILL WE DO IN THIS CHAPTER?

- Calculate net pay after making deductions from gross pay
- Calculate taxable income after making allowable deductions from income
- Calculate income tax and Medicare levy
- Calculate tax payable or refund owing after PAYG tax has been paid
- Calculate goods and services tax (GST)

HOW ARE WE EVER GOING TO USE THIS?

- Checking the amount of tax deducted from our pay
- Making sure we don't pay too much tax
- Keeping accurate records and collecting receipts to calculate allowable tax deductions
- Calculating whether we're entitled to a tax refund

Payday

5.01 Gross and net pay

Everyone who's had a job knows that the amount of money they've earned isn't what they end up getting. Most workers pay tax under the **PAYG** (Pay As You Go) system, where your employer estimates and takes out the tax from your **gross pay** before they give you your **net pay**. Some people have other amounts deducted from their pay, such as health insurance payments, superannuation contributions or union fees. What's left of their pay after tax and other **deductions** is called their 'take-home pay'.

> Gross pay is the total amount earned.
>
> Net pay = gross pay – tax
>
> Take-home pay = gross pay – tax – other deductions

EXAMPLE 1

Chris earns $155 per shift in his part-time job stacking supermarket shelves late at night. He works three shifts per week and each fortnight his employer deducts $45.60 for PAYG tax.

a What is Chris' fortnightly gross pay?

b Calculate Chris' fortnightly net pay.

Solution

a Chris works $3 \times 2 = 6$ shifts per fortnight.

Gross pay = $6 \times \$155$

= $930

b Net pay = gross pay – tax.

Net pay = $930 – $45.60

= $884.40

Exercise 5.01 Gross and net pay

Example 1

1 Dylan earns $680 per week for working 4 days per week and his employer deducts $84.50 for PAYG tax.

   a Calculate Dylan's net weekly pay.

   b How much PAYG tax does Dylan pay per year?

2 Complete the missing values in the table.

Gross weekly income	Weekly tax	Net weekly income
$720	$32	**a**
$860	**b**	$822
c	$75	$1100
$1258	$137	**d**
$1450	**e**	$1139

3 Sarah earns $17.65 per hour for a 35-hour week. Each week, she has $54 deducted from her pay for PAYG tax and $11 for health insurance.

a How much does Sarah earn per week?

b Calculate Sarah's weekly take-home pay.

4 Grant's annual tax bill is $9704. How much tax should he pay per week to cover his annual bill?

5 Aisha's gross salary is $80 000 and she calculated that her annual tax bill is $17 547. Aisha is paid fortnightly.

a How much PAYG tax should she pay each fortnight?

b Calculate her fortnightly net pay.

6 Juan's PAYG payment summary from his employer showed that he paid $5845 in PAYG tax last year. His gross salary was $35 700.

a How much PAYG tax was deducted from Juan's pay each week?

b What was his net weekly pay?

7 Xander is constructing a spreadsheet to calculate annual net pay.

	A	B	C	D	E
1	Net annual Income				
2	Enter gross fortnightly pay in D2				
3	Enter fortnightly PAYG tax in D3				
4		Annual net pay			
5					
6					

What formula will he need for cell D4?

8 Boun is a junior barrister in a city law firm. His gross fortnightly pay is $3692 and his net fortnightly pay is $2656. How much tax is deducted from Boun's pay each year?

5.02 Taxable income and tax refunds

PAYG tax is deducted from your pay and is only an estimate of the **income tax** you need to pay. The actual amount of tax that you're supposed to pay is calculated at the end of the financial year (30th June) and is based on your total income over the previous 12 months.

After July, you are required to complete a **tax return** to tell the Australian Tax Office (ATO) how much money you've earned and how much PAYG tax you've paid. The ATO then calculates the actual income tax payable and decides whether you've paid too much tax or not enough. If you've paid too much, you will receive a **tax refund**. If you haven't paid enough, you will be sent a bill to pay the **tax debt**.

If you donate some of your income to charities or spend it on work-related items and activities, then you do not have to pay tax on those amounts. These amounts are 'tax deductible' and are called **allowable tax deductions**. We deduct these amounts from our annual income and then income tax is calculated on the remaining amount, called our **taxable income**.

> Taxable income = total income – allowable tax deductions

EXAMPLE 2

Last financial year, Rowan's gross salary was $65 000 and he earned $420 in interest from his bank account. He had allowable tax deductions of his $290 union membership fees and an $80 donation to the RSPCA. How much is Rowan's taxable income?

Solution

Total income = gross salary + bank interest.	Total income = $65 000 + $420 = $65 420
Find total tax deductions.	Deductions = $290 + $80 = $370
Taxable income = total income – deductions.	Taxable income = $65 420 – $370 = $65 050

NCM 11. Mathematics Standard (Pathway 1)

ISBN 9780170413503

EXAMPLE 3

Last financial year, Rowan paid $302 each week in PAYG tax instalments. The tax office calculates that on an income of $65 000 his income tax should be $15 095. Will Rowan receive a tax refund or a tax debt? How much?

Solution

Calculate the total PAYG tax paid.	PAYG tax paid = $302 × 52 = $15 704
Compare this to the amount Rowan has to pay ($15 095).	Rowan paid more than was required ($15 704 > $15 095) so he will receive a tax refund.
Calculate the difference to find the refund.	Tax refund = $15 704 − $15 095
	= $609
	Rowan will receive a $609 tax refund.

Exercise 5.02 Taxable income and tax refunds

1 Nabil's gross annual pay is $69 600 and he earns $1280 in bank interest. He has the following allowable tax deductions: trade tools $370, mobile phone $413, home office equipment $177.

   a Calculate Nabil's gross annual income.

   b How much is Nabil's taxable income?

2 Last year, Sandy paid $11 865 in PAYG tax. The tax office informed Sandy that her income tax payable is $11 450. How much will Sandy receive as a tax refund?

3 Madeleine's weekly net pay is $890 and her weekly gross pay is $995.

   a How much does Madeleine pay in PAYG tax each week?

   b Madeleine's income tax payable is $4790.50. Will she receive a tax refund or a tax bill? What amount is involved?

4 Every fortnight, Sunny earns $1775 in net pay after paying $260 in PAYG tax.

   a Calculate his gross pay.

   b What is Sunny's annual gross pay?

   c Sunny's income tax payable is $7897.75. Will he receive a refund or a bill? Justify your answer.

Example
3

5 Abdul earns $406 per week. During the year, he paid $160 in union fees and $65 in work-related expenses.

   a   Calculate his gross annual income.

   b   What is Abdul's annual taxable income?

   c   Each week, Abdul paid $47 in PAYG tax. How much tax did Abdul pay during the year?

   d   The tax office calculated that Abdul's income tax was $2250. Will Abdul receive a refund or will he have to pay more tax? How much?

6 Mrs Durham pays $178.20 PAYG tax every week. At the end of the year, the ATO calculated that the actual tax she was required to pay was $9320.40. Did she receive a refund or did she have to pay more tax?

7 Peter's salary was $77 500 and he earned $165.20 from his shares. His allowable taxation deductions were conference fees of $310 and a donation of $60 to the Salvation Army. How much was his taxable income?

8 Each week, Gianna earned $1635 and paid $496.18 in PAYG tax. Her work-related travelling expenses of $199 were tax-deductible.

   a   Calculate Gianna's gross annual income.

   b   How much PAYG tax did she pay for the year?

   c   How much was her annual taxable income?

   d   The ATO calculated Gianna's tax to be $19 113.84. Did she receive a refund? Explain your answer.

5.03 Income tax and Medicare levy

This table shows the tax rates for different income brackets (2017 rates).

Income tax
tables

Percentage
calculations

Income tax

Taxable income	Tax on this income
0 – $18 200	Nil
$18 201 – $37 000	19c for each $1 over $18 200
$37 001 – $87 000	$3572 plus 32.5c for each $1 over $37 000
$87 001 – $180 000	$19 822 plus 37c for each $1 over $87 000
$180 001 and over	$54 232 plus 45c for each $1 over $180 000

© Australian Taxation Office for the Commonwealth of Australia

There is also a tax that goes towards funding Medicare, the public health system, called the **Medicare levy**.

For most people, the Medicare levy is 2% of their taxable income.

> Note: The Medicare levy is due to increase to 2.5% in July 2019.

EXAMPLE 4

Rodney's gross salary is $46 248 and he has allowable tax deductions totalling $7310. Calculate Rodney's income tax and Medicare levy.

Solution

Calculate Rodney's taxable income.

$$\text{Taxable income} = \$46\,248 - \$7310$$
$$= \$38\,938$$

$38\,938 is in the $37 001 to $87 000 row of the tax table.

$3572 plus 32.5c for each $1 over $37 000 (which means 32.5% or 0.325).

$$\text{Income tax} = \$3572 + 0.325 \times (\$38\,938 - \$37\,000)$$
$$= \$4201.85$$

Medicare levy is 2% of taxable income.

$$\text{Medicare levy} = 0.02 \times \$38\,938$$
$$= \$778.76$$

Exercise 5.03 Income tax and Medicare levy

1 Calculate the income tax, Medicare levy and the total tax for each taxable income.

  a $45 200

  b $36 960

  c $49 625

  d $122 500

  e $186 000

2 Marianne's gross salary is $46 230 and her total allowable deductions are $7520.

  a   What is her taxable income?

  b   Calculate her income tax.

  c   How much is Marianne's Medicare levy?

  d   Marianne's PAYG summary from her employer shows that she has paid $4520 in PAYG tax. Will this cover her income tax and Medicare levy?

  e   How much extra will she have to pay?

3 Muspah's gross salary is $126 450 and he has $11 200 of allowable deductions.

    a How much income tax is he required to pay?

    b Because Muspah has a high taxable income and he doesn't have any private health insurance he has to pay a Medicare levy surcharge. His Medicare levy is increased to 3.25% of his taxable income. Calculate Muspah's Medicare levy, including the surcharge.

    c How much extra does Muspah have to pay for Medicare because he doesn't have private health insurance?

4 Kait works in a hospital. She recently received a $1600 pay rise that took her taxable income from $75 200 to $76 800.

    a How much will her income tax increase as a result of her pay rise?

    b How much will her Medicare levy increase as a result of her pay rise?

5 Jacob's taxable income is $50 000 and his friend Garth's taxable income is $150 000. Garth claimed that because his income is three times Paul's income he pays three times as much tax as Paul. Use calculations to determine whether Garth's statement is correct.

6 Last financial year, Tina earned $56 800 from her full-time building job and $26 000 from her weekend work. Her allowable deductions for the year totaled $5600.

    a Calculate Tina's income tax and Medicare levy.

    b During the year, Tina's full-time employer deducted $10 300 in PAYG tax. How much does Tina owe the tax office?

7 Single people with a taxable income over $90 000 and no private health insurance are required to pay the Medicare levy surcharge which starts at 1% of their taxable income.

    a Calculate the Medicare levy plus surcharge; that is, a 3% Medicare levy on a taxable income of $90 000.

    b How much extra is this compared to a similar person who *does* have private health insurance?

    c 'Basic plus' private health insurance for young singles costs $16.05 per week. How much more than the Medicare levy surcharge does this health insurance cost a person with a taxable income of $90 000?

8 This year, Jules' Medicare levy was $1285, which was 2% of his taxable income. Calculate Jules' taxable income.

INVESTIGATION

TAXATION, MEDICARE AND WORKING FAMILIES

Ask your teacher to download the Income tax spreadsheet from NelsonNet for this investigation. We will see how having children affects income tax and the Medicare levy.

Imran and Michelle's data

Imran and Michelle are a young married couple. Imran's gross pay is $50 000 and he has allowable deductions of $6300. During the year, he has paid $8000 in PAYG tax. Michelle is pregnant with their first child and is not working.

1 On the spreadsheet, enter the data for Imran and Michelle in income 1.

2 Determine Imran's tax refund.

3 Enter the same income data in income 2 and increase the number of children to one.

4 Determine Imran's tax refund when the couple has one child.

5 What tax refund is Imran entitled to with different numbers of children?

6 Determine why the tax refund changes.

7 Why does the refund size change for up to three children but not for a greater number?

8 How does the size of Imran's income affect the maximum number of children who change his Medicare levy?

COMPLETING A TAX RETURN

In Chapter 4, you looked at your possible career in the 'My future career' investigation. Imagine you have just completed a financial year working in one of the careers recommended for you.

Part 1: Your financial details

1 Enter your 9-digit tax file number if you have one or make one up if you haven't.

2 Research the income you can expect to earn per year in your recommended career.

3 Calculate 20% of the value of your income and assume you have paid that amount in PAYG tax.

4 List any expenses you could incur that are work-related.

5 Decide whether you will have private medical insurance and research the annual insurance cost.

6 Record any bank interest or dividends from shares that you think you might receive.

Australian
Tax Office

Part 2: Completing the tax return form

1 Log onto the Australian Taxation Office website www.ato.gov.au and download a tax return form for individuals.

2 Print the form, then open the 'Individual tax return' instructions document. Printing this document is not required.

3 Enter the information from Part 1 into the tax return form. If you don't know how to complete any section, ask your teacher for help or look at the relevant pages of the tax return instructions.

4 Determine your taxable income, the income tax you are required to pay, and your Medicare levy.

5 Determine the amount of your tax refund or any extra tax you have to pay.

6 You can use the tax calculators on the ATO website to check your calculations.

When you have a real job and tax file number, you can complete your tax return online.

5.04 Goods and services tax (GST)

Most countries have some form of **goods and services tax (GST)** or value-added tax (VAT). In Australia, 10% GST is included in the price of all non-essential goods and services. Some countries use the Australian convention of including the GST in the marked price, whereas other countries show prices before the GST is added.

EXAMPLE 5

The price of a Persian carpet was $3250 before GST. Calculate the final price of the carpet.

Solution

Add 10% GST.

$$GST = 0.10 \times \$3250 = \$325$$
$$Final\ price = \$3250 + \$325$$
$$= \$3575$$

OR: To increase a price by 10%, multiply the price by 1.10 or 1.1.

$$Final\ price = \$3250 \times 1.1$$
$$= \$3575$$

EXAMPLE 6

How much GST is included in the price of a shirt that is on sale for $62?

Solution

First, work out the price without GST by solving an equation.

Let P = the price without GST.

$$P \times 1.1 = \$62$$

Divide both sides of the equation by 1.1.

$$P = \$62 \div 1.1$$
$$= \$56.36$$

The GST is the difference between $62 and $56.36.

$$GST = \$62 - \$56.36$$
$$= \$5.64$$

OR: $62 is 110% of the price without GST.

Let P = the price without GST.

110% of P = $62

So, 10% of the price without GST is $62 divided by 11.

10% of P = $62 ÷ 11

$$= \$5.64$$

Exercise 5.04 Goods and services tax (GST)

1 Without GST, the price of a pair of shoes is $46.

    a Calculate the GST (at 10%) that has to be added to the price of the shoes.

    b What is the price of the shoes including GST?

2 When Anthony the builder issues invoices to customers, he is required to add 10% GST to the invoice. How much GST will Anthony have to add to an invoice for $3800?

3 Rhondell calculates the price of items and completes price tags in a fashion store. Find the missing values **a** to **f** in the table below.

Item	Price before GST	GST	GST-inclusive price
Jeans	$75	**a**	**b**
Jumper	**c**	$8.50	$93.50
Coat	$230	**d**	$253
Scarf	$15	**e**	**f**

4 GST is included in a plumber's invoice for $290.40.

    a How much is the GST?

    b What is the value of the invoice without GST?

5 Find the missing values **a** to **j** in the following table.

Item	Price including GST	GST	Price without GST
Basket of flowers	$55	**a**	**b**
Bottle of perfume	$169.40	**c**	**d**
Restaurant meal	$105.60	**e**	**f**
1 litre of petrol	$1.65	**g**	**h**
A haircut	$22	**i**	**j**

6 Businesses registered for GST can receive a refund of the GST included in items they purchase. Sally's new company car cost $57 200 including GST. Calculate the value of her GST refund.

7 Thanh designed a spreadsheet to help him calculate GST-free prices quickly.

What formulas could Thanh have used in cells C4 and C5?

	A	B	C	D
1	GST Calculator			
2				
3	GST inclusive price		$78.00	
4		GST	$7.09	
5	Price without GST		$70.91	
6				

8 Walid is registered to charge 10% GST on all services his IT business provides. At the end of every 3 months, he has to send the GST collected to the Australian Tax Office. The total value of Walid's invoices is $82 350, including GST. How much GST does Walid have to send the tax office?

9 When Stuart was in New Zealand he bought a jacket for $138, GST-inclusive.

a The New Zealand GST rate is 15%. Explain why solving the equation $P \times 1.15 = 138$ will give the value of P, the price of the jacket before GST was added.

b Calculate the price of Stuart's jacket without GST.

10 Some countries use value-added tax (VAT) in a similar way to GST. When Linda was on holiday in Estonia she bought a piece of jewellery for 3364 euros which included 20% VAT.

a Calculate the price of Linda's jewellery without VAT.

b When Linda left the European Union, she was able to claim the VAT back. How much was her VAT refund?

11 Klaus paid 6000 Danish krone for some business supplies. In Denmark, the rate of VAT is 25%. Calculate the amount of VAT that is included in the price.

INVESTIGATION

WHICH SUPERMARKET ITEMS DO NOT HAVE A GST?

At the supermarket, a hot BBQ chicken has a GST but a cold BBQ chicken does not. Why?

1 Collect a supermarket docket that includes many items.

2 How much GST is included in the final sale amount?

3 Classify the items on the docket into two lists: items that are GST-free and items that include GST.

4 What do all the items that include GST have in common?

TAX CROSSWORD

Earning and
taxation
crossword

Copy this crossword and complete it using the clues given in the paragraphs below.
NelsonNet also contains a different crossword called 'Earning and taxation'.

The Government needs money to provide public facilities such as hospitals, **2**_____,
roads, public transport and **5**_____ services. Part of the funding for these facilities comes
from income tax. The **3**_____ **8**____ helps to pay for public health and medical services.

Income tax is charged on your taxable **6**_____ which is your total income less your
allowable **10**_____. If your taxable income is lower than the tax-**1 across**_____
threshold you don't have to pay any tax. Each **11**____, fortnight or month when
employees receive their pay, the tax has already been taken out as part of the
4_____ (PAYG) system.

At the end of the **1 down**_____ year, you complete a tax return to inform the ATO
about the total income you have earned and the **9**_____ you have paid. If you have paid
too much tax, you will receive a tax **7**_____, but if you haven't paid enough tax you will
receive a bill for more.

SOLUTION TO THE CHAPTER PROBLEM

Problem

Hamish is preparing his income tax return. The annual payment summary shown is for his day job only. Hamish also earned $3640 and paid $1040 tax in his part-time job. He calculated that he has $460 in allowable tax deductions.

Will Hamish receive a tax refund?

PAYG payment summary		
Employee	Hamish Donald	
Employment period	1 July 2017–30 June 2018	
	Total income	$48 230
	Total tax withheld	$8461

Solution

Hamish's total income = $48 230 + $3640

$$= \$51\ 870$$

Hamish's taxable income = $51 879 – $460

$$= \$51\ 410$$

According to the income tax table on page 128, the income tax payable on taxable incomes between $37 001 and $87 000 is:

$3572 plus 32.5c for each $1 over $37 000.

Income tax = $3572 + 0.325 × ($51 410 – 37 000)

$$= \$8255.25$$

PAYG tax already paid = $8461 + $1040

$$= \$9501$$

Hamish has paid too much tax.

Tax refund = $9501 – $8255.25

$$= \$1245.75$$

- What new things did you learn in this chapter?

- How do members of your family pay tax?

- Make a list of the terminology used in this chapter.

- What is the difference between PAYG tax, income tax payable and taxable income?

Copy and complete this mind map of the topic, adding detail to its branches and using pictures, symbols and colour where needed. Ask your teacher to check your work.

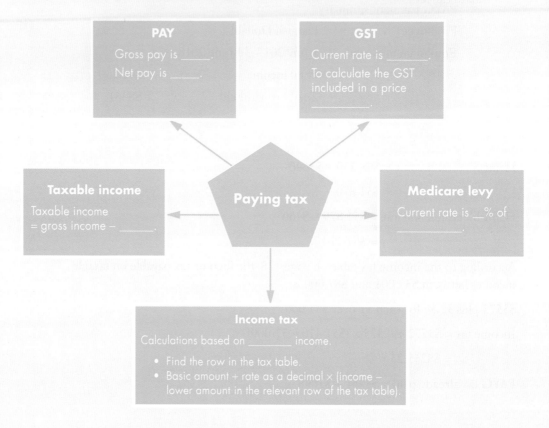

PAY

Gross pay is _____.

Net pay is _____.

GST

Current rate is _____.

To calculate the GST included in a price

_____.

Taxable income

Taxable income
= gross income − _____.

Paying tax

Medicare levy

Current rate is __% of

_____.

Income tax

Calculations based on _____ income.

- Find the row in the tax table.
- Basic amount + rate as a decimal × (income − lower amount in the relevant row of the tax table).

5. TEST YOURSELF

1 Ava earns $19.50 per hour for a 36-hour week. Each week, she has $84 deducted from her pay for PAYG tax and $14 for private health insurance.
   a How much does Ava earn per week?
   b Calculate Ava's weekly take-home pay.

2 Jim's salary was $87 500 and he received $2965 interest on his investment account. His allowable tax deductions were $340 for uniform and shoes, $460 for tools and a donation of $80 to cancer research. How much was his taxable income?

3 Pia's gross annual income is $65 850 and she has allowable deductions totalling $1250. Calculate her taxable income and Medicare levy.

4 Use the income tax table on page 128 to calculate each person's income tax.
   a Claire's taxable income is $82 500.
   b Will's annual taxable income is $26 500.

5 Suzanne's taxable income is $89 000. Each year, she makes $25 000 concessional contributions ('salary sacrifice') to her superannuation account, which are allowable tax deductions. How much income tax does Suzanne save by paying the $25 000 contribution?

6 Part of Lee's job in a clothing shop is calculating GST-inclusive prices for all items in the shop. Find the missing values **a** to **d** in the following table.

Item	Price without GST	Price including GST
Pair of trousers	$48	**a**
Shirt	$36	**b**
Pajamas	**c**	$27.50
Jacket	**d**	$198

7 The rate of VAT in Ireland is 23%.
   a Calculate the selling price of some Irish linen with a VAT-free price of 40 euros.
   b When she was in Ireland, Jade bought a hand-knitted jumper for 124 euros. How much VAT was included in the price?

Section A: Multiple-choice questions

For each question, select the correct answer A, B, C or D.

1 Which of the following is *not* numerical data?

   A   The postcode of a town or suburb

   B   The average temperature for each month in Cobar

   C   The number of shoes each student owns

   D   The daily rainfall for Merimbula

2 What is the solution to $3x = -12$?

   A $x = -36$ **B** $x = -15$ **C** $x = -9$ **D** $x = -4$

3 Walid earns $3672 per month. What is his weekly pay?

   A $847.38   **B** $918   **C** $1694.77   **D** $1836

4 For which study would you use a census rather than a sample to investigate?

   A   Testing coffee for the best taste

   B   Finding the number of migrants from Japan to Australia

   C   Checking for drink drivers on the road

   D   Finding out people's views on whether Australia should become a republic

5 One hamburger with chips at the local cafe costs $8.50 plus 10% GST. Calculate the selling price.

   A $8.60   **B** $7.50   **C** $9.35   **D** $9.50

6 A boat travels 24 km in 90 minutes. What is its average speed?

   A 12 km/h   **B** 16 km/h   **C** 18 km/h   **D** 36 km/h

7 Sienna sells cars and is paid a weekly retainer of $320 plus 3.5% of the value of the cars she sells. Calculate her income for a week when she sells cars worth a total of $87 000.

   A $3365   **B** $3045   **C** $1120   **D** $331.20

8 Sophie is paid $45 for each scarf she knits. One week she is paid $270. How many scarves did she knit? Let N represent the number of scarves. Choose the correct equation required to solve this problem.

   A $45 \times 270 = N$ **B** $45N = 270$ **C** $N + 45 = 270$ **D** $270 - N = 45$

9 According to this table, which of the following will use 3500 kJ?

Exercise 3.01

Activity	Time required to use 1000 kJ
Sleeping	4 hours
Eating	3 hours
Working in class	2 hours 30 minutes
Studying	
Watching TV	
Walking	1 hour
Bike riding	50 minutes
Swimming	30 minutes

A Sleeping for 12 hours and 30 minutes

B Studying for 7 hours and 30 minutes

C Bike riding for 2 hours and 55 minutes

D Swimming for 1 hour and 35 minutes

10 Jeremy works 12 hours at a normal hourly wage of $22.80 and 3 hours overtime at time-and-a-half. Which calculation could be used to find his weekly wage?

Exercise 4.02

A $22.80 × 15

B $22.80 × 15 × 1.5

C $22.80 × 12 + $22.80 × 3 × 1.5

D $22.80 × 12 + $22.80 × 15 × 1.5

11 For the formula $A = \dfrac{h(a+b)}{2}$, evaluate A when $h = 4.5$, $a = 6$ and $b = 8$.

Exercise 2.04

A 9.25   **B** 18.5   **C** 31   **D** 31.5

12 Reece earns $1538.50 per week. He claims deductions of $715 for travel expenses and $218 for membership of a professional association. He also gives $50 a month to charity. Calculate his taxable income for one year.

Exercise 5.02

A $80 002   **B** $79 069   **C** $79 019   **D** $78 469

Section B: Short-answer questions

Exercise 2.01

1 Simplify each expression.

   a $7ab + 2bc - 3ab + bc$ **b** $6x^2 - 4x - 4 - 9x + 6$

   c $9d \times 4e \div 6d$ **d** $\dfrac{15n^2 p}{3n^2}$

Exercise 1.04

2 Melinda wants to survey student views on the school uniform.
Describe how she could choose each type of sample.

   a   random sample   **b**   stratified sample

   c   self-selected sample   **d**   systematic sample

Exercise 2.07

3 A number is doubled and then has 4 added to it. The result is 38.
Write an equation to represent this problem and solve it to find the number.

Exercise 5.03

4 Rajesh has a taxable income of $91 300. Calculate the 2% Medicare levy payable.

Exercise 1.09

5 A die was rolled 20 times and these were the results:

   5   2   3   4   2   2   6   2   6   3

   5   1   4   1   2   5   6   2   4   1

   a   Draw a dot plot for this data.

   b   How many times was a number greater than 3 rolled?

   c   What percentage of rolls resulted in a 1 or a 2?

Exercise 4.01

6 **a**   Janine is paid $109 000 annual salary. How much is this each fortnight?

   b   Michael is paid $33.50 per hour. How much is he paid for a 35-hour week?

Exercise 4.02

7 Simon is paid $276 for working 6 hours on a Sunday at double time.
What is his usual hourly rate of pay?

Exercise 1.07

8 In order to improve customer service, the NCM Bank recorded the waiting times in
minutes per customer at one of its biggest branches:

   14   11   12   7   6   13   12   21   11   6   13

   11   7   3   8   11   10   8   9   7   9   11

   10   5   14   12   12   10   13   14   9   15

   9   12   7   22   8   9   12   7   6   9

  NCM 11. Mathematics Standard (Pathway 1)   ISBN 9780170413503

a Copy and complete this frequency table for the data.

Waiting time	Tally	Frequency
1–5		
6–10		
11–15		
16–20		
21–25		

b On how many occasions was the waiting time from 11 to 15 minutes?

c What percentage of customers waited more than 20 minutes? Answer correct to one decimal place.

9 Draw a frequency histogram and polygon for the data in Question **8**.

10 Annabel's total income is $79 500 and she receives $220 in interest from her bank account. Annabel has allowable tax deductions of $680 for union membership fees, $485 for tools and $1345 for donations to charities. How much is Annabel's taxable income?

11 Annabel has $334.32 deducted each week in PAYG tax.

a Calculate how much tax Annabel has paid in one year.

b Use this table to calculate how much tax is payable on her taxable income.

Taxable income	Tax on this income
0–$18 200	Nil
$18 201–$37 000	19c for each $1 over $18 200
$37 001–$87 000	$3572 plus 32.5c for each $1 over $37 000
$87 001–$180 000	$19 822 plus 37c for each $1 over $87 000
$180 001 and over	$54 232 plus 45c for each $1 over $180 000

c Does Annabel receive a refund or does she owe money to the tax office? Find the amount she receives or owes.

12 Solve each equation.

a $4p + 3 = 23$

b $\dfrac{a}{5} - 8 = 4$

Exercise
3.01

Exercise
3.02

13 Jane has lunch consisting of 2 slices of ham, 1 slice of cheese and 1 slice of buttered bread, followed by a banana.

a Use this table to calculate how many kilojoules are in Jane's meal.

Food	kJ	Food	kJ	Food	kJ
Roast lamb, gravy	1064	Apple pie	1380	Apple	243
Potato bake	1175	One slice of buttered bread	520	Large tomato	120
				Small tomato	30
Mixed grill	2600	Yoghurt	315	Chocolate biscuit	493
Steak	3900	Muesli	1470	Can of soft drink	372
Bacon (2 slices)	640	Eggs (2)	735	Orange juice (glass)	206
Fish	340	Cheese	115	Milk (glass)	628
Grilled chicken breast	1264	Ham (2 slices)	224	Milk for cereal	400
Chips	1425	Mixed nuts (100 g)	2640	Coffee with milk and sugar	295
Ice cream	810	Broccoli	98	Banana	546
Sauce for steak or chicken	246	Sauce for fish	265	Beef sausage	176

b In the afternoon, Jane goes swimming. Use the table from Question **9** on p.141 to determine how long Jane must swim to use all the energy from her lunch.

Exercise
2.04

Exercise
2.06

Exercise
2.08

14 A catering company calculates its charges using the formula $C = 350 + 49n$, where n is the number of people attending the function and C is the total cost.

a Calculate the cost to cater for a party of 200 people.

b How many people attended a party where the catering cost was $7210?

c Change the subject of this formula to n.

Exercise
5.04

15 Grant's new car costs $47 990 including GST. How much GST is included in the cost of Grant's car?

Exercise
3.04

16 Manu is driving at 80 km/h on a wet asphalt road. What distance will he travel after he applies the brakes? Use the formula $D = \dfrac{V}{1000}(210 + 97R)$ where $R = 1.6$ for wet asphalt.

17 Anilha earns $970 per week. Each year, she is paid 17.5% annual leave loading on 4 weeks pay. Calculate:

4.04

   a her pay for 4 weeks

   b the annual leave loading she will be paid

   c her total gross pay.

18 Judy and Sue are paid royalties on the sale of their books. Each author is paid 7.5% of the selling price of the book. Calculate the royalties each one will receive on the sale of 2400 books priced at $49.95 each.

4.05

19 Nelson Software has developed a new business program. This Pareto chart shows the reasons for calls to technical support in the first 3 days after customers purchase the software.

1.09

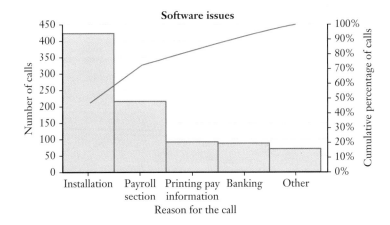

   a How many calls concerned the payroll section?

   b What are the two major problems that customers encountered with the software?

   c What percentage of calls were concerned with these two major problems?

6.

TAKING CHANCES

Chapter problem

Based on past records, the school sports coordinator claims that the probability of weekly winter sport being cancelled due to wet weather in Terms 2 and 3 is 0.15.

a Calculate the expected number of cancelled sports days over the 20 weeks of winter sport.

b Charlie thinks that this probability value is wrong because school sport was actually cancelled 5 times in winter this year. Is the sports coordinator incorrect in claiming a probability of 0.15?

CHAPTER OUTLINE

WHAT WILL WE DO IN THIS CHAPTER?

- Use the language of probability
- Calculate the probability of a simple event and complementary events
- Perform simple experiments and calculate relative frequency
- Use tree diagrams and grids to determine the sample space and calculate probabilities
- Compare relative frequency and theoretical probability
- Calculate probabilities involved in simple games of chance

HOW ARE WE EVER GOING TO USE THIS?

- Determine the chance that something will or won't happen
- Make predictions about chance events
- Know whether common beliefs about chance events are true or false
- Make sensible decisions if we are going to gamble on games of chance

6.01 Probability of simple events

The language
of chance

Sample space

WS

Games of
chance

The **probability** of an **event** occurring, where all outcomes are **equally likely**, is given by the formula:

$$P(\text{event}) = \frac{\text{number of ways the event can happen}}{\text{total number of possible outcomes}}$$

This can also be written:

$$P(\text{event}) = \frac{\text{number of favourable outcomes}}{\text{total number of outcomes}}$$

Because probability is a fraction, the smallest value for probability is 0 and the largest value is 1. An **impossible event** has a probability of 0, whereas a **certain event** has a probability of 1. Unlikely events have a probability close to 0 whereas **likely** events have a probability close to 1.

EXAMPLE 1

A group of 7 female and 2 male whales are swimming along the Australian coast. One of the whales has water coming out of its spout. What is the probability that this whale is female?

Solution

There is a total of $7 + 2 = 9$ whales.
There are 7 females out of the 9 whales.

$$P(\text{whale is female}) = \frac{7}{9}$$

EXAMPLE 2

There are four major blood types A, B, AB and O, and each blood type can be positive or negative. This table shows the percentage of the Australian population with each blood type.

What is the probability that a person selected at random from the Australian population has blood type A?

	+	−
A	31%	7%
B	8%	2%
AB	2%	1%
O	40%	9%
Totals	81%	19%

'at random' means every person has an equal chance of being selected.

Solution

Read the A row of the table and add the values.

Type A blood $= 31\% + 7\%$
$= 38\%$

Express the probability as a fraction and simplify.

$P(\text{blood type A}) = \dfrac{38}{100}$

$= \dfrac{19}{50}$

Exercise 6.01 Probability of simple events

1 There are 19 adult and 11 young seals in a colony. A photographer selected at random a seal from the colony to photograph. What is the probability that she selected an adult seal?

2 A dingo has 3 male and 2 female pups. Jemma chooses one of the pups at random. What is the probability that she chooses a:

a female pup?

b male pup?

3 One of Paula's 4 cockatoos can talk. What is the probability that she randomly selects a cockatoo that can't talk?

4 A pregnant female whale is expecting a single calf. What is the probability that the calf is male?

5 Voula has 4 yellow and 5 blue budgerigars for sale. Her customer chooses one of them at random. What is the probability that it is blue?

6 Members of a wildlife rescue group have specialist training. Five members are trained to rescue reptiles, 8 are trained to rescue birds and 7 are trained to rescue marsupials.

a One member is selected at random to give a talk to the local primary school. What is the probability that the selected person:

i is trained to rescue reptiles?

ii is not trained to rescue birds?

b Anu is joining the rescue group and she is being assigned to a training group at random. What is the probability that she will be placed in the group trained to rescue marsupials?

7 Cameron works at a call centre that operates 7 days a week. He is allocated 2 random days each week to answer the phone. What is the probability that Cameron will be working on Friday?

8 A turtle laid 60 eggs but 24 of them weren't fertilised and won't hatch. Also, predators will eat half of the fertilised eggs.

    a On average, how many of the 60 eggs will survive and hatch?

    b What is the probability that an egg selected at random was not fertilised?

    c Calculate the probability that a newly laid egg will result in the birth of a baby turtle.

9 Use the table in Example 2 (p.148) to find the probability that a student in Year 11 selected at random has:

    a O+ blood   **b** AB– blood   **c** any positive blood group.

10 This table shows the percentage of the Bangladeshi population with each blood type.

What is the probability that a person selected at random in Bangladesh has:

    a type B blood?

    b a blood type that's negative?

	+	–
A	21%	1%
B	35%	1%
AB	9%	1%
O	31%	1%
Totals	96%	4%

11 A migrating adult whale has a 95% probability of returning to Antarctica by Christmas. Write this probability as a simple fraction.

12 Explain why it is impossible to have a probability of 110%.

13 Aman has created a frog-friendly environment in her backyard with 9 brown frogs and 6 green frogs living there. She checks the frogs every night before she goes to bed.

    a Calculate the probability that the first frog she sees tonight is green.

    b Calculate as a decimal the probability that the first frog she sees tonight is brown.

14 Toby works for a fisheries research company. He is researching the incidence of killer whale attacks on humpback whales. In a group of 40 whales, Toby counted 6 whales that had scars from killer whale attacks from which they had escaped.

    a What is the probability that a whale selected at random from the group has scars from a killer whale attack?

    b Calculate the probability that a randomly selected whale from the group doesn't have scars.

6.02 Relative frequency

Sometimes, we can't calculate the theoretical probability of an event. Instead, we can use past records, an experiment or simulation to calculate a **relative frequency** or **experimental probability**.

A page of spinners

Coins probability

Dice probability

Relative frequency of an event $= \dfrac{\text{number of times the event happens}}{\text{total number of trials}}$

The **relative frequency** of an event is the frequency of the event as a fraction of the total frequency.

EXAMPLE 3

Emily is a park ranger. This table shows the data she recorded about kangaroo deaths in the park.

a   Calculate the probability that a kangaroo was killed by a motor vehicle. Express your answer as a decimal, correct to three places.

b   Use Emily's data to estimate the probability that a kangaroo in her area will die from old age. Express the answer correct to the nearest percentage.

Cause of death	Number of deaths
Hit by a motor vehicle	78
Shot	12
Caught in a fence or trap	11
Old age	25
Starvation	3
Other	7
No known cause	4
Total	140

Solution

a   The relative frequency of the number of deaths due to motor vehicles is $\dfrac{78}{140}$.

$P(\text{killed by motor vehicle}) = \dfrac{78}{140}$

$= 0.55714\ldots$

≈ 0.557

b   The relative frequency of the number of deaths due to old age is $\dfrac{25}{140}$. Convert this to a percentage.

$P(\text{death from old age}) = \dfrac{25}{140} \times 100\%$

$= 17.8571\ldots\%$

$\approx 18\%$

EXAMPLE 4

Souraya counted the contents of
10 boxes of matches that were each
labelled as containing 50 matches.
Her results were:

53 49 50 48 52

51 50 49 50 51

a What is the relative frequency of a
box containing exactly 50 matches?

b What is the probability that a box
contains more than 50 matches?

Solution

a In Souraya's data there are 3 boxes
that contain 50 matches. That's 3
boxes out of 10 boxes.

The relative frequency is $\dfrac{3}{10}$.

b Four of the 10 boxes contain more
than 50 matches.

$P(\text{more than 50 matches}) = \dfrac{4}{10}$

$= \dfrac{2}{5}$

Exercise 6.02 Relative frequency

Example
3

1 Renee measured the tail lengths of a sample of
adult quokkas on Rottnest Island, WA.

Tail length (cm)	Frequency
24	2
25	7
26	9
27	10
28	8
29	7
30	4
31	3
Total	50

a What is the probability that an adult quokka will have a 27 cm tail?

b Renee's wildlife manual says that adult quokka tails range from 25 to 30 cm long.
What is the probability that an adult quokka will have a tail that is outside this range?

c What is the most likely length for an adult quokka's tail?

2 Renee was surprised at how friendly the quokkas were. She gave each animal a friendliness rating and displayed the data in a pie chart. Calculate the experimental probability that a randomly selected quokka will be:

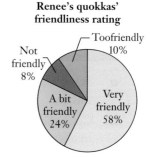

Renee's quokkas' friendliness rating

a not friendly (give your answer as a decimal)

b very friendly or too friendly (give your answer as a fraction).

3 In the past 12 months, when Jason visited his favourite restaurant, he noted that he had to wait for a table 8 times and he got a table straight away 16 times.

Example
4

a How many times did Jason go to the restaurant?

b What is the relative frequency of 'not having to wait'?

c What is the experimental probability that the next time Jason goes to the restaurant he will have to wait for a table?

4 Murphy's Law states that if anything can go wrong, then it will! Libby decided to test this theory by dropping a piece of toast and seeing whether it landed buttered-side up (good) or buttered-side down (bad). She performed 40 trials of her experiment:

Buttered-side up: 5

Buttered-side down: 35

> Try this experiment at home but make sure that you first cover the floor to avoid staining.

a What is the relative frequency of the toast landing buttered-side up?

b Use Libby's data to determine the probability that a dropped piece of toast will land buttered-side down.

5 Mitchell visited the old whaling station in Albany, WA. He asked a random selection of people some questions about whaling and presented his results below.

Question	Yes	No
Do you agree with the international ban on killing whales?	19	1
Is it OK to make and sell souvenirs made from whale bones?	13	7
Are there too many whales in the oceans around WA?	2	18

a What is the probability that a person selected at random agrees with the international ban on killing whales?

b Calculate the probability that a person selected at random does not agree with making and selling souvenirs from whale bones. Express your answer as a decimal.

c What is the probability that a randomly selected person thinks there are too many whales in the oceans around WA?

d Predict the answers that a person who is a member of the 'Save our whales' group would give to each of Mitchell's questions.

e Suggest a factor that could create bias in Mitchell's results.

6 Latu rolled a die 75 times and displayed the results in a table.

> 'Die' is the singular word for 'dice': one die, two dice.

Number	Frequency
1	21
2	12
3	10
4	11
5	12
6	9

a What is the relative frequency of rolling a 3 with this die?

b Copy and complete the table to show the relative frequency of rolling each number, correct to 2 decimal places.

Number on the die	1	2	3	4	5	6
Relative frequency			0.13			

c To 2 decimal places, the theoretical probability of tossing each number on a normal die is 0.17. Latu believes his die is biased. Do his results support this belief? Justify your answer.

7 Lauren loves to collect 'Sports heroes' cards. There are 10 different cards in the set and one of them is placed randomly in each packet of chewing gum.

Sports heroes

a Predict the number of packets of chewing gum Lauren will need to buy to get all 10 cards.

b Ask your teacher to download the 'Sports heroes' spreadsheet from NelsonNet.

c Run the spreadsheet 20 times to simulate buying 20 packets of gum. Record the number of packets required to obtain all 10 cards.

d Find the relative frequency of obtaining a full set of cards from fewer than 15 packets of chewing gum.

PRACTICAL ACTIVITY

THE GAME SHOW PROBLEM

For this activity, each pair of students will need three cards from a deck of playing cards.

You are the contestant in a TV game show. Behind one of three doors there is a car that you could win. The host asks you to choose a door and you choose Door 1. The host then opens *Door 3* to show you that the car isn't behind it. This means that the car is behind either Door 1 or Door 2. Should you stick with Door 1 or change to Door 2?

To simulate this problem, decide which of your three cards will be the 'CAR' card. Then you are going to work out the relative frequency that the car is behind the door that wasn't your first choice.

ISBN 9780170413503

1 Decide who will be the game show host and who will be the contestant.

2 Copy this frequency table.

Outcome	Tally	Frequency
Car was first choice		
Car wasn't first choice		

3 The host shuffles the three cards, looks at them, then places them face down on the table.

4 The contestant chooses one card and moves it down as shown in the photo to the right.

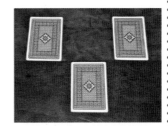

5 The host displays one of the two remaining cards that is *not* the car.

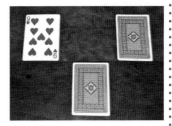

6 Turn over both the remaining cards. Record in your frequency table whether the car was the card first chosen by the contestant or whether it was the other card.

7 Perform the simulation at least 24 times.

8 Determine the relative frequency that the car is behind the door that wasn't the contestant's first choice.

9 Decide on the best strategy. Should the contestant stick with their original choice or switch?

6.03 Complementary events

When flipping a coin, getting heads and getting tails are examples of **complementary events**. The **complement** of an event is every other thing that could happen.

When two events are complementary, the sum of their probabilities is 1.

EXAMPLE 5

On Penguin Island, penguins that have been orphaned or rejected by their parents are raised in the penguin centre. The probability that a penguin from the penguin centre will survive in the wild is $\frac{3}{40}$. What is the probability that a penguin will die in the wild?

Solution

Surviving and dying are complementary events. The sum of their probabilities is 1.

$$P(\text{die}) = 1 - P(\text{survive})$$
$$= 1 - \frac{3}{40}$$
$$= \frac{37}{40}$$

Write the answer.

The probability that a penguin will die is $\frac{37}{40}$.

Exercise 6.03 Complementary events

Example
5

1 The probability that an adult male sperm whale will live to 70 is $\frac{3}{5}$. What is the probability that an adult male sperm whale will die before it is 70 years old?

2 The probability that Isabella will have homework tonight is $\frac{11}{12}$. What is the probability that Isabella won't have homework tonight?

3 The probability that the African painted dog will be extinct in the next 10 years is 0.35. What is the probability that the African painted dog won't be extinct in the next 10 years?

4 Zac doesn't like getting out of bed. The probability that he will sleep in and be late for school on any day is $\frac{1}{3}$. What is the probability that Zac won't sleep in and be late for school?

5 For leafy sea dragons, it is the *father* that gives birth to baby sea dragons.

 a Only 1 in 20 baby sea dragons lives to become an adult. What is the probability that a baby sea dragon won't live to become an adult?

 b A male leafy sea dragon laid 250 eggs, of which 110 were female. What is the probability that a randomly selected egg from the 250 eggs laid will be male?

6 The probability that the traffic lights at the T-intersection will be green is 25%. What is the probability that they won't be green?

7 Groper fish change sex as they grow. Small gropers are male. The probability that an 89-centimetre-long groper will be female is $\frac{19}{20}$. What is the probability that an 89-centimetre-long groper will be male?

Shutterstock.com/Ian Scot

8 The east coast of NSW is experiencing a very long spell of dry, hot weather. The weather reporter on TV says that there is an 80% chance of bushfires tomorrow. What is the probability that there won't be bushfires tomorrow?

6.04 Using tables to list outcomes

A list of all the possible outcomes in a chance situation is called the **sample space**.

Listing all possible outcomes can be difficult when probability problems become more complex. For more complicated situations such as rolling two dice together, drawing a grid or a table is a useful method.

EXAMPLE 6

Zoe tosses a coin and a die together. What is the probability that she tosses:

a a 5 and a head? **b** a 5 or a head?

Solution

For the coin, there are two possible outcomes: heads or tails.

For the die, there are six possible outcomes: 1 to 6.

The table shows the sample space for a coin and a die tossed together.

		Die					
		1	**2**	**3**	**4**	**5**	**6**
Coin	**Head**	H1	H2	H3	H4	H5	H6
	Tail	T1	T2	T3	T4	T5	T6

a The table shows that there are 12 possibilities. $P(5 \text{ and a head}) = \frac{1}{12}$
Only one of the outcomes (H5) shows a 5 and a head.

b Count every outcome in the Head row and the 5 column $P(5 \text{ or a head}) = \frac{7}{12}$
(but don't count H5 twice).

Exercise 6.04 Using tables to list outcomes

1 Ziad tosses a coin and an eight-sided die numbered 1 to 8.

a Copy and complete the table to show all possible outcomes.

b What is the probability that Ziad tosses:

		Die							
		1	**2**	**3**	**4**	**5**	**6**	**7**	**8**
Coin	Head								
	Tail								

   i a 7 and a tail?   **ii** a 7 or a tail?

   iii a number less than 5 and a tail?   **iv** a head and an odd number?

2 Jemma rolls a pair of dice and adds the 2 numbers that come up.

a Copy and complete this 2-dice grid to list all possible totals.

b How many outcomes are in the sample space?

c Find the probability of rolling a total of:

   i 5   **ii** 8

   iii 10   **iv** 12

		2nd die					
	+	**1**	**2**	**3**	**4**	**5**	**6**
1st die	**1**			4			
	2						
	3				7		
	4						
	5	6				10	
	6						12

d Which is more likely: a total of 9 or a total greater than 10?

e How many times more likely is a total of 7 than a total of 4?

3 Alyssa uses a pair of unusual dice in a board game she is designing. She numbers one die 0, 1, 2, 3, 4 and 5, and the other die 1, 1, 3, 3, 4 and 6. Players roll the dice and add the two numbers to determine their score.

a Copy and complete this grid for Alyssa's dice.

b What is the probability of rolling the following scores on Alyssa's dice?

   i 5   **ii** 10   **iii** 3 or 4

c What is the most likely score on Alyssa's dice?

+	**0**	**1**				
1						
1						
3						
3	3	4				
4						
6						

4 Stefan uses a normal pair of dice in his board game, but his rules require players to *subtract* the smaller number from the larger number.

a Copy the grid and show all the possible scores using Stefan's rules.

b What is the probability of scoring a 3?

c What is the probability of scoring 6?

d What is the most likely score with Stefan's rules?

–	**1**	**2**	**3**	**4**	**5**	**6**
1						
2		0				
3					2	
4						
5						
6			3			

5 Julianne uses a normal pair of dice in her board game, but her rules require players to use just the *larger* number on the two dice as the score.

   a Copy and complete the table for Julianne's dice.

   b What is the probability of scoring 3 with Julianne's rules?

   c What is the most likely score?

   d What score has a probability of 0.25?

	1	2	3	4	5	6
1				4		
2						
3						
4						
5						5
6	6					

6 Christina made up an interesting special rule for a board game she is designing. Players roll a pair of dice, but they can choose to move:

- the sum of the two numbers on the dice, or
- either of the individual numbers showing.

Jayden is playing Christina's game and he needs 6 to win. What is the probability that Jayden can move 6 on his next roll of the dice?

6.05 Using tree diagrams

Tree diagrams are another way of listing all possible outcomes systematically. Tree diagrams start from a point on the left side of the page and grow sideways to the right. It's a good idea to leave space above the starting point so that the 'tree' has room to grow!

Tables and tree diagrams

Tree diagrams

EXAMPLE 7

Caleb has these three cards. He chooses two cards at random to make a two-digit number.

a How many different two-digit numbers can he make?

b What is the probability that he makes a number greater than 40?

Matching probabilities

Solution

Draw a tree diagram that lists all possible two-digit numbers. For the first digit, Caleb can select 8, 4 or 1. For the second digit, he can choose one of the remaining two cards (he can't choose the same card twice).

Probability review

First digit	Second digit	Outcomes
8	4	84
	1	81
4	8	48
	1	41
1	8	18
	4	14

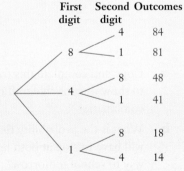

a Count the outcomes.   Caleb can make 6 different two-digit numbers.

b Four of the numbers listed (in red) are greater than 40. $P(\text{number} > 40) = \dfrac{4}{6} = \dfrac{2}{3}$

Exercise 6.05 Using tree diagrams

Example
7

1 Britney selects two cards at random from these three cards to make a two-digit number.

a Copy and complete the tree diagram to show all possible two-digit numbers.

First digit   **Second digit**   **Outcomes**

1

6   —

1

6

b What is the probability that Britney will make a number less than 35?

2 Every morning on her way to school, Gordana drives through two sets of traffic lights at which she is equally likely to have to stop or not stop.

Shutterstock.com/ktsdesign

a Copy and complete this tree diagram to show the possible traffic light combinations.

b What is the probability that Gordana will have to stop at both lights on her way to school tomorrow?

c Calculate the probability that she will have to stop for at least one set of traffic lights tomorrow morning.

First set   **Second set**   **Outcomes**

stop

stop

not stop

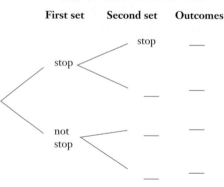

3 Claire has a bag that contains 3 highlighter pens: one red, one blue and one yellow.

a She selects a pen from the bag at random and then another without putting the first one back. Copy and complete the tree diagram to show the sample space.

> When we choose a second item without putting the first item back we call it 'without replacement'. If we put the first one back before we choose the second it's called 'with replacement'.

First pen	Second pen	Outcomes

yellow — red, —

red — yellow, —, —

— — —, —, —

b What's the probability Claire chooses a red pen followed by a blue pen?

c What is the probability she chooses a yellow and a blue pen (in any order)?

d Construct another tree diagram to determine the sample space if Claire *replaces* the first pen before she selects the second.

e When Claire replaces the first pen, what's the probability that she will select the same pen twice?

4 A tennis squad has three boys, Mark, Wang and Peter, and three girls, Sonia, Delta and Aniela. The coach chooses a boy and a girl from the squad to represent the school in a mixed doubles competition.

a Construct a tree diagram to determine all the possible mixed doubles pairs.

b What is the probability that Wang and Delta will be selected to represent the school?

5 Seth selects two cards at random from these four cards.

a Construct a tree diagram to list the sample space.

b How many different two-card pairs can he select?

c What is the probability that Seth chooses two cards that are the same colour?

Scout Kozakiewicz

6.06 Comparing relative frequency and theoretical probability

If the probability of rolling a 4 on a die is $\frac{1}{6}$, it does not mean that, in the next 60 rolls, a 4 will come up *exactly* 10 times. Theoretical probability tells you what will happen in the long run. Over many rolls, a 4 should come up in approximately $\frac{1}{6}$ of the rolls.

EXAMPLE 8

Kevin rolled an 8 sided die 120 times and recorded the results in a frequency table.

a Determine the relative frequency and theoretical probability of rolling a 6 on this die. Express the values as decimals correct to three places.

b Comment on the statement: 'The law of averages says that the next roll will probably be a 4 because there haven't been enough 4s rolled so far.'

Number	Frequency
1	17
2	10
3	11
4	9
5	20
6	22
7	13
8	18

Solution

a In the experiment, the number 6 occurred 22 times in 120 rolls.

$$\text{Relative frequency} = \frac{22}{120}$$
$$= 0.1833 \ldots$$
$$\approx 0.183$$

Theoretically, there are 8 possible outcomes and the number 6 is one of them.

$$\text{Theoretical probability} = \frac{1}{8}$$
$$= 0.125$$

b With each new roll, the probability of each number is the same, so 4 is not more likely.

The statement is incorrect. Each time the die is rolled, there is a $\frac{1}{8}$ chance that a 4 will be rolled.

A common mistake that gamblers make is thinking that if a number hasn't come up much in the past it's got a higher chance of coming up next. The die doesn't have a memory or remember the numbers rolled in the past. Each number has the same chance.

ISBN 9780170413503

Exercise 6.06 Comparing relative frequency and theoretical probability

1 a What is the theoretical probability of getting a head when you toss a coin?

Example
8

b Toss a coin 40 times and record the number of heads you get.

c Use the data you obtained in part **b** to determine the relative frequency of getting a head.

d Calculate the difference between the theoretical probability and the relative frequency.

e Josie has been playing a game that involves tossing a coin. The coin was tossed 10 times and heads came up only three times. Josie thinks that there is something wrong with the coin or someone is cheating.

Heads and
tails

 i Ask your teacher to download the 'Heads and tails' spreadsheet from NelsonNet.

 ii Run the simulation 40 times and calculate the relative frequency of getting three or fewer heads when you toss a coin 10 times.

 iii Josie expected to get heads about half of the time. Run the simulation numerous times and concentrate on the percentage of heads. In which group, 10, 50, 100, 200 or 300 tosses, does the percentage of heads change by the biggest amount?

 iv Write a sentence to explain how the percentage of heads changes as you increase the number of tosses.

 v Are Josie's concerns about the coin, or the people she is playing with, justified? Explain your answer.

2 When you roll a pair of dice, which event do you think is more likely to happen:

- a 1 or a 2 (or both) will show, or
- neither a 1 nor a 2 will show?

a Roll a pair of dice 40 times and record how many times a 1 or a 2 (or both) shows.

b Calculate the relative frequency that a 1 or a 2 (or both) shows.

c Repeat your experiment to check your results.

The table below lists all possible outcomes when a pair of dice is rolled.

One die → ↓The other die	1	2	3	4	5	6
1	1, 1	1, 2	1, 3	1, 4	1, 5	1, 6
2	2, 1	2, 2	2, 3	2, 4	2, 5	2, 6
3	3, 1	3, 2	3, 3	3, 4	3, 5	3, 6
4	4, 1	4, 2	4, 3	4, 4	4, 5	4, 6
5	5, 1	5, 2	5, 3	5, 4	5, 5	5, 6
6	6, 1	6, 2	6, 3	6, 4	6, 5	6, 6

d What is the theoretical probability of rolling two numbers that are the same?

e Determine the probability that at least one of the numbers showing will be a 1 or a 2.

f Why is it more likely that a 1 or a 2 will show than neither a 1 nor a 2 will show?

g Write a sentence to compare and contrast the relative frequency and theoretical probability of obtaining a 1 or a 2 when you roll a pair of dice.

INVESTIGATION

ROULETTE WHEEL

Roulette is a popular gambling game. There are 37 slots on a roulette wheel. The slots are numbered from 0 to 36. The zero slot is green and there are an equal number of red and black slots for the numbers from 1 to 36. There is a variety of ways people can gamble at roulette. Some people like to bet on the colour of the slot where the ball finishes.

Shutterstock.com/Galushko Sergey

Mike believes that he has a winning roulette strategy. He believes that it's unlikely for the ball to land on a black number 4 times in a row. Mike watches the game and when the ball has landed on black 3 times in a row he then bets on red for the next roll. He claims that he always wins when he uses this strategy.

Roulette

1 Ask your teacher to download the 'Roulette' simulation spreadsheet from NelsonNet.

2 Run the spreadsheet and find occasions in the simulation where the ball lands on black three times in a row and record the colour for the *next* time.

3 Determine whether Mike's strategy is successful.

6.07 Probability tree diagrams

Tree diagrams can become large and complicated for more complex chance situations. Writing probability values on the branches simplifies this problem.

Tree diagrams

Rolling a die

Greedy pig game

EXAMPLE 9

This weekend, the probability of rain on any particular day is 0.15.
Find the probability that:

a it won't rain on Saturday

b it will rain on both Saturday and Sunday

c it will rain on Saturday but not on Sunday.

Solution

a $P(\text{rain})$ and $P(\text{no rain})$ are complementary events.

$$P(\text{no rain}) = 1 - P(\text{rain})$$
$$= 1 - 0.15$$
$$= 0.85$$

b Draw a probability tree diagram to list the possible outcomes for Saturday and Sunday, with the probabilities listed on the branches.

	Saturday	Sunday	Outcomes
		0.15 rain	rain, rain
	0.15 rain	0.85 no rain	rain, no rain
	0.85 no rain	0.15 rain	no rain, rain
		0.85 no rain	no rain, no rain

To find the probability that it rains on both days, we follow the top branches (rain, rain), and we multiply the probabilities on the branches.

$$P(\text{rain, rain}) = 0.15 \times 0.15$$
$$= 0.0225$$

c Follow the 'rain, no rain' branches and multiply.

$$P(\text{rain, no rain}) = 0.15 \times 0.85$$
$$= 0.1275$$

6. Taking chances

EXAMPLE 10

A box contains 6 soft-centred and 4 hard-centred chocolates. Ruby selected a chocolate at random and ate it. Then she selected another chocolate.

a What is the probability that the first chocolate had a soft centre?

b Calculate the probability that both chocolates had soft centres.

c What is the probability that one chocolate was soft-centred and the other was hard-centred?

Solution

This is a 'without replacement' situation because once Ruby eats a chocolate she can't replace it. When she chooses the second chocolate there are only 9 for her to choose from and the numbers of hard- and soft-centred chocolates remaining depend on the first chocolate she ate.

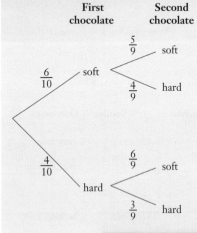

Note that on a probability tree, branches from the same point always have probabilities that add to 1; for example, $\frac{6}{9}$ and $\frac{3}{9}$.

a There are 6 soft-centred chocolates from a total of 10.

$$P(\text{first has soft centre}) = \frac{6}{10}$$
$$= \frac{3}{5}$$

b Multiply the probabilities along the 'soft, soft' branches.

$$P(\text{both soft}) = \frac{6}{10} \times \frac{5}{9}$$
$$= \frac{1}{3}$$

c Multiply the probabilities along the 'soft, hard' and 'hard, soft' branches, then add them together.

$$P(\text{different centres}) = P(\text{soft, hard}) + P(\text{hard, soft})$$
$$= \frac{6}{10} \times \frac{4}{9} + \frac{4}{10} \times \frac{6}{9}$$
$$= \frac{8}{15}$$

In a **probability tree diagram**:

- branches from the same point have probabilities that add to 1.
- to calculate the probability of an outcome, *multiply* the probabilities along that branch.
- to calculate the probability of two or more outcomes, *add* their calculated probabilities.

Exercise 6.07 Probability tree diagrams

1 Rachel is a keen pistol shooter. During competitions, the probability that she will hit the target is 0.9.

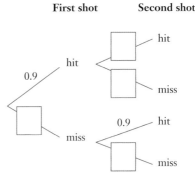

First shot   Second shot

a What is the probability that Rachel will miss the target in any competition shot?

b Rachel fires two shots in a competition. Copy and complete the probability tree.

c Calculate the probability that Rachel will hit the target with both shots.

d What is the probability she will miss the target with her first shot and hit it with her second shot?

2 Nick drives through two sets of traffic lights on his way to work. The probability that the first light is red is 0.5 and the probability that the second light is red is 0.4.

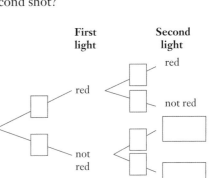

First light   Second light

a Copy and complete this probability tree.

b What is the probability that both sets of lights will be red?

c Calculate the probability that neither light will be red.

d Find the probability that Nick will get a red light followed by a light that isn't red.

3 7% of adults have a phobia (fear) of dentists.

a What is the probability that a randomly selected adult is not afraid of dentists?

b A journalist chooses two adults at random to interview about health issues. Copy and complete the tree diagram.

c Calculate the probability that one of the adults has a dentist phobia but the other doesn't.

d What is the probability that at least one of the adults doesn't have a dentist phobia?

> 'At least one' means one or more.

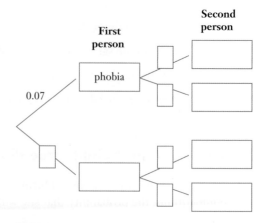

First person

Second person

0.07

phobia

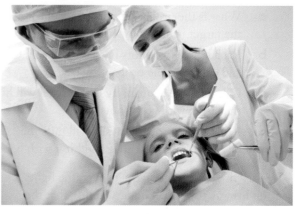

Shutterstock.com/pressmaster

4 Lance plays in a soccer competition on the weekend. The probability of rain on Saturday is 0.25 and on Sunday is 0.4. Draw a probability tree to represent this situation.

a Calculate the probability that it will rain on both days.

b What is the probability that it will rain on only one of the days?

c Calculate the probability that it will not rain all weekend.

5 A butcher has a large jar of jelly beans to give to well-behaved children. He knows that 60% of the jelly beans in the jar are red. The butcher lets Samir choose two jelly beans from the jar at random. What is the probability that:

a the first jelly bean he chooses is red?

b both jelly beans he chooses are red?

c at least one of the jelly beans he chooses is red?

> When there are a large number of items, or the probability is given as a percentage, we don't have to change the probability for selecting the second item.

6 In Nelson Waters, the probability of being tested by a random breath-testing unit late on Friday night is 0.3. On the next two Friday nights, Leo is meeting friends and will be driving home late.

a Find the probability that he won't be random breath-tested on the first Friday.

b What is the probability that he will be random breath tested at least once on the two Friday nights?

7 Sarah and Rhys are sitting the driving theory test to get L-plates. The probability that Sarah will pass is 0.8 and that Rhys will pass is 0.6.

   a What is the probability that they will both pass the test?

   b Calculate the probability that Rhys will pass but Sarah won't.

   c Calculate the probability that at least one of them will pass.

8 On any day, the chance that Chloe will be late to school is 0.3. For the next two days, find the probability that Chloe will be:

   a late both days   **b** on time both days

   c late for school on at least one of the days.

9 Research shows that 85% of Australian haemophiliacs lack 'factor 8' in their blood, one of the factors that makes blood clot. If two Australian haemophiliacs are selected at random, what is the probability that they both lack 'factor 8' in their blood?

10 Ella flipped a biased coin that shows tails 60% of the time.

   a What is the probability that the coin shows heads?

   b Ella flips the coin twice. Calculate the probability that she flips a head and a tail in any order.

INVESTIGATION

FREEWAY ACCIDENTS

Accidents are a serious problem on the M1, the freeway between Sydney and Newcastle. The probability that a severe accident (where a person is killed or seriously injured) on the M1 involves a truck or other heavy vehicle is 0.12.

Freeway accidents

To complete this activity, ask your teacher to download the 'Freeway accidents' spreadsheet from NelsonNet. This spreadsheet simulates 50 accidents on the M1 and counts the number of accidents involving trucks and other heavy vehicles.

1 Over 50 accidents, what is the expected number of times that a truck is involved?

2 Run 40 simulations of 50 accidents using the spreadsheet and record the number of times a truck is involved.

3 Calculate the percentage of simulations in which six trucks were involved in accidents.

4 Calculate the percentage of simulations in which the number of accidents involving trucks was:

   a 5, 6 or 7   **b** within the range 3 to 9.

5 What conclusion could you make from this simulation?

FACT OR FALLACY?

Scott is attempting to swim from Newcastle to Sydney. There are two possibilities; either he will swim the distance or he won't. Therefore, the probability that Scott will swim the distance is $\frac{1}{2}$.

The claim is a fallacy (false) because the two outcomes, 'swimming the distance' and 'not swimming the distance' are *not* equally likely.

In your group, discuss each of the following probability statements and decide whether they are fact (true) or fallacy (false). Be ready to justify your group's opinion when other groups disagree!

1 I will either die when I'm 99 or I won't die when I'm 99. That's 1 possibility out of 2. The probability that I will die on my birthday is $\frac{1}{2}$.

2 The probability of flipping 2 heads on a pair of coins is $\frac{1}{3}$, because there are 3 possible outcomes: 2 heads, 2 tails or a head and a tail.

3 There are 10 runners in a race.

   a The name of each runner is in a hat on a separate piece of paper. The chance of picking the name of the winner out of the hat at random is $\frac{1}{10}$.

   b Each runner in the race has a probability of $\frac{1}{10}$ of winning the race.

4 Melissa has 4 sons and she would like to have a daughter. She correctly calculated that the probability of a family of 5 children all being boys is $\frac{1}{32}$. She says that she is going to have another baby because the probability that her next baby will be a girl is $\frac{31}{32}$ which is close to a certainty.

5 There is a 60% chance of rain on Saturday and a 40% chance of rain on Sunday. There is a 100% chance of rain on the weekend.

6 Tony plays table tennis. He wins 3 out of 5 matches he plays, making the probability that he will win any match $\frac{3}{5}$. He is playing in a 5-match competition and he has won the first 3 games. It is likely that he will lose the next 2 matches.

7 In Lotto, players select six numbers from the numbers 1 to 45.

   a The numbers 3, 11, 15, 16, 25 and 31 are more likely to be the six winning numbers than the numbers 1, 2, 3, 4, 5 and 6.

   b If you include some or all numbers bigger than 30 in your six numbers and your six numbers come up, you will win more money than if the six winning numbers included no number bigger than 30.

8 Martin is interested in buying a house that is in a 1-in-a-100-years flood zone. The house flooded in March this year. It won't flood again for another 100 years.

PROBABILITY OF $\frac{1}{2}$

What does a theoretical probability of $\frac{1}{2}$ really mean?

Group activity: The lines of text in the following passage are in the wrong order. Your group's challenge is to reassemble the lines in the correct order. To make the task easier, ask your teacher to print a copy of this activity from NelsonNet to cut up.

row, then the next toss of the coin is more likely to be a head. Coins can't
happen. The theoretical probability of getting a head when we toss a coin
of a head being $\frac{1}{2}$ does mean is that if we toss the coin thousands of times, about
get a head. Neither does it mean that when we toss a coin 100 times we
will get a head 50 times. It also doesn't mean that if we get 8 tails in a
is $\frac{1}{2}$. This doesn't mean that every second time we toss a coin we will
what is going to happen.
half of the time we'll get a head, but on no individual future occasion can we know
Theoretical probability is about the long-term chance that something will
remember what's happened in the past and the chance of getting a head in the
future doesn't change because we've had lots of tails. What the probability

Probability of $\frac{1}{2}$

Probability crossword

SOLUTION TO THE CHAPTER PROBLEM

Problem

Based on past records, the school sports coordinator claims that the probability of weekly winter sport being cancelled due to wet weather in Terms 2 and 3 is 0.15.

a Calculate the expected number of cancelled sports days over the 20 weeks of winter sport.

b Charlie thinks that this probability value is wrong because school sport was actually cancelled 5 times in winter this year. Is the sports coordinator incorrect in claiming a probability of 0.15?

Solution

a Expected number of cancelled sports days $= 20 \times 0.15$

$$= 3$$

We expect, on average, to have three wet sports days in Terms 2 and 3.

b Expectation is only a long-term average and not a guaranteed amount. The number of cancelled sports days in winter will not be the same every year, and having five cancelled sports days this year is possible when the probability is 0.15. So, the sports coordinator is not wrong in claiming this probability.

ISBN 9780170413503

- What was the most interesting part of this chapter?

- List some real-life examples where you've been involved in chance.

- List some jobs where chance plays a significant part.

- Are there any parts of this chapter that you don't understand fully? If yes, ask your teacher for some additional assistance.

Copy and complete this mind map of the chapter, adding detail to its branches and using pictures, symbols and colour where needed. Ask your teacher to check your work.

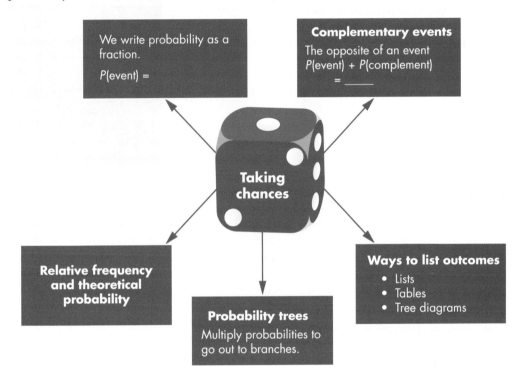

We write probability as a fraction.

$P(\text{event}) =$

Complementary events

The opposite of an event
$P(\text{event}) + P(\text{complement})$

$= \_\_\_$

Taking chances

Relative frequency and theoretical probability

Probability trees

Multiply probabilities to go out to branches.

Ways to list outcomes

- Lists
- Tables
- Tree diagrams

Exercise
6.01

1 I have an 8-sided die with the numbers 0, 1, 2, 3, 4, 5, 6, and 7 on it. When I roll this die, what is the probability I will get:

a a 4?

b a number less than 3?

c an odd number more than 2?

d an 8?

Exercise
6.02

2 Billie is a keen bird watcher. She spent 48 nights looking for a Tawny Frogmouth and she spotted one on 16 of the nights.

What is the relative frequency that Billie will see a Tawny Frogmouth when she goes looking for one at night?

iStock.com/CraigRJD

Exercise
6.03

3 Koalas are nocturnal animals. The probability that a koala will be awake at lunchtime is $\frac{1}{20}$. What is the probability that a koala won't be awake at lunchtime?

Exercise
6.04

4 Li Jing rolls a 4-sided die numbered 1 to 4 and a normal 6-sided die together. She adds the two numbers shown on the dice.

a Construct a table showing all possible outcomes and determine the total number of possibilities.

b What is the probability of rolling a sum of 5?

Exercise
6.05

5 Ngoc is setting up her loom to weave a scarf that has four stripes running along the length. She has green, blue, white and yellow wool for the stripes. Use a tree diagram to show that there are 24 different orders in which she can arrange the stripes.

6 The members of an extended family were told that the probability that any child born in the family will be severely short-sighted is $\frac{1}{4}$. There are 15 children in the extended family and seven of them are severely short sighted.

Exercise
6.06

   a Using the theoretical probability, how many of the 15 children would you expect to be severely short-sighted?

   b How can you explain the difference between the expected frequency and the reality?

   c Another baby is due to be born into the extended family. The family believes that this baby won't be severely short-sighted because they already have seven severely short-sighted children. Is this thinking correct? Explain your answer.

7 The tiny corroboree frog is Australia's most endangered frog. The frogs are dying from the effects of an environmental fungus. The probability that any living corroboree frog has the fungus is 60%. Environmentalists found two corroboree frogs. Use a probability tree diagram to calculate the probability that:

Exercise
6.07

   a both frogs have the fungus

   b neither frog has the fungus

   c at least one of the frogs has the fungus.

Getty Images/Torsten Blackwood

7.

MEASUREMENT

Chapter problem

Jake is going to Scotland to 'bag' (climb) as many Munros as he can. Munros are Scottish mountains that are more than 3000 feet (914 metres) high. At 1.34 km above sea level, Ben Nevis is Scotland's highest Munro. How many metres higher than the qualifying height of 914 metres is the top of Ben Nevis?

CHAPTER OUTLINE

WHAT WILL WE DO IN THIS CHAPTER?

- Measure, calculate and convert with metric units, including mass
- Determine absolute error, limits of accuracy and percentage error of a measured value
- Use significant figures and scientific notation
- Express quantities in scientific notation
- Calculate perimeters of triangles, quadrilaterals, circles, sectors and composite shapes, including the use of Pythagoras' theorem

HOW ARE WE EVER GOING TO USE THIS?

- Measurement skills are an essential component of many trades; for example, shopfitting, carpentry, upholstery, building, interior decorating and dressmaking
- Measurement skills are needed when building or renovating a home; for example, pool fencing, garden edging, making curtains and replacing gutters
- Measurement skills are used in cooking when we weigh the amount of flour or sugar required in a recipe

7.01 Metric units

In Australia, we measure most things using the **metric system**.

Unit	Relationships
Length	
micrometre (μm)	
millimetre (mm)	
centimetre (cm)	1 cm = 10 mm
metre (m)	1 m = 100 cm = 1000 mm = 1 000 000 μm
kilometre (km)	1 km = 1000 m
Mass	
milligram (mg)	
gram (g)	1 g = 1000 mg
kilogram (kg)	1 kg = 1000 g
tonne (t)	1 t = 1000 kg
megatonne (Mt)	1 Mt = 1 000 000 t
Time	
second (s)	
minute (min)	1 min = 60 s
hour (h)	1 h = 60 min = 3600 s
day (day)	1 day = 24 h
Capacity	
millilitre (mL)	
litre (L)	1 L = 1000 mL
kilolitre (kL)	1 kL = 1000 L
megalitre (ML)	1 ML = 1 000 000 L

The metric system makes it easy to convert from one unit to another, as we usually multiply or divide by a power of 10, often by 1000.

When converting between metric units:
- divide if you are converting to a bigger unit
- multiply if you are converting to a smaller unit.

ISBN 9780170413503

EXAMPLE 1

Convert:

a 2500 m to km   **b** 1.2 kL to mL.

Solution

a There are 1000 m in 1 km. Because we   2500 m = 2500 ÷ 1000
are converting from a smaller unit (m)   = 2.5 km
to a larger unit (km), we divide.

b First, convert kL to L, and then L to mL.   1.2 kL = 1.2 × 1000 × 1000
There are 1000 L in a kL and 1000 mL   = 1 200 000 mL
in a L. Because we are converting to a
smaller unit (mL), we multiply.

Exercise 7.01 Metric units

1 Write the unit that is most suitable for measuring:

   a your height

   b the time it takes to eat lunch

   c your friend's mass

   d the distance from your home to the next town or city

   e how long it takes for a plant seed to shoot

   f the amount of medicine in a teaspoon

   g the height of your main school building

   h the amount of water in a bucket

   i the length of an ant

   j the time it takes to drive from Sydney to Melbourne

   k the mass of one egg

   l the amount of water in an Olympic-sized swimming pool

   m the mass of a truck

   n how long it takes to tie your shoelace

   o the mass of a feather.

2 Convert each length to metres.

   a 1800 mm   **b** 9.5 km   **c** 1365 cm   **d** 19.8 km

3 Convert each capacity to litres.

   a 12.2 kL   **b** 6340 mL   **c** 550 mL   **d** 0.98 kL

4 Copy and complete each statement.

a 32 min = _____ s   **b** 550 g = _____ kg   **c** 9050 m = _____ km

d 45 g = _____ mg   **e** 15.2 L = _____ mL   **f** 0.8 h = _____ min

g 96 h = _____ days   **h** 18 cm = _____ mm   **i** 192 m = _____ mm

j 65 700 L = _____ kL   **k** 16 h = _____ min   **l** 12 300 kg = _____ t

5 Convert:

a 3.2 t to grams   **b** 3 days to min

c 765 000 cm to km   **d** 56 800 mg to kg.

6 Copy and complete each statement.

a 0.5 kg = _____ mg   **b** 3 h = _____ s

c 45 000 cm = _____ km   **d** 3 194 500 g = _____ t

e 0.98 ML = _____ L   **f** 5 460 000 mL = _____ kL

g 7200 s = _____ h   **h** 1.2 km = _____ mm

i 5040 min = _____ days

7 For each set of measurements, select the largest one **A**, **B** or **C**.

a   **A**   345 mm   **B**   3.45 m   **C**   34 cm

b   **A**   4345 g   **B**   440 000 mg   **C**   4.5 kg

c   **A**   0.055 kL   **B**   55 L   **C**   550 000 mL

d   **A**   25 200 s   **B**   450 min   **C**   6 h

8 Hannah recorded how much her vegetable plants grew in one week. She measured the growth in millimetres. Her results were:

- runner beans   66 mm
- corn   130 mm
- tomato   88 mm
- silver beet   35 mm.

a What would Hannah's measurements be if they were in centimetres?

b Assuming each plant kept growing at this rate, how high would each plant be at the end of 12 weeks? Give your answers in metres.

iStock.com/cjp

9 Nathan delivers 6750 kg of grapes to the market. How many tonnes is this?

NCM 11. Mathematics Standard (Pathway 1)   ISBN 9780170413503

10 A large egg weighs 60 g on average. A carton of eggs contains 12 eggs.

    a What is the mass of eggs in one carton?

    b For delivery to the supermarket, the eggs are packaged in containers holding 24 cartons each. How many kilograms of eggs are there in a package?

    c The vehicle transporting the eggs can carry 20 of these containers. What is the mass of eggs transported in kilograms and in tonnes?

11 Lizzie has some pet rabbits which she feeds with a special milk supplement. Each rabbit is given 125 mL of the supplement each day.

    a How many litres of the supplement will she need per day to feed 12 rabbits?

    b Lizzie buys the supplement in 4.5 L containers. How many 125 mL are in one container?

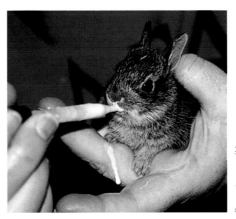

12 It takes Yestin $3\frac{1}{2}$ minutes to sow a row of corn. Today, he intends to plant 48 rows. How many hours will it take him?

7.02 Mass

When we talk about the weight of an object we really mean **mass**. When Louise says that she *weighs* 85 kg, she should say that her *mass* is 85 kg.

The **gram** is the basic unit for mass and all other mass units are based on the gram.

Unit	Relationships
milligram (mg)	
gram (g)	1 g = 1000 mg
kilogram (kg)	1 kg = 1000 g
tonne (t)	1 t = 1000 kg

EXAMPLE 2

Express 750 kg in:

  a grams   **b** tonnes.

Solution

a There are 1000 g in 1 kg. Changing from kg to g is changing to a smaller unit, so multiply by the conversion factor.

$$750 \text{ kg} = 750 \times 1000 \text{ g}$$
$$= 750\,000 \text{ g}$$

b There are 1000 kg in 1 t. Changing from kg to t is changing to a larger unit, so divide by the conversion factor.

$$750 \text{ kg} = 750 \div 1000 \text{ t}$$
$$= 0.75 \text{ t}$$

People in the USA use **pounds**, which is a non-metric unit of mass.

Rachel is catching a flight from Mexico City to
Monterey, California. She wants to take her baby
stroller on the plane but the American airline only
allows strollers with a mass of 20 pounds or less.
Rachel's stroller has a mass of 8.5 kg. Will she
be able to take the stroller on the flight?
(1 kg = 2.2 pounds.)

Solution

Convert 8.5 kg to pounds.

$8.5 \text{ kg} = 8.5 \times 2.2$ pounds

1 kg = 2.2 pounds.

$= 18.7$ pounds

Changing from kg to pounds is changing to
a smaller unit, so multiply by the conversion
factor.

Rachel's stroller is 18.7 pounds which is
less than the airline's limit of 20 pounds.
She can take it on the flight.

OR Convert 20 pounds to kg by dividing by 2.2.

$20 \text{ pounds} = 20 \div 2.2 \text{ kg}$

$\approx 9.09 \text{ kg}$

The airline's limit is 9.09 kg and Rachel's
stroller is 8.5 kg which is less than this.
She can take it on the flight.

Exercise 7.02 Mass

1 Copy and complete each conversion.

   a 3 kg = ___ g   **b** 12 t = ____ kg   **c** 1500 g = ___ kg

   d 2400 kg = ___ t   **e** 850 kg = ____ g   **f** 900 g = ____ kg

   g 2.5 g = _____ mg   **h** 500 mg = ___ g

2 A hospital pharmacist ordered 2000 tablets. Each tablet has a mass of 5 mg.

   a Calculate the total mass of the tablets in mg.

   b What is the total mass in grams?

3 Vitamin C powder contains 90% ascorbic acid and 10% calcium.

   a   What mass of calcium is in 40 milligrams of vitamin C?

   b   What mass of ascorbic acid is in 60 milligrams of vitamin C?

   c   Calculate the number of milligrams of ascorbic acid in 2.4 grams of vitamin C.

4 The gross mass of a bottle of 500 tablets is 155 g. The mass of the bottle only is 20 g.

   a   Calculate the net mass of the tablets.

   b   What is the net mass of the tablets in mg?

> gross mass = total mass including bottle
> net mass = mass of tablets only

   c   What is the mass of one tablet in mg?

5 How many 50 mg injections can a nurse make from a 1 g container of streptomycin medicine?

123rf/sakpols

6 List three items whose mass you would measure in:

   a   tonnes   **b**   kilograms   **c**   grams   **d**   milligrams.

7 We measure the size of precious stones in carats. Erin's engagement ring contains a 1.8-carat diamond. What is the mass of the diamond in mg? (1 carat = 200 mg.)

iStock.com/AplTone

8 Nelsonlink Airlines has a carry-on luggage limit of 12 pounds. Karen's bag is 5 kg.

   a   Calculate the mass of Karen's bag in pounds. 1 kg = 2.2 pounds.

   b   Is Karen's bag light enough to take on the flight? Justify your answer.

Example
3

9 Jettison Air has two sets of restrictions on the size of bags it allows on flights.

- The mass of the bag must be 50 pounds or less.

- The sum of the bag's dimensions (length + width + height) must be less than 62 inches.

Orlando's bag is 50 cm long, 19 cm high, 32 cm wide and has a mass of 24 kg. 1 kg = 2.2 pounds and 1 inch = 2.5 cm

Is Orlando's bag allowed on the flight? Justify your answer.

10 A standard house brick has a mass of 2.7 kg.

a A pallet of bricks contains 500 bricks. Calculate the mass of one pallet of bricks.

b A truck carries 8 pallets of bricks. Calculate the mass of the bricks in tonnes.

11 In China, the mass of tea leaves is measured in 'jins'. One jin = 500 g. Calculate in grams the mass of a packet of tea that is 3.2 jin. Answer correct to one decimal place.

12 We measure the mass of precious metals in troy ounces (1 troy ounce = 31.103 g). Gazi bought a 1 kg gold bar as an investment. How much was Gazi's gold bar worth on a day when gold was valued at $1331 per troy ounce?

PRACTICAL ACTIVITY

ESTIMATING MASS

You will need a set of kitchen scales (for measuring small masses) and a set of bathroom scales (for larger masses). You also need some items so that you can estimate and measure their mass.

This table shows some common items and their approximate mass.

Mass	Items with this approximate mass
1 mg	Grain of sand Tiny insect such as a sandfly
1 g	Paper clip
2 g	5-cent coin
100 g	iPhone without a case
1 kg	1 litre of water 5 medium-sized oranges
71 kg	Average Australian woman
84 kg	Average Australian man

NCM 11. Mathematics Standard (Pathway 1)   ISBN 9780170413503

1 Choose an item; for example, a library card.

2 Select an item in the right column of the table that has a similar mass. For example, a library card is similar in mass to a 5c coin.

3 Compare the two items and estimate the mass of the unknown item. A library card is about twice as heavy as a 5c coin, so it should be about 4 g.

4 Use the scales to check the accuracy of your estimate.

7.03 Error in measurement

We make many types of errors when measuring.

- **Parallax error** results from looking at the measuring scale from an angle instead of from directly in front.

- **Zero error** happens if the measuring device is not properly set to zero beforehand or if the zero mark on a ruler is not placed exactly at the start of the object being measured.

- **Reaction error** occurs when measuring time, with a stopwatch for example, when we take a short time to react at the start of the measurement.

We can avoid most of these errors with a bit of care. Another way we can reduce their effect is to take a number of measurements and average them.

Even if we reduce these human errors, our measurements are only as good as the **precision** allowed by the measuring device we are using. The ruler's scale shown has a precision of 1 cm (the size of one unit), so a measurement anywhere in the shaded zone will be recorded as 12 cm but can involve an error of up to half a centimetre either side of the 12-cm mark.

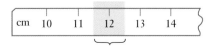

This type of error due to the precision of the measuring device is called the **absolute error**.

The absolute error is half of the **precision** of the measuring device being used. We record the absolute error as ± 0.5 of one unit.

The percentage error of a measurement is the absolute error as a percentage of the measurement.

$$\text{Percentage error} = \frac{\text{absolute error}}{\text{measurement}} \times 100\%$$

EXAMPLE 4

While repairing her boundary fence, Shannon measured the distance between the strands of wire to be 24 cm using the ruler shown.

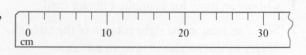

a What is the precision shown on the ruler scale?

b What is the absolute error of Shannon's measurement?

c What are the limits of accuracy of Shannon's measurement?

d What is the percentage error correct to one decimal place?

Solution

a There are five units between 0 and 10 cm on the ruler, so each unit is $10 \div 5 = 2$ cm.

Precision = 2 cm.

b The absolute error is half of the precision.

Absolute error = $\pm 0.5 \times 2$

$= \pm 1$ cm

c So, the shortest length that the measurement could be is $24 - 1 = 23$ cm.

The longest length that the measurement could be is: $24 + 1 = 25$ cm.

The actual distance is between 23 cm and 25 cm.

The limits of accuracy are 23 cm and 25 cm.

d The percentage error is the absolute error as a percentage of the measurement.

Percentage error $= \dfrac{\text{absolute error}}{\text{measurement}} \times 100\%$

$= \dfrac{\pm 1}{24} \times 100\%$

$= \pm 4.1666...\%$

$\approx \pm 4.2\%$

The **limits of accuracy** of a measurement are the lower and upper limits of the range where the actual measurement lies:

- lower limit = measurement − absolute error
- upper limit = measurement + absolute error.

EXAMPLE 5

Gina measured her mass on a set of bathroom scales to be 58 kg.

a   What are the limits of accuracy of this measurement?

b   What is the percentage error correct to two decimal places?

Solution

a   Because the measurement is given to the nearest kilogram, we assume that the precision of the scales is 1 kg.

Absolute error = $\pm 0.5 \times 1$ kg = ± 0.5 kg

Lower limit = $58 - 0.5 = 57.5$ kg

Upper limit = $58 + 0.5 = 58.5$ kg

The limits of accuracy are 57.5 kg and 58.5 kg.

b   The percentage error is the absolute error as a percentage of the measurement.

$$\text{Percentage error} = \frac{\text{absolute error}}{\text{measurement}} \times 100\%$$

$$= \frac{\pm 0.5}{58} \times 100\%$$

$$= \pm 0.8620\ldots\%$$

$$\approx \pm 0.86\%$$

Exercise 7.03 Error in measurement

1   For each measuring instrument shown next page, state:

   i   its precision

   ii   the measurement reading to the nearest unit

   iii   the absolute error of the measurement

   iv   the percentage error correct to one decimal place.

Example
4

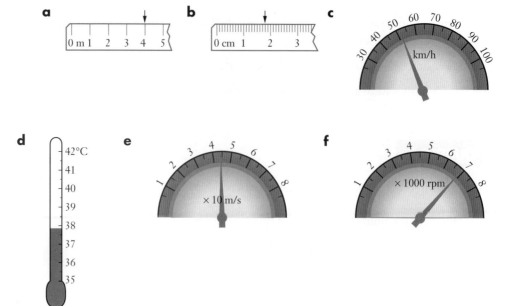

2 The time it takes Saxon to drive his tractor from one end of his property to the other was measured as 7 minutes.

    a What is the absolute error for this measurement?

    b What is this error expressed in seconds?

    c What is the percentage error correct to two decimal places?

3 Find the missing values in this table.

	Measurement	Absolute error	Lower limit of accuracy	Upper limit of accuracy
a	18 L correct to the nearest litre	**i**	**ii**	**iii**
b	78 cm correct to the nearest cm	**i**	**ii**	**iii**
c	35 kg correct to the nearest 5 kg	**i**	**ii**	**iii**
d	9.4 seconds correct to 1 decimal place	**i**	**ii**	**iii**
e	730 km correct to the nearest 10 km	**i**	**ii**	**iii**

4 Lucy has decided to divide one of her existing paddocks into two by erecting a new fence. She has measured the width of the field to be 240 metres (to the nearest 10 metres).

    a What is the absolute error of the measured width?

    b What are the limits of accuracy of the width?

    c What length of fencing wire should Lucy buy to ensure she has enough to do the job?

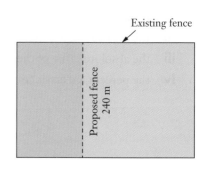

Existing fence

Proposed fence 240 m

5 James has listed the stock of a number of the materials he has in his farm shed. Copy and complete the table.

	Item	Quantity	Recorded to the nearest	Absolute error	Limits of accuracy
a	Fuel	280 L	5 L		
b	Fence posts	350	50		
c	Fencing wire	540 m	20 m		
d	Fertiliser	870 kg	10 kg		

6 We can obtain a more reliable measurement by taking repeated measurements and averaging them. Find the average value of each set of repeated measurements.

a Length of a shed: 12.3 m, 11.9 m, 12.2 m, 12.3 m, 12.1 m

b Mass of a carton of fruit: 15.9 kg, 15.7 kg, 16.1 kg, 15.5 kg

c Capacity of a water trough: 75 L, 77 L, 75 L, 72 L, 79 L, 77 L, 76 L

PRACTICAL ACTIVITY

TAKING ACCURATE MEASUREMENTS

This group activity involves measuring some common items around the school, and will require the following instruments:

- 30 cm ruler
- tape measure
- trundle wheel
- stopwatch
- kitchen scales that measure in grams and kilograms
- a 1 litre measuring cylinder

Each group will measure:

a the longest width of the classroom

b the width of a classroom desk

c the thickness of a $2 coin

d the length of the longest building in the school

e the mass of 4 pencils

f the mass of 5 textbooks

g the volume of water in a glass tumbler

h the time it takes for a ball to hit the ground when dropped from the second floor.

Before dividing into groups to take the measurements:

1 Copy this table.

Measurement	Instrument	Units	Tolerance	Results of measurement					Average
				Group 1	Group 2	Group 3	Group 4	Group 5	
a									
b									
c									
d									
e									
f									
g									
h									

2 As a class, agree on:

- the instrument that should be used for each measurement
- the units (e.g. cm, mL) that should be used for the measurement
- the absolute error for the measurement.

These should be entered into your table.

3 In your groups, take the measurements. Take care to avoid parallax and zero errors. Write your group's result for each measurement in your table.

4 When all measuring is complete, obtain the results of the other groups and enter them in your table as well.

5 Using the data for all of the groups, calculate the average for each measurement and enter this in the last column.

6 Discuss the results as a class and answer the following questions.

a What problems were encountered in taking the measurements?

b Do you think that averaging all the groups' results gave a more reliable outcome?

c How sure are you that your results were as accurate as they could have been?

d What errors other than absolute error may have occurred in any of the measurements?

ISBN 9780170413503

7.04 Significant figures

Significant figures

When we're talking about large quantities, we often use an approximate amount. In a disaster, for example, the media could report that a bushfire burned out 3 000 000 hectares, or that 20 000 people were stranded because of floods. Just as numbers can be rounded to decimal places, they can also be rounded to **significant figures**.

With large numbers, we start counting significant figures from the first digit. With small decimals, we start counting from the first digit that is not 0.

EXAMPLE 6

Round each number correct to two significant figures.

a 28 315 008   **b** 32 724   **c** 0.006 952

Solution

a The first two significant figures are 28. Decide whether 28 315 008 is closer to 28 000 000 or 29 000 000. The third digit is 3, which is less than 5, so round down.

28 315 008 ≈ 28 000 000

b The first two significant figures are 32. The third digit is 7, which is bigger than 5, so round up.

32 724 ≈ 33 000

c With 0.006 952, the first two significant figures are 69 because the 0s are not significant. Decide whether 0.006 952 is closer to 0.0069 or 0.0070. The third digit is 5, so round up.

0.006 952 ≈ 0.0070

- When we round to two decimal places, we have two digits following the decimal point.

- When we round to two significant figures, we put only the two most important digits in the amount.

Exercise 7.04 Significant figures

1 The city of Nelson Springs has a population of 372 413. Write this population correct to two significant figures.

Example 6

2 Ian's superannuation account contains $105 496. Round this amount to:

a 2 significant figures   **b** 1 significant figure   **c** 3 significant figures.

3 Express $482 356 correct to:

   a 1 significant figure   **b** 2 significant figures   **c** 3 significant figures.

4 Express each decimal correct to two significant figures.

   a 0.004 623   **b** 0.025 73   **c** 0.019 623   **d** 0.000 466

5 a In 2011, the population of Melbourne was 3 999 982. Express this amount correct to three significant figures.

   b By 2014, Melbourne's population had grown to 4 250 000. Write a possible population that can be rounded to 4 250 000.

   c Melbourne's population is predicted to reach 7 739 000 by 2051. By how much is Melbourne's population predicted to increase between 2011 and 2051? Express your answer correct to three significant figures.

6 Write each number correct to one significant figure.

   a 2914.23   **b** 3.2548   **c** 11 950

   d 0.005 134   **e** 20.46   **f** 0.6517

7 Evaluate each expression correct to two significant figures:

   a $0.2 \div 0.3$ **b** $11 \div 1990$ **c** $16 \div 12$

   d $\sqrt{0.0075}$ **e** $9\,300\,000 \times 0.085$ **f** 2.7^2

   g $\sqrt{560}$ **h** $\sqrt{5.6}$ **i** 3.4×9.9

8 Express each of the following measurements correct to three significant figures.

   a mass of an aeroplane, 351 540 kg

   b average depth of the Pacific Ocean, 4188 m

   c speed of a greyhound, 67.14 km/h

   d capacity of a tablespoon, 14.79 mL

   e distance from the Earth to the Sun, 149 573 881 km

7.05 Scientific notation

Scientific notation
puzzle

Scientific notation (or **standard form**) is a way of writing very large numbers and very small numbers.

It shows numbers as a decimal between 1 and 10 multiplied by a power of 10.

Australia's national debt was $516 000 000 000 in 2017. In scientific notation, this amount is 5.16×10^{11}.

The diameter of an atom is 0.000 000 03 m. In scientific notation, this length is 3×10^{-8}. Very small numbers have a negative power of 10.

EXAMPLE 7

Write each number in scientific notation.

a   190 000   b   43 000 000   c   0.0054   d   0.000 087 4

Solution

a   Make the decimal 1.9, then count how many $190\ 000 = 1.9 \times 10^5$
    places the decimal point has moved to
    the left.

    1.9 0 0 0 0   Five places to the left: 10^5

b   Make the decimal 4.3, then count the places $43\ 000\ 000 = 4.3 \times 10^7$
    the point has moved.

    4.3 0 0 0 0 0 0   Seven places to the left: 10^7

c   Make the decimal 5.4, then count places to the $0.0054 = 5.4 \times 10^{-3}$
    right.

    0.0 0 5.4   Three places to the right: 10^{-3}

    > Large numbers have positive powers
    > of 10, small numbers have negative
    > powers of 10.

d   Make the decimal 8.74. $0.000\ 087\ 4 = 8.74 \times 10^{-5}$

    0.0 0 0 0 8.7 4   Five places to the right: 10^{-5}

EXAMPLE 8

Express each number in normal decimal form.

a 3.29×10^4 b 9.1×10^6 c 2.5×10^{-3} d 5.8×10^{-5}

Solution

a   Move the decimal point 4 places to the right. $3.29 \times 10^4 = 32\ 900$
    Add 0s where necessary.

b   Move the decimal point 6 places to the right. $9.1 \times 10^6 = 9\ 100\ 000$

c   The negative power of 10 means this is a small $2.5 \times 10^{-3} = 0.0025$
    number. Move the decimal point 3 places to
    the left, adding zeros where necessary.

d   The negative power of 10 means this is a small $5.8 \times 10^{-5} = 0.000\ 058$
    number. Move the decimal point 5 places to
    the left, adding zeros where necessary.

You can enter numbers in scientific notation into your calculator using the $\boxed{\times 10^x}$ or $\boxed{\text{EXP}}$ key.

EXAMPLE 9

Evaluate $(6.5 \times 10^6) \times (2.2 \times 10^{-2})$.

You don't need to enter the brackets on the calculator.

Solution

On your calculator, enter:

6.5 $\boxed{\times 10^x}$ 6 $\boxed{\times}$ 2.2 $\boxed{\times 10^x}$ –2 $\boxed{=}$

$(6.5 \times 10^6) \times (2.2 \times 10^{-2})$

$= 143\ 000$ or 1.43×10^5

Exercise 7.05 Scientific notation

1 Write each number in scientific notation.

   a   860 000   **b**   9 140 000 000   **c**   2010

   d   0.000 36   **e**   0.000 000 0018   **f**   0.000 101

2 Write each number in normal decimal form.

   a 3.52×10^6 **b** 3.5×10^4 **c** 8.7×10^3 **d** 6.45×10^8

   e 6.1×10^{-3} **f** 1.93×10^{-5} **g** 5×10^{-7} **h** 1×10^{-2}

3 A gigametre is a billion (thousand million) metres. Use scientific notation to express 1 gigametre in metres.

4 The thickness of a soap bubble is 0.000 000 1 metres. Write this thickness in scientific notation.

5 Which has the larger value 7×10^6 or 6×10^7, and by how much?

6 Evaluate each expression.

   a $(5 \times 10^7) \times (2 \times 10^4)$ **b** $(4 \times 10^4) \times (1.6 \times 10^{-2})$

   c $(3.8 \times 10^6) \times (5.7 \times 10^3)$ **d** $(7.8 \times 10^7) \times (4.5 \times 10^{-3})$

   e $(4.4 \times 10^8) \div (4 \times 10^5)$ **f** $(9.6 \times 10^3) \div (1.6 \times 10^{-4})$

   g $(7.2 \times 10^6) \div (2.4 \times 10^{-3})$ **h** $(4.7 \times 10^8) \div (1.25 \times 10^{-2})$

7 The Horsehead Nebula, one of the sky's most distinctive features, is 1500 light years from Earth. One light year equals 9.46×10^{12} km. How many kilometres is it from Earth to the Horsehead Nebula? Express your answer in scientific notation.

NCM 11. Mathematics Standard (Pathway 1)   ISBN 9780170413503

8 How many times larger is the diameter of the Universe (6×10^{22} km) than the diameter of the Earth (1.3×10^4 km)? Express your answer in scientific notation correct to one decimal place.

9 A bacteria is approximately 3.5×10^{-3} mm wide and an influenza virus is approximately 2×10^{-7} mm wide. Which of them is wider and by how much?

10 The diameter of a HIV virus is approximately 9×10^{-6} mm. Approximately how many HIV viruses placed side by side would be required to make a line 1 metre long? Answer to the nearest whole number.

11 Express each number in scientific notation correct to two significant figures.

a 53 467 892	**b** 146 089	**c** 2453	
d 0.000 457 3	**e** 0.002 652	**f** 0.102 05	

12 In 2017, Australia's population was 24 379 972 and the average debt per person was $21 188. Calculate the total personal debt in Australia in scientific notation correct to three significant figures.

7.06 Pythagoras' theorem

Pythagoras was an ancient Greek mathematician who lived from 580 to 500 BCE. His famous theorem relates the sides of a right-angled triangle.

Pythagoras' theorem time trial

Pythagoras' theorem

In a right-angled triangle, the square of the **hypotenuse** is equal to the sum of the squares of the other two sides:

$$(\text{hypotenuse})^2 = (\text{side 1})^2 + (\text{side 2})^2.$$

$$c^2 = a^2 + b^2.$$

The **hypotenuse** is the longest side of a right-angled triangle.

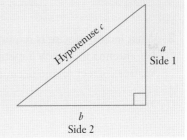

EXAMPLE 10

Use Pythagoras' theorem to find the height (h) of this roof truss. Answer correct to one decimal place.

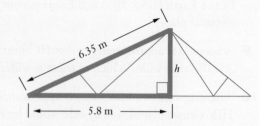

Solution

The hypotenuse is 6.35 and one side is 5.8.

Write Pythagoras' theorem.

Solve the equation for h.

$$c^2 = a^2 + b^2$$

$$6.35^2 = 5.8^2 + h^2$$

$$h^2 = 6.35^2 - 5.8^2 = 6.6825$$

$$h = \sqrt{6.6825}$$

$$= 2.585$$

$$\approx 2.6 \text{ m}$$

Exercise 7.06 Pythagoras' theorem

1 The diagram represents a wheelchair ramp. The 36 m long ramp covers a horizontal distance of 35 m. Use Pythagoras' theorem to calculate the rise h m, of the ramp. Answer correct to one decimal place.

Example
10

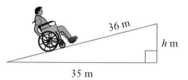

2 What length of wire is required to connect the top of a 23 metre TV antenna to a hook 6 metres from the base of the antenna? Answer correct to one decimal place.

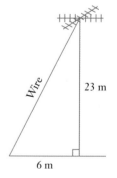

3 Saravanan installs a new section of pipe to join two existing pipes. Calculate the length of the new section of pipe. Express your answer in metres correct to the nearest millimetre.

When a measurement is in metres, the nearest millimetre means 3 decimal places.

4 Ivana is flying a small plane at 160 km/h against a 50 km/h wind, as shown in the diagram. Calculate the plane's ground speed correct to the nearest km/h.

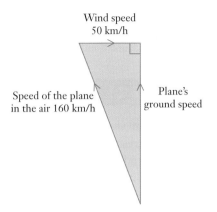

5 John is laying a concrete slab 12 m by 3 m in front of his shed. He uses Pythagoras' theorem to check that the corners of the slab are right angles. How long should the diagonal be? Express your answer in metres correct to the nearest centimetre.

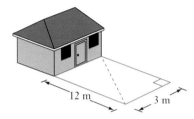

6 The school cross-country course is in the shape of a right-angled triangle. The first leg is 1200 m and the second leg is 950 m. The third leg, through thick scrub, is difficult to measure. Calculate its length correct to the nearest metre.

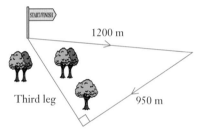

7 Belle had to cross the Paroo River to get from Mud Springs to Eulo. The new bridge across the river was closed for repairs so she had to use the old bridge. How much further did she have to travel using the old road and old bridge compared to the direct route across the new bridge? Answer correct to one decimal place.

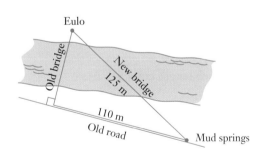

8 David is the pilot of a small plane. He planned to fly 200 km from Mount Surprise to Cairns. Because of poor weather conditions between Mount Surprise and Cairns, David flew 135 km due north to Chillagoe, then turned due east and flew to Cairns. Calculate the straight line distance from Chillagoe to Cairns correct to the nearest kilometre.

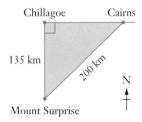

9 Sketch two different right-angled triangles that have a hypotenuse of length 24 m, writing the lengths of the other two sides correct to two decimal places.

7.07 Perimeter

When we measure the distance around the outside of a shape, we are measuring its **perimeter**.

The diagram shows the dimensions of a
roof in metres. Stuart is going to install gutters
around the edge of the roof.
The guttering costs $11.67 per metre and
it comes in 3 metre lengths.

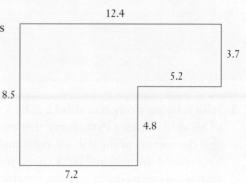

a   Calculate the perimeter of the roof.

b   How many lengths of guttering will
Stuart need?

c   Calculate the cost of the guttering.

Solution

a   Add the sides for the perimeter.

Perimeter = 8.5 + 12.4 + 3.7 + 5.2 + 4.8 + 7.2

= 41.8 m

b   Guttering comes in 3 m lengths, so
divide the perimeter by 3.

Number of lengths = 41.8 ÷ 3

= 13.9333 …

≈ 14

> Always round *up* when calculating
> amounts to purchase.

Stuart needs to buy 14 lengths.

c   Each length costs $11.67.

Total cost = 14 × 11.67

= $163.38

Exercise 7.07 Perimeter

1 Calculate the perimeter of each shape.

a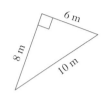

b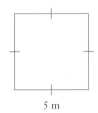

c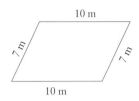

2 To decorate the nursery for her new baby, Menhal is putting a wallpaper frieze around the tops of the walls. The rectangle shows the dimensions of the room.

Example **11**

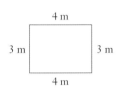

A **frieze** is a small pattern that goes around walls.

a What is the perimeter of the room?

b The frieze is available in 5 m rolls for $18.50 and in 10 m rolls for $29.95. How many of each size roll do you recommend Menhal buy?

c Calculate the cheapest cost for the wallpaper frieze.

3 The sides of a rectangle are 4 m and 11 m. Calculate its perimeter.

4 Measure the sides of each quadrilateral accurately, then calculate its perimeter.

a

b

5 The perimeter of a rectangle is 36 m. What could the lengths of the sides of the rectangle be? Suggest two possible sets of sides.

6 All the sides in a **regular polygon** are the same size. Calculate the perimeter of each regular polygon below.

a

5 m

b

4 cm

c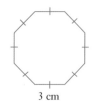

3 cm

7 Chen is replacing the skirting boards around the floor in his TV room.

   a Calculate the length of skirting boards Chen will need.

   b The boards cost $14 per 3 metre length. How much will the boards cost?

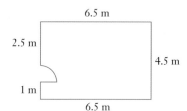

8 Roulla is installing new guttering to a shed and she is connecting the guttering to a rainwater tank.

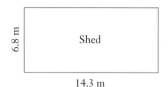

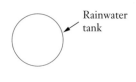

Rainwater tank

HARDWARE COSTS
Guttering $11.67 per metre
in 3 m lengths
Pipes $44.90 per 6 m length

To complete the job, Roulla will need guttering and 9 m of pipe. Calculate the cost of the guttering and extra piping.

7.08 Perimeter of circular shapes

The perimeter of a circle is called its **circumference**.

> There are two formulas for the **circumference of a circle**:
>
> - $C = \pi \times d$ where d is the **diameter** of the circle
>
> - $C = 2\pi \times r$ where r is the **radius** of the circle.

EXAMPLE 12

To stop joins from leaking,
Roulla wraps plumbers'
tape around pipes before she
screws them together.
How many metres of tape
(rounded up to one
decimal place) will she use
when she puts 6 wraps
around a pipe of diameter 150 mm?

Shutterstock.com/rukawajung

Solution

$C = \pi \times d$ where $d = 150$.

Do not round the partial answer.

$$\text{Circumference of pipe} = \pi \times 150$$
$$= 471.2388 \ldots \text{ mm}$$

For 6 wraps, multiply by 6.

$$\text{Total amount of tape} = 471.2388 \ldots \times 6$$
$$= 2827.4333 \ldots \text{ mm}$$

Convert to metres.

$$= 2827.4333 \ldots \div 1000 \text{ m}$$
$$= 2.8274333\ldots \text{ m}$$
$$\approx 2.9 \text{ m}$$

Answer the question.

Roulla will need 2.9 m of tape to wrap the pipe.

EXAMPLE 13

This logo's shape is a sector with a central angle of 270°
and a radius of 8 mm. Calculate its perimeter.

270°

8 mm

Solution

A sector is a fraction of a circle and
there are 360° in a circle.

$\text{Perimeter} = \dfrac{270}{360}$ of the circumference + 2 radii.

$$\text{Perimeter} = \dfrac{270}{360} \times 2\pi \times 8 + 2 \times 8$$
$$= 53.6991 \ldots$$
$$\approx 53.7 \text{ mm}$$

Radii is the plural form of
radius. One radius, two radii.

Exercise 7.08 Perimeter of circular shapes

1 Calculate correct to one decimal place the circumference of a circle with:

   a   radius 12 mm   **b**   diameter 15 cm

   c   radius 4.75 m   **d**   diameter 5.75 mm.

2 The Central Coast Lions run around the outside of a circular field as part of their training. The diameter of the field is 180 m.

   a   How far do they run (to the nearest metre) when they complete 8 laps of the field?

   b   The coach wants the team to run 2 km. How many laps of the oval will the team have to run to complete this distance? Answer correct to one decimal place.

3 Calculate correct to one decimal place the perimeter of each shape.

   a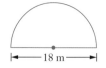
   b
   c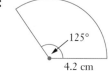

4 The diagram shows a belt around a pair of pulleys. Calculate the length of the belt correct to three significant figures.

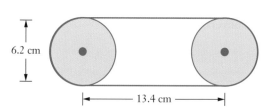

5 A traditional softball field is in the shape of a quadrant of a circle. What is the perimeter of the field? Answer to the nearest metre.

> A **quadrant** is a quarter of a circle.

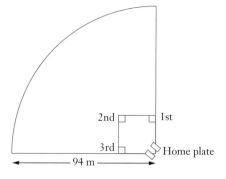

COMPLETE THE BLANKS

Copy and complete this summary of the chapter.

Throughout history, people have used different ways to measure items and it was quite confusing. In the 18th century, the French developed a measuring system called the **1**_____ system. In this system, the metre is the basic unit for measuring **2**_____. For longer lengths, we use **3**_____ which are equivalent to 1000 m. To measure small lengths, we can use millimetres which are **4**_____ of a metre.

We use length units when we measure the distance around the outside of a shape, which is called the **5**_____.

The basic unit for measuring **6** _____ is the gram. One kilogram is equivalent to 1000 g and 1000 kg is called one **7** _____.

SOLUTION TO THE CHAPTER PROBLEM

Problem

Jake is going to Scotland to 'bag' (climb) as many Munros as he can. Munros are Scottish mountains that are more than 3000 feet (914 metres) high. At 1.34 km above sea level, Ben Nevis is Scotland's highest Munro. How many metres higher than the qualifying height of 914 metres is the top of Ben Nevis?

Solution

Convert 1.34 km to metres to compare with 914 m.

$$1.34 \text{ km} = 1.34 \times 1000 \text{ m}$$

$$= 1340 \text{ m}$$

$$\text{Height difference} = 1340 - 914$$

$$= 426 \text{ m}$$

Ben Nevis is 426 metres higher than the qualifying height for Munros.

- What parts of this chapter were new to you?

- Give examples of jobs where you would use the metric system and make calculations involving perimeter.

- Write any difficulties you had with the work in this chapter. Ask your teacher to help you.

Copy and complete this mind map of the chapter, adding detail to its branches. Use pictures, symbols and colour where needed. Ask your teacher to check your work.

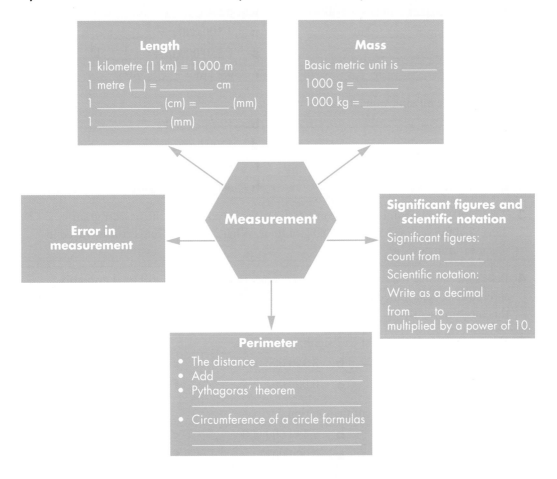

Length
1 kilometre (1 km) = 1000 m
1 metre (__) = _____ cm
1 _____ (cm) = _____ (mm)
1 _____ (mm)

Mass
Basic metric unit is _____
1000 g = _____
1000 kg = _____

Measurement

Error in measurement

Significant figures and scientific notation
Significant figures:
count from _____
Scientific notation:
Write as a decimal
from ___ to _____
multiplied by a power of 10.

Perimeter
- The distance _____
- Add _____
- Pythagoras' theorem _____
- Circumference of a circle formulas _____

7. TEST YOURSELF

Exercise 7.01

1 Copy and complete each statement.

a 2 min = _____ s   **b** 750 g = _____ kg

c 8500 m = _____ km   **d** 25 g = _____ mg

e 5.6 L = _____ mL   **f** 0.3 h = _____ min

g 48 h = _____ days   **h** 11 cm = _____ mm

i 266 m = _____ mm   **j** 95 700 L = _____ kL

k 0.9 kg = _____ mg   **l** 8 h = _____ s

m 85 000 cm = _____ km   **n** 4 500 000 g = _____ t

o 0.98 ML = _____ L   **p** 7 550 000 mL = _____ kL

q 10 800 s = _____ h   **r** 4.2 km = _____ mm

Exercise 7.02

2 Calculate the missing value in each statement.

a 5.6 kg = ___ g   **b** 18 t = ____ kg

c 4500 g = ___ kg   **d** 6200 kg = ___ t

e 930 kg = ____ g   **f** 400 g = ____ kg

g 4.1 g = _____ mg   **h** 600 mg = ___ g

Exercise 7.03

3 Tina is making bread. Her electronic scales measure correct to the nearest gram. Tina used the scales to measure 180 g of flour. What is the absolute error involved in this measurement?

Exercise 7.03

4 Bella's thermomix measures ingredients correct to the nearest 5 g. The thermomix showed that Bella added 65 g of sugar to the mixture. What are the limits of accuracy of this measurement?

Exercise 7.04

5 Write each number correct to 2 significant figures.

a 741 600   **b** 8 340 000 000   **c** 4030

d 0.000 722   **e** 0.000 003 495   **f** 0.000 206

Exercise 7.05

6 Write each number in scientific notation.

a 740 000   **b** 8 340 000 000   **c** 4030

d 0.000 722   **e** 0.000 0034   **f** 0.000 206

Exercise 7.05

7 Write each number in normal decimal form.

a 5.6×10^6 **b** 9.1×10^4 **c** 4.1×10^{-3} **d** 1×10^{-4}

8 Find the value of *h* correct to one decimal place.

a

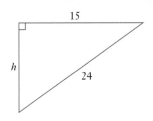

15

h

24

b

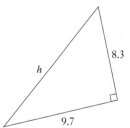

8.3

h

9.7

9 Calculate the perimeter of each shape.

a

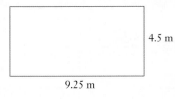

4.5 m

9.25 m

b

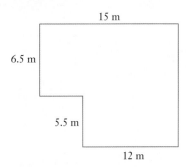

15 m

6.5 m

5.5 m

12 m

c

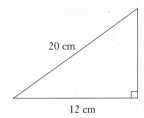

20 cm

12 cm

Find the height of the triangle first!

10 Calculate the perimeter of each shape correct to one decimal place.

a

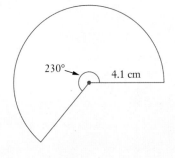

230°   4.1 cm

b

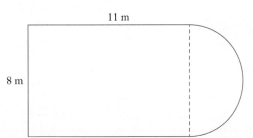

11 m

8 m

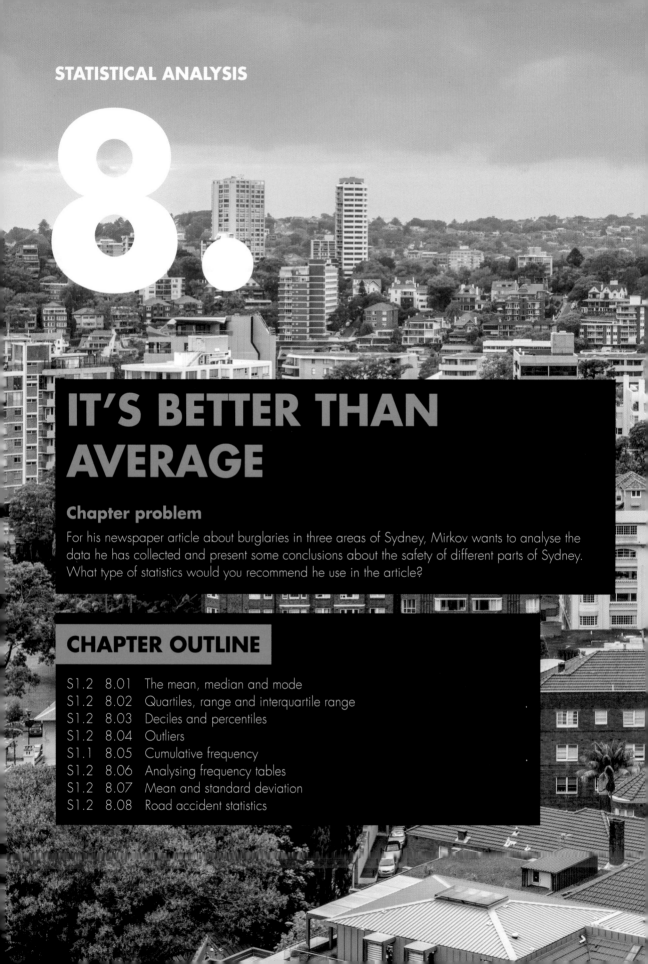

STATISTICAL ANALYSIS

8.

IT'S BETTER THAN AVERAGE

Chapter problem

For his newspaper article about burglaries in three areas of Sydney, Mirkov wants to analyse the data he has collected and present some conclusions about the safety of different parts of Sydney. What type of statistics would you recommend he use in the article?

CHAPTER OUTLINE

WHAT WILL WE DO IN THIS CHAPTER?

- Find the mean, median and mode of a data set, including data presented in a frequency table and data grouped into class intervals
- Calculate the range, the quartiles and the interquartile range of a data set
- Interpret quartiles, deciles and percentiles from a graph
- Identify outliers in a data set and their effects on the mean and median
- Construct cumulative frequency histograms and polygons
- Use cumulative frequency graphs to find the median and interquartile range
- Use the statistics mode on a calculator to find the mean and standard deviation
- Examine statistics involving road accidents

HOW ARE WE EVER GOING TO USE THIS?

- When calculating and examining statistics to compare data
- When interpreting different 'averages' and deciding which is best in different situations
- When deciding whether data has been analysed correctly in order to support opinions

8.01 The mean, median and mode

Statistical
measures

Sets of **data** are often too big to understand just by looking at them. We can analyse them by finding a typical or central value to represent all of the data. There are three types of averages or **measures of central tendency**.

Mean,
median and
mode

The **mean** is calculated by adding all the scores and dividing by the number of scores. This is what most people call the 'average'.

$$\text{Mean} = \frac{\text{sum of scores}}{\text{number of scores}}$$

The **median** is the middle score when the scores are arranged from smallest to largest. If there is an even number of scores, it is the average of the two middle scores.

The **mode** is the most common or frequent score(s). | If all scores occur the same number of times, there is no mode.

	Features	When is it best to use?
Mean	Depends on all the scores. Is affected by extreme scores (outliers).	When the **data set** does *not* have many outliers. Not suitable for categorical data.
Median	Can be one of the scores. Is not affected by outliers.	When the data set has outliers. Not suitable for categorical data.
Mode	May be more than one mode, or no mode at all. Is not affected by outliers.	When the most common score is needed. Suitable for categorical data.

EXAMPLE 1

The daily maximum temperatures (in °C) for 15 days in Cairns in July were:

32	30	31	32	31	30	31	31

31	31	29	25	28	27	29

For this set of data find:

a   the median   **b**   the mode   **c**   the mean correct to two decimal places.

Solution

a   Put the scores in order.

There are 15 scores, so the middle one will be the 8th score.

25, 27, 28, 29, 29, 30, 30, ③1, 31, 31, 31,
31, 31, 32, 32

The median is 31.

b 31 occurs 6 times, more often than any other score.

Mode is 31.

c $\text{Mean} = \dfrac{\text{sum of scores}}{\text{number of scores}}$

$$\text{Mean} = \dfrac{448}{15}$$

$$= 29.86666\ldots$$

$$\approx 29.87$$

EXAMPLE 2

This stem-and-leaf plot shows the ages of people enrolled in a computer course.

For the data shown find:

a the median

b the mode

c the mean.

Which measure best represents the data?

Stem	Leaf
1	7 7 8
2	1 4 4 4 8 9
3	0 4 5
4	4 6 7
5	2

Solution

a The scores are already in order. There are 16 scores, so the middle will be between the 8th and 9th scores.

Stem	Leaf
1	7 7 8
2	1 4 4 4 ⑧ ⑨
3	0 4 5
4	4 6 7
5	2

$$\text{Median} = \dfrac{28+29}{2}$$

$$= 28.5$$

b 24 occurs most often.

The mode is 24.

c $\text{Mean} = \dfrac{\text{sum of scores}}{\text{number of scores}}$

$$\text{Mean} = \dfrac{490}{16}$$

$$= 30.625$$

These measures are quite different, as the ages are spread out from 17 to 52. In this case, the median best represents the data as it is in the middle.

Exercise 8.01 The mean, median and mode

1 This data shows the number of people who breached their bail conditions in a large city each month for one year.

| 14 | 12 | 16 | 14 | 17 | 12 | 14 | 14 | 10 | 13 | 11 | 19 |

For this set of data find:

a the median   **b** the mode

c the mean (correct to one decimal place).

2 This is the data for the same offence in a country region over one year.

| 7 | 9 | 7 | 6 | 3 | 4 | 5 | 3 | 2 | 1 | 1 | 4 |

For this set of data find:

a the median   **b** the mode

c the mean (correct to one decimal place).

3 What are the differences in the statistics for the data in Question **1** and in Question **2**? Suggest a reason for these differences.

4 For this stem-and-leaf plot find:

a the median

b the mode

c the mean.

Which measure of central tendency best represents the data?

Stem	Leaf
1	5 6 8 9
2	1 3 4 5 8
3	6

5 The stem-and-leaf plot shows the ages of all employees at the CHATPHONE Company.

a Find:

    i the median   **ii** the mode   **iii** the mean.

b The three measures are different. Explain why this happens. Which measure best represents the data?

Stem	Leaf
1	7 8 9
2	6 8 9
3	1 2 6 6 7 8
4	2

6 Find the mean, the median and the mode for each set of data.

a The distance (in kilometres) a salesperson travels each working day for 2 weeks:

| 45 | 80 | 65 | 50 | 45 | 41 | 50 | 40 | 45 | 50 |

b The number of seats filled on an aircraft over seven journeys:

| 56 | 104 | 78 | 100 | 89 | 125 | 100 |

c The ages of the Maths teachers in a school:

| 49 | 39 | 37 | 41 | 27 | 25 | 41 | 47 | 39 |

7 For each set of data in Question **6**, which measure best represents the data? Why?

8 In Year 10, students are awarded a grade (A, B, C, D or E) for their achievement in English. One class obtained the following results:

Grade	A	B	C	D	E
Number of students	4	9	8	6	3

  a What is the mode of the class results?

  b Is it possible to find the mean or the median for this data? Explain your answer.

9 Twelve people work in a small company. Their annual salaries are:

  $71 000 $67 000 $76 000 $64 000 $61 000 $63 000

  $72 000 $66 000 $73 000 $70 000 $275 000 $890 000

  a Calculate the mean salary for this company.

  b Find the median salary for this company.

  c Which measure (mean or median) is the better reflection of a typical salary for this company? Explain your answer.

  d If you wanted to attract new employees to the company, which would you advertise as the average salary? Why?

  e For this company, why would the mode be an inappropriate measure to use to describe the typical salary?

10 Darryl is a market gardener. In the last five years, his annual profits were:
$2000, $57 000, $61 000, $62 000 and $65 000.

  a Find the mean and the median for this data.

  b Which measure gives the more accurate impression of Darryl's profit? Why?

  c In what situation might Darryl use the other measure to describe his usual profit?

11 A developer has 40 new apartments in a new building for sale. The 20 apartments on the first 5 floors are priced at $330 000 each. The next 8 apartments are priced at $380 000 and then the next 8 are priced at $425 000. The 3 apartments on the second highest floor are $835 000 and the penthouse apartment is priced at $1.7 million.

  a Determine the mean and the median price of the apartments.

  b When the developer is advertising the apartments for sale, which measure should he use? Explain your answer.

  c The developer will be speaking to potential investors in his company. What measure do you think he will use to make his company look profitable?

12 Simon and Daniel's batting scores for their past seven innings of cricket are:

Simon: 45   48   53   38   32   40   51

Daniel: 23   44   6   125   60   15   37

   a   Calculate the mean score for each player correct to one decimal place.

   b   Which player is better if the mean is used to judge them?

   c   Find the median score for each player.

   d   Which player would you rather have in your cricket team? Why?

8.02 Quartiles, range and interquartile range

The **range** and the **interquartile range** (**IQR**) are both **measures of spread** (how the data is spread out).

Range

$$\textbf{Range} = \text{highest score} - \text{lowest score}$$

It can give a false impression of the data if there are outliers (very high or very low scores).

Quartiles

When a set of data is arranged from smallest to largest, **quartiles** divide the data into four equal groups. 25% of the data is in each quartile.

Lowest score	1st quartile (Q_1)	2nd quartile (Q_2)	3rd quartile (Q_3)	Highest score

Q_1 is called the **lower quartile** or 1st quartile.

Q_3 is called the **upper quartile** or 3rd quartile.

Q_2 is the 2nd quartile and is the same as the **median**.

Interquartile range

$$\textbf{Interquartile range} = \text{upper quartile} - \text{lower quartile}$$

$$\text{IQR} = Q_3 - Q_1$$

The interquartile range is not as affected by outliers and covers the middle 50% of the scores.

EXAMPLE 3

The ages of the 23 people at a cafe are shown here.

33 23 28 36 27 15 32 18 13 13 38 38

27 7 34 27 12 26 33 21 24 39 20

Find:

a each quartile **b** the range **c** the interquartile range.

Solution

a Place the ages in ascending order and find the middle score; it is the median but it is also the 2nd quartile.

7, 12, 13, 13, 15, 18, 20, 21, 23, 24, 26, (27), 27, 27, 28, 32, 33, 33, 34, 36, 38, 38, 39

$Q_2 = 27$

The median splits the scores into two halves. Find the middle of each half of the scores. These are the 1st and 3rd quartiles.

7, 12, 13, 13, 15, (18), 20, 21, 23, 24, 26, (27), 27, 27, 28, 32, 33, (33), 34, 36, 38, 38, 39

$Q_1 = 18$, $Q_3 = 33$

The quartiles are 18, 27 and 33.

b Range = highest score – lowest score

Range = 39 – 7
= 32

c Interquartile range = $Q_3 - Q_1$

Interquartile range = 33 – 18
= 15

Exercise 8.02 Quartiles, range and interquartile range

Keep your answers to this exercise because you will need them for Exercise 14.01 Boxplots.

1 The number of fish sold each day at lunchtime in a fish and chip shop during the month of August is shown below. (Example 3)

17 27 28 18 18 17 19 19 25 27 17 19 20 19 21 26

28 18 19 20 17 19 23 24 20 18 17 20 19 27 28

For this data find:

a the range **b** each of the quartiles **c** the interquartile range.

2 The following data shows the daily maximum temperatures (in °C) for 15 days in Cairns in July.

32 30 31 32 31 30 31 31 31 31 29 25 28 27 29

For this data find:

a   the range   **b**   the upper and lower quartiles   **c** the interquartile range.

3 The heights of 25 Year 11 students in centimetres are:

151	167	181	172	179	155	159	162	169	174
178	180	158	166	171	168	157	160	175	172
150	169	163	170	176					

Find:

a   the range   **b**   each quartile   **c** the interquartile range.

4 A Year 11 class was surveyed to find the number of hours each student spent on homework each week. These are the results:

4	5	7	8	3	6	9	9	8	4
7	5	3	3	4	9	3	4	7	8

> If there is an even number of scores, then the quartile is the average of the two middle scores.

Find:

a   the range   **b**   each quartile   **c** the interquartile range.

5 The following data is a record of the number of thefts from retail stores in two regions.

Inner Sydney: 13, 13, 13, 12, 16, 25, 23, 20, 33, 25, 27, 25,

55, 20, 27, 33, 28, 26, 38, 24, 33, 55, 42, 48

South Coast: 62, 60, 52, 62, 52, 63, 60, 65, 74, 61, 36, 66

36, 69, 70, 47, 39, 64, 69, 55, 40, 60, 58, 52

a   For each set of data find:
  i the range   **ii** each quartile   **iii** the interquartile range.

b   Comment on the differences between these two sets of data.

DECILES AND PERCENTILES

When we divide an ordered group into 4 equal subgroups, we are making **quartile** groups.

When we sort a large, ordered group into 10 equal subgroups we are making **decile** groups.

When we sort a large ordered group into 100 equal subgroups, we are making **percentile** groups.

Now imagine that all the students in your school are lined up on the school oval from the shortest to the tallest. Find out how many students there are in your school.

1   If the students were divided into 100 approximately equal groups, how many would be in each group?

2   Each of the 100 groups contains 1% of the whole school.

•   The height of the tallest person in the shortest group is called the 1st percentile.

•   The height of the tallest person in the second group is called the 2nd percentile.

•   The height of the tallest person in the tallest group is called the 100th percentile.

3   Which group do you think you would be in?

4   If the students were divided into 10 approximately equal groups, how many would be in each group?

5   Each of the 10 groups is 10% of the whole group.

   •   The height of the tallest person in the shortest group is called the 1st decile.

   •   The height of the tallest person in the second group is called the 2nd decile.

   •   The height of the tallest person in the tallest group is called the 10th decile.

6   Which group do you think you are in?

7   In which groups would you expect most of the Year 8 students to fall?

8   In which groups would you expect most Year 12 students to fall?

9   How many percentile groups are in each decile group?

10   How many percentile groups are in one quartile group?

11   Why is the 10th percentile the same as the 1st decile?

12   What percentile is the same as the 9th decile?

8.03 Deciles and percentiles

Previously, we divided data into four equal parts using quartiles. Another way of sorting large sets of data into groups is to divide it using **deciles** and **percentiles**.

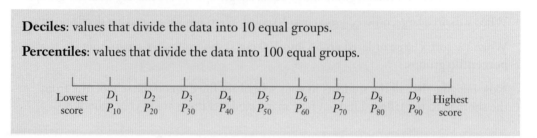

Deciles: values that divide the data into 10 equal groups.

Percentiles: values that divide the data into 100 equal groups.

Deciles

Deciles (D_1, D_2, D_3, D_4, D_5, D_6, D_7, D_8 and D_9) separate data into tenths. 'Deci-' means $\dfrac{1}{10}$.

- D_1 cuts off the lowest 10% of scores

- D_4 cuts off the lowest 40% of scores

- D_9 cuts off the lowest 90% of scores (or the top 10% of scores)

EXAMPLE 4

This graph shows the changes in annual income over 20 years using the median and 3rd and 7th deciles.

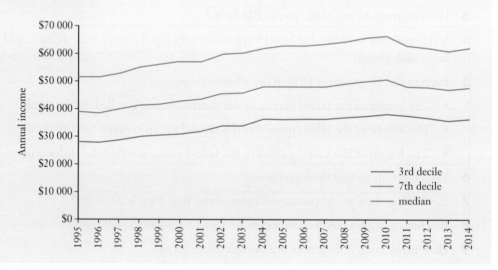

---

a What decile is the median?

b In 2002, what income was on the 7th decile?

c In 2008, Grant's annual income was above 70% of the population. What was his income?

d In 1995, between what two values were the middle 40% of incomes?

Solution

a The median is the middle. Half of 10 is 5.

The median is the 5th decile.

b Read from the graph to the top line for 2002.

From the graph, approximately $60 000.

c Top 30% means above the 7th percentile. Read from the graph to the top line for 2008.

From the graph, Grant's income was approximately $65 000.

d The middle 40% is between the 3rd and 7th deciles.

In 1995:

- the 3rd decile was approximately $28 000

- the 7th decile was approximately $52 000.

So, the middle 40% of incomes were between $28 000 and $52 000.

Percentiles

Percentiles ($P_1, P_2, P_3, \ldots P_{99}$) separate data into hundredths. 'Percent' means $\dfrac{1}{100}$.

- P_{24} cuts off the lowest 24% of scores

- P_{60} cuts off the lowest 60% of scores

- P_{87} cuts off the lowest 87% of scores (or the top 13% of scores)

EXAMPLE 5

This graph shows the annual income percentiles for last year.

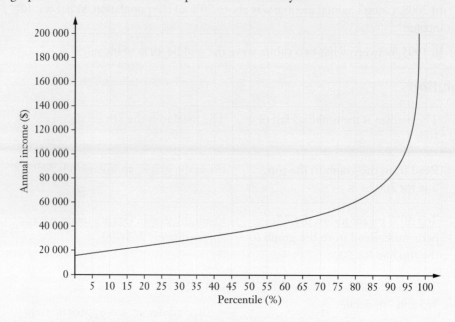

a What income is at the 20th percentile?

b Georgina earns $40 000 per year. What percentile is this?

c Give the approximate difference in income between the 80th and 90th percentiles.

d The graph only goes up to the 99th percentile. Suggest a possible reason.

Solution

a Find the 20th percentile on the horizontal axis and read off the value.

The income is about $26 000.

b Find $40 000 on the vertical axis and read off the value.

It is the 54th percentile.

c Find the incomes for the 80th and 90th percentiles.

Difference means subtract.

80th percentile: $64 000

90th percentile: $83 000

Difference: $83 000 − $64 000 = $19 000

d The 100th percentile would have very high incomes.

Including very high incomes makes it difficult, or nearly impossible, to have an appropriate scale on the vertical axis.

Exercise 8.03 Deciles and percentiles

1 Use the graph in Example 4 to answer these questions.

   a In 1999, what income was at the 3rd decile?

   b In 2000, what income was at the 5th decile?

   c Indira has an annual income of $50 000. In what year(s) is this at the 5th decile?

   d Between what two values were the middle 40% of incomes in 2014?

   e Harry earns $50 000 per year. In what year(s) does he drop out of the top 50% of the population for income?

2 Use the graph in Example 5 to answer these questions.

   a What income is at the 40th percentile?

   b Sue earns $60 000. What percentile is this?

   c What is the approximate difference in income between the 30th and 50th percentiles?

   d What is another name for the 50th percentile?

   e Approximately what percentage of the population earns more than $80 000?

3 This graph shows the results of a group of students on an exam out of 100. The results have been divided into 10 groups.

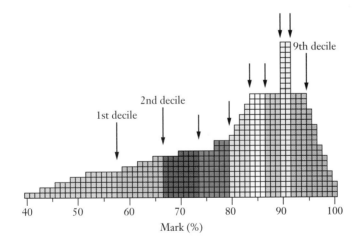

   a What percentage of students scored 80 or more?

   b Find the value that separates the bottom 70% of students from the top 30%.

   c Krystal's score was on the second decile. What did she score in the exam?

   d Fatima scored 75. Between which two deciles did she score?

   e Do you think this exam was easy or difficult? Give reasons for your answer.

4 This graph shows the rainfall in South East Queensland for the 2014–2015 financial year compared to long-term deciles and averages.

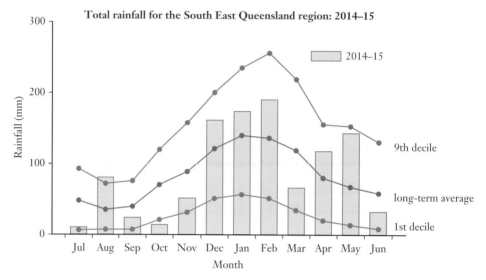

Total rainfall for the South East Queensland region: 2014–15

© Commonwealth of Australia Bureau of Meteorology CC BY 3.0 http://creativecommons.org/licenses/by/3.0/au/

a For how many months of the year was the rainfall below average?

b State the month in which the rainfall was below the 1st decile.

c State the month in which the rainfall was above the 9th decile.

d How often was the rainfall between the long-term average and the 9th decile?

e Was the 2014–2015 financial year a good or a poor year for rain? Give reasons for your answer.

ISBN 9780170413503

5 This percentiles chart shows the range of heights for boys aged 2 to 20 years.

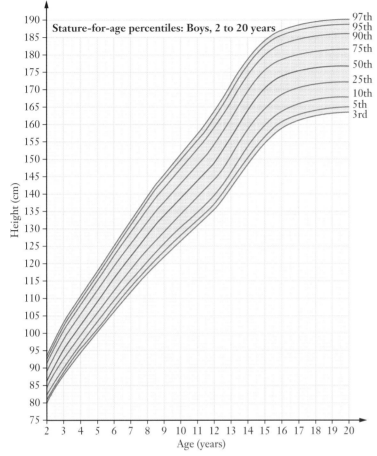

Source: National Center for Health Statistics (US) and National Center for Chronic Disease Prevention and Health Promotion (CDC).

a Adam is aged 9 years and is 129 cm tall. What percentage of boys his age are shorter than him?

b Justin is 11 years old and is 155 cm tall. What percentage of boys his age are shorter than him?

c How tall should Justin be when he turns 18?

d Liong is 103 cm tall, which is at the 1st decile for boys his age. How old is Liong?

e Asam is 16 years and his height is at the 3rd quartile.

　i What is Asam's height now?

　ii Predict Asam's height when he turns 20 years old.

8.04 Outliers

An **outlier** is a score that is much bigger or much smaller than the other scores.

If we see an outlier, we should decide whether the score is reasonable to include or is an error that has been wrongly recorded.

EXAMPLE 6

Peter asked 8 friends about the amount of pocket money they received each week. The results (in dollars) were:

| 20 | 32 | 40 | 18 | 32 | 18 | 75 | 32 |

a Identify the outlier in this data.

b Is the outlier reasonable or likely to be a wrongly recorded score? Explain your answer.

Solution

a Write the data in order.   18, 18, 20, 32, 32, 32, 40, 75

Identify the much bigger or smaller   The outlier is $75 because it is much
score.   bigger than the other scores.

b It is quite possible that one of Peter's friends receives a lot of pocket money, so this outlier is reasonable.

An outlier can also be identified by the use of a formula that uses the quartiles and interquartile range (IQR). A score is considered an outlier if it is a lot less than Q_1 or a lot more than Q_3.

Outliers

An **outlier** is a score that is either:

- less than $Q_1 - 1.5 \times IQR$ or
- greater than $Q_3 + 1.5 \times IQR$

EXAMPLE 7

These are the ages of the 23 people at a cafe from Example 3 on p. 215.

33	23	28	36	27	15	32	18	13	13	38	38
27	7	34	27	12	26	33	21	24	39	20	

We found that $Q_1 = 18$, $Q_3 = 33$ and IQR $= 15$.

a Find the value of:

  i $Q_1 - 1.5 \times \text{IQR}$ **ii** $Q_3 + 1.5 \times \text{IQR}$.

b Are any of the scores outliers? Explain your answer.

Solution

a **i** Substitute $Q_1 = 18$, $Q_3 = 33$, IQR $= 15$ into each formula.

$$Q_1 - 1.5 \times \text{IQR} = 18 - 1.5 \times 15$$
$$= -4.5$$

ii

$$Q_3 + 1.5 \times \text{IQR} = 33 + 1.5 \times 15$$
$$= 55.5$$

b So, any value below -4.5 and above 55.5 is an outlier. Place the data in ascending order.

7, 12, 13, 13, 15, 18, 20, 21, 23, 24, 26, 27, 27, 27, 28, 32, 33, 33, 34, 36, 38, 38, 39

There are no scores below -4.5 or above 55.5. So, none of the scores are outliers.

Exercise 8.04 Outliers

1 For each set of data, identify the outlier without using the formula.

  a 12 15 28 19 15 14 16

  b 7 5 6 8 7 1 8 6 9

  c 32 35 12 40 36 29 38 30

  d 94 49 35 38 31 44

Example
6

2 For each set of data in Question **1**, identify the outlier using the formula.

  i Calculate the quartiles and the interquartile range.

  ii Calculate $Q_1 - 1.5 \times \text{IQR}$ and $Q_3 + 1.5 \times \text{IQR}$

  iii Find the outlier that is beyond the above values and determine whether the outliers identified in Question **1** are also outliers when the formula is applied.

Example
7

3 Data can have more than one outlier:

| 6 | 8 | 0 | 6 | 8 | 16 | 8 | 7 | 2 | 8 | 9 | 9 | 8 |

   a   What are the outliers without using the formula?

   b   What are the outliers using the formula?

4 At a hospital, the triage nurse recorded the masses (in kilograms) of patients.

| 68 | 59 | 63 | 80 | 68 | 54 | 48 |

| 64 | 47 | 48 | 59 | 68 | 30 | 49 |

   a   Place these results in order from lowest to highest.

   b   What are the outliers in this data set (without using the formula)?

   c   Calculate $Q_1 - 1.5 \times IQR$ and $Q_3 + 1.5 \times IQR$.

   d   Are the outliers you identified in part **b** still outliers if you apply the formula test?

5 Katrina measured the heights of a sample of 25 Year 11 males in centimetres.

| 175 | 176 | 185 | 176 | 25 | 184 | 197 | 161 | 186 | 169 | 171 | 170 | 182 | 165 |

| 179 | 180 | 167 | 169 | 198 | 167 | 170 | 180 | 182 | 173 | 230 |

   a   Draw a stem-and-leaf plot of this data.

   b   What are the outliers for this data?

   c   For each outlier, decide if it is reasonable or a wrongly recorded height. Explain your answer.

6 Eleven houses were sold in Keswick Street over two years. The selling prices were:

| $620 000 | $625 000 | $700 500 | $738 000 | $625 000 | $1 800 000 |

| $598 000 | $612 000 | $696 500 | $720 000 | $705 000 |

   a   What is the outlier for this data?

   b   Is this outlier reasonable or likely to be a mistake? Explain your answer.

   c   Find the median sale price for the houses.

   d   Find the mean sale price.

   e   Which measure of central tendency best describes the prices of the houses in Keswick Street? Justify your answer.

   f   Calculate the mean of the remaining prices when this outlier is removed. Is this mean closer to the median you found in part **c**?

7 Twelve people work in a small business which sells electronics equipment. Their annual salaries are:

$71 000 $67 000 $76 000 $64 000 $61 000 $63 000

$72 000 $66 000 $73 000 $70 000 $275 000 $890 000

a What is the outlier(s) for this data?

b Is this outlier reasonable or likely to be a mistake? Explain your answer.

c Calculate the mean salary for this company. Answer to the nearest dollar.

d Find the median salary for this company.

e Which measure (mean or median) is the better reflection of a typical salary for this company? Justify your answer.

f If you wanted to attract new employees to the company, which would you advertise as the average salary? Why?

g If the outliers are removed, calculate the mean salary of the remaining scores. Is this mean closer to the median you found in part **d**?

8 An obstetrician is a doctor who specialises in providing care for expectant mothers and delivering their babies. The data below shows the number of caesarean deliveries performed by a sample of 20 Australian obstetricians in a 12-month period.

22	38	15	204	3	16	21	13	24	32
19	24	22	31	8	21	14	37	28	21

a What is the outlier in this set of data?

b Some people may accuse a doctor who performed a lot of caesarean sections of doing unnecessary operations. Why might one obstetrician need to perform a lot more caesarean deliveries than all the other doctors?

c Calculate the mean and the median of the 20 values.

d Calculate the mean and the median without the outlier.

e What do your answers to parts **c** and **d** show?

8.05 Cumulative frequency

The **cumulative frequency** is a progressive total (subtotal) of the frequency.

EXAMPLE 8

a This frequency table shows the results of a survey on the number of children in a sample of families. Complete the cumulative frequency column.

Number of children	Frequency	Cumulative frequency
1	4	
2	5	
3	8	
4	3	
5	2	
6	1	
7	2	

b Draw a cumulative frequency histogram and polygon for this set of data.

Solution

a

Number of children	Frequency	Cumulative frequency	
1	4	4	
2	5	9	4 + 5
3	8	17	9 + 8
4	3	20	17 + 3
5	2	22	
6	1	23	
7	2	25	

b

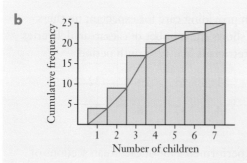

A **cumulative frequency histogram** is a bar chart of cumulative frequency.

A **cumulative frequency polygon** is drawn by joining the top right-hand corner of each column of a cumulative frequency histogram.

NCM 11. Mathematics Standard (Pathway 1)

ISBN 9780170413503

EXAMPLE 9

The ages of people swimming at an indoor pool were recorded and grouped into 6 class intervals. Then the data was graphed on a cumulative frequency histogram and polygon.

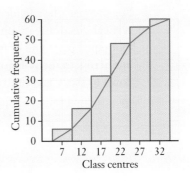

a Copy and complete this table.

Ages	Class centre	Frequency	Cumulative frequency
5–9		6	6
10–14			16
15–19			32
20–24			48
25–29			56
30–34			60

b Use the cumulative frequency polygon to estimate:

i the quartiles   **ii** the median   **iii** the interquartile range.

Solution

a For each class interval, the **class centre** is the average of the ends of the interval. So, for the interval 5–9, the class centre is $\dfrac{5+9}{2} = \dfrac{14}{2} = 7$, halfway between 5 and 9.

Complete the frequency column by subtracting pairs of cumulative frequencies: $16 - 6 = 10$, $32 - 16 = 16$, and so on.

Ages	Class centre	Frequency	Cumulative frequency
5–9	7	6	6
10–14	12	10	16
15–19	17	16	32
20–24	22	16	48
25–29	27	8	56
30–34	32	4	60

b There are 60 scores in the data set.

i For the quartiles, divide the vertical axis of the graph into four equal parts and draw a line across to the polygon, then down to the 'Class centres' axis.

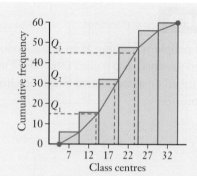

Estimating values on the 'Class centres' axis:

$$Q_1 \approx 14$$

$$Q_2 \approx 18$$

$$Q_3 \approx 23$$

ii The median is Q_2.

The median is 18.

iii The interquartile range is $Q_3 - Q_1$.

$$IQR \approx 23 - 14$$

$$= 9$$

Exercise 8.05 Cumulative frequency

Example 8

1 Copy and complete each table and construct a cumulative frequency histogram and polygon for the data.

a

Numbers of computers in household	Frequency	Cumulative frequency
0	6	
1	8	
2	2	
3	1	
4	2	
5	1	
Total		

b

Hours of sleep	Frequency	Cumulative frequency
6	3	
7	9	
8	11	
9	6	
10	5	
11	2	
Total		

c

Hours spent training	Frequency	Cumulative frequency
1–3	1	
4–6	3	
7–9	6	
10–12	10	
13–15	14	
16–18	6	
Total		

2 The test marks of 90 students in a French exam were grouped into 5 class intervals and summarised in this cumulative frequency graph.

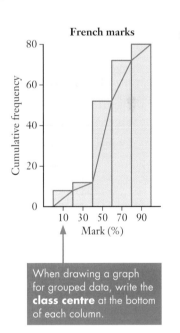

a Copy and complete this table.

Class centre	Frequency	Cumulative frequency
10		8
30		12
50		52
70		72
90		80

b Find an estimate for the median.

c Find an estimate for the interquartile range.

When drawing a graph for grouped data, write the **class centre** at the bottom of each column.

3 The waters around the southern coast of Australia can be very dangerous. This graph shows information about the masses of ships wrecked on the Victorian coast, grouped into class intervals with the class centres shown on the Mass axis.

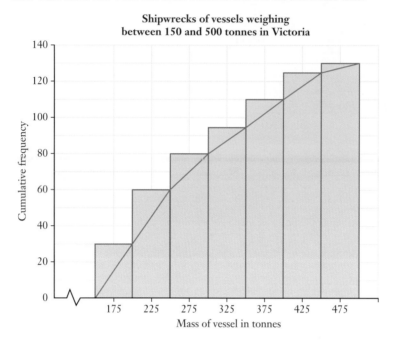

Shipwrecks of vessels weighing between 150 and 500 tonnes in Victoria

a Copy and complete this table.

Class centre	Frequency	Cumulative frequency
175		30
225		
275		
325		
375		
425		
475		130

b Use the graph to estimate the median.

c Use the graph to estimate the interquartile range.

4 Customers complained that they had difficulty contacting the Chatphone company because its helpline was always busy. The company hired Francine to investigate the time the operators took to complete each call. This cumulative frequency graph summarises this data during a 1-hour period.

Find an estimate for:

a the median   **b** the interquartile range.

   ISBN 9780170413503

8.06 Analysing frequency tables

We can analyse information given in a frequency table or presented as grouped data. The **median class** is the class interval that contains the median score. The **modal class** is the class interval with the highest frequency.

Statistical match-up

Statistics from a frequency table

EXAMPLE 10

This frequency table shows the number of letters in each word in a paragraph of text.

For this set of data find the:

a median

b mode

c mean correct to one decimal place

d range.

Number of letters	Frequency
2	2
3	14
4	20
5	7
6	2

Solution

a To find the median, add a cumulative frequency column.

There are 45 scores, so the middle score is the 23rd score.

Reading down the cumulative frequency column, the 23rd score is between the 16th (the last of the 3s) and the 36th (the last of the 4s) scores. So, the 23rd score is 4.

Number of letters	Frequency	Cumulative frequency
2	2	2
3	14	16
4	20	36
5	7	43
6	2	45

The median is 4.

b The mode is the score with the highest frequency.

The mode is 4.

c To find the mean, we add a 'Frequency × score' column to the table. This is a shortcut to adding the 45 scores.

Number of letters	Frequency	Frequency × score	
3	4	12	4 × 3
4	7	28	7 × 4
5	8	40	
6	14	84	
7	12	84	
Total	45	248	

$$\text{Mean} = \frac{248}{45}$$

$$= 5.51111 \ldots$$

$$\approx 5.5$$

d In the first column, the highest number of letters is 7 and the lowest is 3.

$$\text{Range} = 7 - 3$$

$$= 4$$

EXAMPLE 11

Customers of a local coffee shop were asked how many cups of coffee they drank each week, and the responses were grouped into class intervals.

For this set of data find:

a the median class

b the modal class

c an estimate for the mean.

Is it possible to find the range for grouped data? Explain your answer.

Number of cups of coffee	Frequency
1–5	3
6–10	13
11–15	15
16–20	7
21–25	12

Solution

a Add a cumulative frequency column. There are 50 scores, so the middle score will be the average of the 25th and the 26th scores. Reading from the cumulative frequency column, both of these scores are in the 11–15 class.

Cups of coffee	Frequency	Cumulative frequency
1–5	3	3
6–10	13	16
11–15	15	31
16–20	7	38
21–25	12	50

The median class is the 11–15 class.

b The class with the highest frequency is 11–15 (with a frequency of 15).

The modal class is the 11–15 class.

c For an estimate of the mean, we need to add columns for Class centre and
Frequency × class centre.

Cups of coffee	Class centre, cc	Frequency, f	$f \times$ cc
1–5	3	3	9
6–10	8	13	104
11–15	13	15	195
16–20	18	7	126
21–25	23	12	276
	Totals	50	710

The mean $\approx \dfrac{710}{50}$

$= 14.2$

This is only an *estimate* for the mean as we don't have the exact data.

It is not possible to find the range because we only have the class intervals and we don't know the actual scores.

Exercise 8.06 Analysing frequency tables

1 Here are the scores of students on a quick quiz out of 10.

a Copy this table and add columns for cumulative frequency and frequency × score.

b Use the table to find:
- **i** the median
- **ii** the mode
- **iii** the mean (correct to one decimal place)
- **iv** the range.

Score	Frequency
5	4
6	3
7	8
8	4
Total	19

Example 10

2 Andrea surveyed 50 students on how many text messages they sent within the last hour.

a Copy this table and add columns for cumulative frequency and frequency × score.

b Use the table to find:
- **i** the median
- **ii** the mode
- **iii** the mean (correct to one decimal place)
- **iv** the range.

Number of texts	Frequency
0	8
1	4
2	10
3	10
4	15
5	3
Total	50

3 Centrelink officers want to encourage jobseekers to keep applying for jobs. They recorded how many jobs each person applied for before they received their first job interview.

a How many people are in this random sample?

b Find the mode, the median, the mean and the range.

c Which measure of central tendency would you use to encourage people to keep job hunting? Explain.

Score	Frequency
1	2
2	5
3	10
4	6
5	11
6	13
7	3

8. It's better than average 235

Example
11

4 Jordan is looking for a good used car. This table shows the prices of the cars she has considered.

Price ($)	Frequency
5000–<6000	3
6000–<7000	5
7000–<8000	6
8000–<9000	12
9000–<10 000	8

  a How many cars has Jordan looked at?

  b Copy this table and add columns for class centre, cumulative frequency and frequency × class centre.

  c Use the table to find:

     i the median class   **ii** the modal class

     iii an estimate for the mean.

5 Max surveyed the amount of money his friends spent on fuel in the last week.

Fuel cost ($)	Frequency
10–<20	9
20–<30	11
30–<40	5
40–<50	3
50–<60	2
60–<70	1
70–<80	2
80–<90	3

  a How many people did Max survey?

  b Copy this table and add centres for class centre, cumulative frequency and frequency × class centre.

  c Use the table to find:

     i the median class   **ii** the modal class

     iii an estimate for the mean (correct to one decimal place).

6 The percentage marks of a Mathematics Standard assessment task were recorded and grouped into six class intervals, and a cumulative frequency graph was produced.

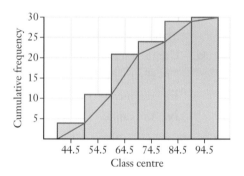

  a How many students are in this class?

  b Copy and complete this frequency table for the data.

Marks (%)	Class centre	Frequency	Cumulative frequency	Frequency × class centre
40–49	44.5		4	
50–59	54.5		11	
	64.5		21	
	74.5		24	
	84.5		29	
	94.5		30	
	Total			

  c For this set of data find:

     i the median class   **ii** the modal class

     iii an estimate for the mean (correct to one decimal place).

8.07 Mean and standard deviation

Standard deviation is another measure of spread, like the range and the interquartile range. It describes how far each score is from the mean. The bigger the standard deviation, the more spread out the scores are. The formula for standard deviation is quite complicated because it involves every score in a data set, so we don't have to learn it because your calculator has a statistics mode that calculates it.

Statistical
calculations

EXAMPLE 12

The net weekly wages of eight casual workers are:

$730 $490 $600 $440 $490 $370 $700 $580

Use your calculator to find the mean and standard deviation.

Statistics
review

Solution

Follow the instructions for the statistics mode (SD or STAT) of your calculator:

Statistics mode:
Graphics
calculator

Operation	Casio scientific	Sharp scientific
Start statistics mode.	MODE STAT 1-VAR	MODE STAT =
Clear the statistical memory.	SHIFT 1 Edit, Del-A	2ndF DEL
Enter data.	SHIFT 1 Data to get table 730 = 490 = , etc. to enter in column AC to leave table	730 M+ 490 M+ , etc.
Calculate the mean. ($\bar{x} = 550$)	SHIFT 1 Var $\bar{x}$ =	RCL $\bar{x}$
Calculate the standard deviation. (s = 117.260 ...)	SHIFT 1 Var $x\sigma_n$ =	RCL σx
Return to normal (COMP) mode.	MODE COMP	MODE 0

On your calculator, there are *two* standard deviations given:

- the population standard deviation σ, shown on your calculator as $x\sigma_n$ or σx

- the sample standard deviation s, shown on your calculator as $x\sigma_{n-1}$ or sx.

In our calculations, we will use σ, the population standard deviation, unless the data of a sample is given.

EXAMPLE 13

Twenty echidnas from Booderee National Park were tagged and returned to their habitat. Rangers later captured several samples of 10 echidnas and recorded the number tagged in each sample.

Find correct to two decimal places:

a the mean number of tagged echidnas per sample

b the sample standard deviation of tagged echidnas.

Echidnas tagged per sample	Frequency
0	8
1	11
2	5
3	4
4	2
5	1

Solution

For data presented in a frequency table, follow the instructions for your calculator as shown in the table below for the following answers:

a mean $\bar{x} \approx 1.48$ **b** standard deviation $s \approx 1.36$.

Operation	Casio scientific	Sharp scientific
Start statistics mode.	MODE STAT 1-VAR SHIFT MODE scroll down to STAT Frequency? ON	MODE STAT =
Clear the statistical memory.	SHIFT 1 Edit, Del-A	2ndF DEL
Enter data.	SHIFT 1 Data to get table 0 = 1 = , etc. to enter in x column 8 = 11 = , etc. to enter in FREQ column AC to leave table	0 2ndF STO 8 M+ 1 2ndF STO 11 M+ etc.
Calculate the mean ($\bar{x} = 1.4838 \ldots$)	SHIFT 1 Var $\bar{x}$ =	RCL $\bar{x}$
Calculate the standard deviation ($s = 1.363 \ldots$)	SHIFT 1 Var $x\sigma_{n-1}$ =	RCL sx

Exercise 8.07 Mean and standard deviation

1 For each set of scores, calculate correct to two decimal places:

    i   the mean   **ii**   the standard deviation.

    a   5, 7, 2, 6, 5, 6, 4, 6, 4

    b   8, 2, 4, 2, 6

    c   14, 13, 17, 14, 13, 16, 16, 17, 16, 12

2 This data shows the number of burglaries each month in a city over a year.

    14   12   16   14   17   12   14   14   10   13   11   19

    Find correct to one decimal place:

    a   the mean   **b**   the standard deviation.

3 This data shows the number of burglaries each month in a country town over a year.

    7   9   7   6   3   4   5   3   2   1   1   4

    Find correct to one decimal place:

    a   the mean   **b**   the standard deviation.

4 Which data set from Questions **2** and **3** is the more spread out?

5 Students' marks in a quiz out of 10 are shown in this frequency table.

    Find correct to one decimal place:

    a   the mean

    b   the sample standard deviation.

Mark	Frequency
5	4
6	3
7	8
8	4
Total	19

6 Andrea surveyed 50 students on how many text messages they sent within the last hour, and the results are shown in the table in Question **2** in Exercise 8.06 on page 235. For this data, find correct to two decimal places:

    a   the mean   **b**   the population standard deviation.

7 Centrelink officers recorded how many jobs each person applied for before they received their first job interview, and the results are shown in the table in Question **3** in Exercise 8.06 on page 235. For this data, find correct to two decimal places:

    a   the mean   **b**   the sample standard deviation.

8.08 Road accident statistics

Speed, alcohol, fatigue, inexperience and reckless driving are major causes of road accidents. By analysing the data relating to road accidents, governments find ways to reduce the number of accidents each year.

EXAMPLE 14

Police recorded the speed of cars travelling through a 40 km/h school zone.

40	42	60	38	42	55	75	35	40	40	39	40
62	55	39	45	38	40	41	34	60	45	44	38
55	40	35	36	46	44	40	60	65	64	38	41

a Sort the data into class intervals of 31–35, 36–40 etc. and calculate an estimate for the mean speed of the cars.

b What was the modal speed class?

c Calculate the range.

d What percentage (correct to one decimal place) of cars was exceeding the 40 km/h speed limit?

Solution

a

Class	Class centre (cc)	Tally	Frequency (f)	Cumulative frequency	$f \times cc$
31–35	33	|||	3	3	99
36–40	38	‖‖ ‖‖ ||||	14	17	532
41–45	43	‖‖ |||	8	25	344
46–50	48			1	26	48
51–55	53	|||	3	29	159
56–60	58	|||	3	32	174
61–65	63	|||	3	35	189
66–70	68		0	35	0
71–75	73			1	36	73
		Total	36		1618

$$\text{Mean} = \frac{1618}{36}$$
$$= 44.944 \ldots$$
$$\approx 44.9$$

The mean speed was 44.9 km/h.

b The class interval 36–40 has the highest frequency of 14.

The modal speed class was 36 to 40 km/h.

c The range must be calculated from the original data, not from the table.

$$Range = 75 - 34$$
$$= 41$$

d 19 of the 36 speeds were over 40 km/h.

Percentage exceeding speed limit

$$= \frac{19}{36} \times 100\%$$
$$= 52.777 \ldots$$
$$\approx 52.8\%$$

Exercise 8.08 Road accident statistics

1 a In one year, there were 365 fatal crashes in NSW ('fatal' means causing death). Use this information to copy and complete this table.

	Number of fatal crashes	Percentage of fatal crashes
Speed	146	40.0%
Alcohol	58	
Driver fatigue	54	
Other factors		

b In what fraction of fatal crashes is alcohol a factor?

c List two things that could be 'other factors' in fatal crashes.

2 Bianca recorded the blood alcohol content (BAC) of each person charged with a driving offence.

0.15   0.09   0.04   0.18   0.00   0.07   0.09   0.00   0.02   0.05   0.12   0.20

a Arrange the data in ascending order.

b Determine the mean, mode, median and range.

c When you arrange the scores into quartiles, how many scores will be in each quarter?

d Calculate the value of the lower quartile and the upper quartile.

e Calculate the interquartile range.

f What fraction of people charged had consumed some alcohol?

This table shows driving offences related to BAC readings.

> PCA means 'prescribed concentration of alcohol'. n/a means not applicable.

Offence	BAC	First offence maximum fine	First offence maximum gaol term
High-range PCA	0.15 or greater	$3300	18 months
Mid-range PCA	0.08 to less than 0.15	$2200	9 months
Low-range PCA	0.05 to less than 0.08	$1000	n/a
Novice-range PCA	Between 0 and 0.05	n/a	n/a

g What percentage of the drivers on trial could be charged with high-range PCA?

h What were the BACs of the three drivers who were given a $2200 fine?

3 During the Christmas season, highway billboards displayed the number of drivers booked for speeding each day, from 24 December to 4 January.

146   46   305   102   183   194   210   98   102   208   168   110

a Arrange the data from smallest to largest.

b Determine the value of the mean, median, mode and range.

c Calculate the interquartile range of the data.

d Calculate the standard deviation for the data. Answer correct to one decimal place.

e A journalist is writing an article about the number of motorists booked for speeding each day. Which value do you think she should use for the 'average'? Give a reason for your answer.

4 Since 1978, there has been a huge reduction in the number of road injuries and deaths in NSW, despite a large increase in the population. In 1978, 1384 people were killed on NSW roads. By 2009, this figure had fallen to 460.

a How many fewer people were killed on NSW roads in 2009 than in 1978?

b Calculate the percentage decrease from 1978 to 2009.

c In 2016, the number of people killed on NSW roads was 384. Calculate the percentage decrease from 2009 to 2016.

d Your answer to part **c** is smaller than your answer to part **b**. Suggest reasons for this.

Example 14

5 Aisha recorded the typical number of kilometres her friends drive each day.

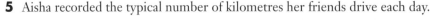

60	54	21	32	18	15	0	25	60	32	12	22
44	51	60	16	31	19	15	23	39	41	50	26
32	45	27	21	15	18	10	62	17	0	11	20

ISBN 9780170413503

a Copy and complete the frequency distribution table.

Class	Class centre, cc	Tally	Frequency, f	Cumulative frequency	$f \times cc$
0–<10	5				
10–<20					
20–<30					
30–<40					
40–<50					
50–<60					
60–<70					
Total					

b Calculate to one decimal place the mean distance Aisha's friends drive.

c In which class is the median distance Aisha's friends drive?

d Calculate the range of the data.

e Which is the modal class?

INVESTIGATION

CHANGING DATA

It is best to use a spreadsheet to complete this investigation.

Score	Frequency
5	11
6	13
7	10
8	8
9	3
Total	45

1 For the set of data shown in the frequency table, find:

  a the mean correct to one decimal place

  b the median

  c the mode

  d the standard deviation correct to one decimal place.

2 In the following challenges, change the frequencies in the table without changing the total frequency of 45.

  a Increase the mean without changing the mode.

  b Decrease the mean without changing the mode.

  c Make the median bigger by 1.

  d Make the standard deviation as small as possible.

  e Make the standard deviation as big as possible.

  f Make the mean equal to 8 in two different ways.

  g Make the mode 7 without changing the mean.

  h Make the mean 7 and the standard deviation as close to 1 as possible.

STATISTICS ON A SPREADSHEET

1 a Enter into a spreadsheet the following data about the daily maximum temperatures in Alice Springs for 1 week.

	A	B	C	D	E
1	**Day**	**Temperature (°C)**			
2	Sunday	29		Mean	
3	Monday	31		Mode	
4	Tuesday	30		Median	
5	Wednesday	33		Standard deviation	
6	Thursday	29		Maximum	
7	Friday	28		Minimum	
8	Saturday	35		Range	
9					
10					

b Enter each formula into the given cell and save the spreadsheet.

Cell E2: =average(B2:B8)   Cell E3: =mode(B2:B8)

Cell E4: =median(B2:B8)   Cell E5: =stdev.p(B2:B8)

Cell E6: =max(B2:B8)   Cell E7: =min(B2:B8)

Cell E8: =E6–E7

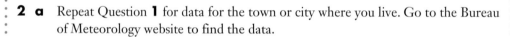

Sometimes, if you type the first couple of letters, the spreadsheet will suggest the correct word.

Bureau of Meteorology

2 a Repeat Question **1** for data for the town or city where you live. Go to the Bureau of Meteorology website to find the data.

b Find data for 1 month instead of for 1 week. Repeat Question **1** for this new data. You will need to adjust the formulas you enter for the statistics.

3 a Enter into a spreadsheet the following data showing the monthly number of motor vehicle thefts in a capital city over two years.

	A	B	C	D	E	F
1	**Motor vehicle theft in inner Sydney**					
2						
3	Jan 2018	20		Jan 2017	15	
4	Feb 2018	19		Feb 2017	19	
5	Mar 2018	26		Mar 2017	17	
6	Apr 2018	17		Apr 2017	21	
7	May 2018	27		May 2017	20	
8	Jun 2018	13		Jun 2017	18	
9	Jul 2018	17		Jul 2017	13	
10	Aug 2018	17		Aug 2017	13	
11	Sep 2018	20		Sep 2017	14	
12	Oct 2018	18		Oct 2017	14	
13	Nov 2018	12		Nov 2017	18	
14	Dec 2018	24		Dec 2017	20	
15						
16	Mean			Mean		
17	Median			Median		
18	Mode			Mode		
19	SD			SD		
20						

b Copy the following formulas into the given cells:

Cell B16: =average(B3:B14) Cell B17: =median(B3:B14)

Cell B18: =mode(B3:B14) Cell B19: =stdev.p(B3:B14)

Cell E16: =average(E3:E14) Cell E17: =median(E3:E14)

Cell E18: =mode(E3:E14) Cell E19: =stdev.p(E3:E14)

c Save your results.

d Comment on the differences between the two years.

4 Repeat Question **3** for Melbourne or Sydney. You will need to use the Internet to find the data. Try searching 'crime statistics and motor vehicle theft'.

WORD MATCH

average	central tendency	cumulative frequency	histogram
interquartile range	mean	mode	number
often	order	outlier	percentiles
polygon	quartiles	range	spread ten

Copy and complete each sentence using a word or phrase from the list.

1 The _____ is the most frequent score.

2 The mean is the sum of scores divided by the _____ of scores.

3 The difference between the highest score and the lowest score is called the _____.

4 _____ divide the data into four equal parts.

5 Another word for the mean is _____.

6 The mode is the score that occurs most _____.

7 A score that is very different to the other scores is called an _____.

8 Mean, median and mode are measures of _____ _____.

9 The difference between the upper quartile and the lower quartile is called the _____ _____.

10 To find the median, you must first put the scores in _____.

11 The standard deviation is a measure of _____.

12 Deciles divide the data into _____ equal parts.

13 The _____ _____ is a progressive total of the frequencies.

14 When a large amount of data is divided into 100 equal parts, they are called _____.

15 When we have calculated the cumulative frequency, we can draw two graphs: a _____ and a _____.

16 A measure of central tendency the value of which depends on all the scores is the _____.

SOLUTION TO THE CHAPTER PROBLEM

Problem

For his newspaper article about burglaries in three areas of Sydney, Mirkov wants to analyse the data he has collected and present some conclusions about the safety of different parts of Sydney. What type of statistics would you recommend he use in the article?

Solution

People reading the article will be interested in knowing the typical or 'average' number of burglaries in each area. Mirkov could use the *mean* or the *median* to describe the typical number of burglaries. The mode is less likely to be relevant as there might not be a single score that is the mode. However, if there are one or two months with very high numbers of burglaries, the mean would give a false impression of the typical number of burglaries, so, in this case, Mirkov should use the *median*.

People reading the article might also be interested in how the number of burglaries per month has changed. Mirkov could include the data over a number of years. This might show that one area has had fewer burglaries over time while another area has had more.

8. CHAPTER SUMMARY

- Have you learned anything new in this topic? If so, what?

- What jobs would use what you have studied in this chapter?

- Is there anything you didn't understand? If so, ask your teacher for help.

Copy and complete this mind map of the topic, adding detail to its branches and using pictures, symbols and colour where needed. Ask your teacher to check your work.

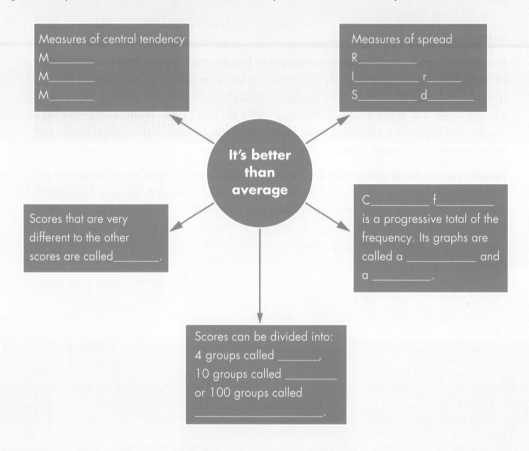

Measures of central tendency
M_____
M_____
M_____

Measures of spread
R_____
I_____ r_____
S_____ d_____

It's better than average

Scores that are very different to the other scores are called_____.

C_____ f_____ is a progressive total of the frequency. Its graphs are called a _____ and a _____.

Scores can be divided into:
4 groups called _____,
10 groups called _____
or 100 groups called
_____.

1 For each set of data, find:

    i the mode   **ii** the median   **iii** the mean.

  a computer frauds per year: 808, 1126, 1003, 913, 300

  b student incomes ($): 32, 29, 41, 34, 29, 40, 40, 37, 39, 40, 33

  c temperatures at Cloncurry (°C): 25, 24, 23, 20, 16, 12, 11, 12, 17, 20, 23, 25

  d rainfall at Tewantin (mm): 152, 227, 202, 124, 149, 127, 83, 90, 59, 88, 96, 143

Exercise 8.01

2 The following percentage scores are the room occupancy rates for NSW hotels per quarter (3 months) for the past 5 years.

57	62	61	56	58	65	60	59	56	66
64	59	58	70	69	62	61	73	68	

  a Find the median for this data.

  b What is the mode for this data?

  c Calculate the mean for this data. Answer correct to one decimal place.

  d Which measure of central tendency most accurately reflects the data? Justify your answer.

3 For each set of data in Question **1**, find:

    i the range   **ii** the upper and lower quartiles   **iii** the interquartile range.

Exercise 8.02

4 Using your answers to Question **3**, decide for which data sets in Question **1** is:

  a the range the best measure of spread

  b the interquartile range the best measure of spread.

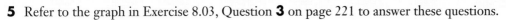

5 Refer to the graph in Exercise 8.03, Question **3** on page 221 to answer these questions.

  a What percentage of students scored more than 51?

  b What is the 4th decile?

  c Joanna scored 53 out of 60. Between which two deciles did she score?

Exercise 8.03

6 Refer to the graph in Exercise 8.03, Question **4** on page 222 to answer these questions.

  a For how many months of the year was the rainfall above average for the month?

  b State the month(s) where the rainfall was between the 10th percentile and the long-term average.

  c What is the difference between the 10th percentile and the 90th percentile in February?

Exercise 8.03

7 For the computer fraud data in Question **1a**:

  a what is the outlier for this data?

  b calculate the mean of the data without the outlier.

  c what effect does the outlier have on the mean calculated in Question **1**?

Exercise 8.04

8 For the rainfall data in Question **1d**:

  a what are the outliers for this data?

  b find the mean and the median for the data without the outliers included.

  c compare the mean with all scores included to the mean without the outliers included. What effect do the outliers have on the mean?

  d compare the median with all scores included to the median without the outliers included. What effect do the outliers have on the median?

  e which measure of central tendency best describes the rainfall in Tewantin? Justify your answer.

9 This frequency table records the number of cars in a sample of households in Nelson Waters Drive.

  a Copy and complete this table.

  b Draw a cumulative frequency histogram and polygon for this set of data.

Number of cars	Frequency	Cumulative frequency
0	2	
1	7	
2	8	
3	2	
4	1	
Total	20	

10 A group of students was asked how long, in minutes, it took them to travel to school. The data was grouped into 6 class intervals and then graphed on a cumulative frequency histogram and polygon as shown.

  a Copy and complete this table.

  b Use the cumulative frequency graph to find an estimate for:

  i the median

  ii the interquartile range.

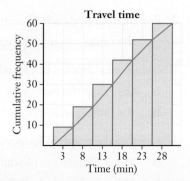

Class centre	Frequency	Cumulative frequency
3		9
8		19
13		30
18		42
23		52
28		60

ISBN 9780170413503

11 Carmelina recorded the number of drinks bought by people visiting her shop during one hour.

Number of drinks	Frequency
0	7
1	5
2	13
3	4
4	1
5	2
6	1
Total	33

a Copy this table and add columns for cumulative frequency and frequency × score.

b For this data find:

   i the median

  ii the mode

 iii the mean (correct to one decimal place)

  iv the range.

12 Flower seeds were sown in boxes. This table shows the number of seedlings that sprouted in each box.

Number of seedlings	Frequency
1–5	3
6–10	6
11–15	6
16–20	9
21–25	7
26–30	5
31–35	4

a Copy this table and add columns for class centre, cumulative frequency and frequency × class centre.

b For this set of data find:

   i the median class

  ii the modal class

 iii an estimate for the mean.

13 Use your calculator to find, correct to two decimal places, the standard deviation of these temperatures in Cloncurry (°C) from Question **1c**:

25   24   23   20   16   12   11   12   17   20   23   25

14 For the table in Question **11**, find correct to two decimal places:

a the mean

b the standard deviation.

9.

MEASURING AREA AND VOLUME

Chapter problem

On a farm, Stuart has to supply and install a septic tank with a capacity of 5000 to 6000 L. The diagram shows the top view of a tank that he is considering. The tank will be 1.7 m deep. Will this tank be suitable for the job?

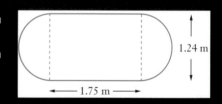

1.24 m

1.75 m

CHAPTER OUTLINE

WHAT WILL WE DO IN THIS CHAPTER?

- Calculate the area of rectangles, triangles, trapeziums, circles, sectors and composite shapes
- Use the trapezoidal rule to estimate the area of irregular shapes
- Calculate the surface area, volume and capacity of prisms, cylinders, spheres and composite solids

HOW ARE WE EVER GOING TO USE THIS?

- Building, renovating or landscaping a property
- Undertaking craft activities including weaving, making stained glass, patchwork and quilting

9.01 Area

Area calculations are involved in many practical, everyday tasks, such as painting, carpeting, tiling and paving. Some area formulas are shown below.

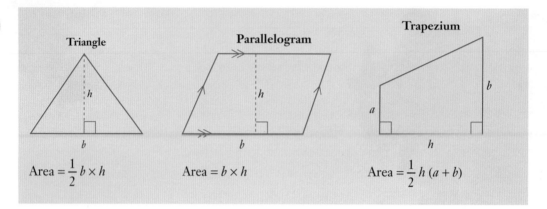

Triangle

Area $= \dfrac{1}{2} b \times h$

Parallelogram

Area $= b \times h$

Trapezium

Area $= \dfrac{1}{2} h\,(a + b)$

EXAMPLE 1

Calculate the area of each triangle.

a

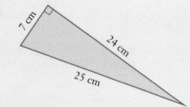

b

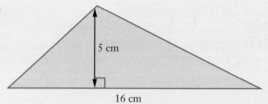

Solution

a The area of a triangle is $A = \dfrac{1}{2}$ base $\times$ height. The base and the height must be at 90° to each other. The base = 24 cm and the height = 7 cm.

$\text{Area} = \dfrac{1}{2} b \times h$

$= \dfrac{1}{2} \times 24 \times 7$

The 25 cm length is not used in this calculation.

$= 84 \text{ cm}^2$

b In this triangle, the base is 16 cm and the height is 5 cm.

$\text{Area} = \dfrac{1}{2} b \times h$

$= \dfrac{1}{2} \times 16 \times 5$

$= 40 \text{ cm}^2$

EXAMPLE 2

Find the area of each shape.

a

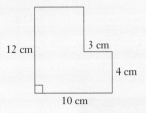

b

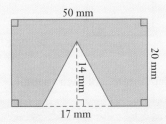

Solution

a Divide the shape into two rectangles as shown.

Width of rectangle 1 = 10 – 3 = 7 cm

7 cm

12 cm **1**

3 cm

2 4 cm

Calculate the area of each rectangle and add them together.

> There is more than one way to divide this shape. All methods result in the same answer.

Area of rectangle 1 = 7 × 12

$$= 84 \text{ cm}^2$$

Area of rectangle 2 = 3 × 4

$$= 12 \text{ cm}^2$$

Total area = 84 + 12 = 96 cm²

b Find the area of the rectangle and subtract the area of the triangle.

> Sometimes, it's easier to find the area by subtraction.

$$\text{Area} = 50 \times 20 - \frac{1}{2} \times 17 \times 14$$

$$= 1000 - 119$$

$$= 881 \text{ mm}^2$$

Exercise 9.01 Area

1 Calculate the area of each quadrilateral.

a

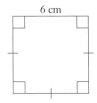

6 cm

b

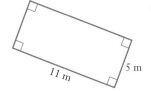

11 m

5 m

c

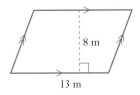

8 m

13 m

d

14 m

6 m

10 m

e

7 m

10 m

f

9 m

8 m

15 m

Example
1

2 Calculate the area of each triangle.

Make sure that you use the two dimensions that are at right angles to each other.

a

23.1 mm

16 mm

28.1 mm

b

36 cm

77 cm

85 cm

c

5 m

22 m

d

10 m

9 m

e

7.4 m

5.6 m

f

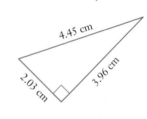

4.45 cm

2.03 cm

3.96 cm

3 Bats are the most common mammal on Earth. The end sections of their wings are triangular. Calculate the area of each triangular section of the bat's wings.

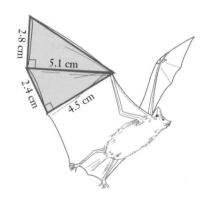

2.8 cm

5.1 cm

2.4 cm

4.5 cm

4 Draw two possible right-angled triangles that each has an area of 24 cm², showing values for the base and height of the triangle.

5 Isabella is using parallelograms to make a star pattern on a quilt.

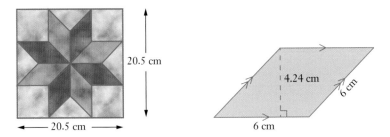

a Determine the area covered by each parallelogram.

b What is the area covered by the star?

c How much of the area of the square is *not* covered by the star?

6 Find the area of each shape correct to two decimal places where necessary.

Example 2

a
20 cm
6.5 cm
9 cm
5.2 cm

b
4.9 m
3.2 m
7.5 m

c
20 mm
25 mm

d
40 cm
20 cm
10 cm   10 cm

7 Calculate the area of each shaded region in the shapes below.

a

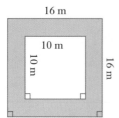

b

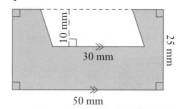

c

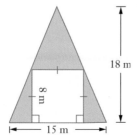

d

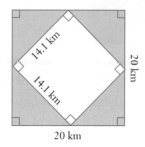

8 A children's playground has an L-shape as shown.

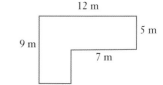

    a   What is the area of the playground?

    b   For safety reasons, the playground is to be covered with woodchips. How much will the woodchips cost if they are sold for $7.20 per square metre?

INVESTIGATION

LAYING GRASS

Keira is planning to use rolls of grass to cover her backyard which is 15.4 m wide by 12 m long. Each roll of grass is 50 cm wide and 2 m long.

a Draw a diagram to represent Keira's backyard.

b Calculate the area of Keira's backyard and the area covered by one roll of grass.

c How many rolls of grass will Keira need to buy to cover her backyard?
We can buy only whole rolls of grass.

d On the diagram you drew in part **a**, decide whether it will be better to lay the grass in rows or in columns or both. Find the best way.

e Describe the size and shape of the pieces of grass that will be left over after Keira has finished covering her backyard.

9.02 Areas of circular shapes

Perimeter
and area of
a sector

Perimeter
and area of
a sector

A page of
circular
shapes

Area of a circle

$A = \pi r^2$ where r = radius.

Area of a sector

$A = \dfrac{\theta}{360} \times \pi r^2$ where θ = central angle, r = radius.

EXAMPLE 3

Calculate correct to two decimal places the area of:

a   this circle

7 cm

b   this sector.

145°   4 cm

Solution

a   Radius $r = 7$

$A = \pi r^2$

$= \pi \times 7^2$

$= 153.9380 \ldots$

$\approx 153.94 \text{ cm}^2$

b   The sector is $\dfrac{145}{360}$ of the circle
(there are 360° in a full circle).

Radius $r = 4$

$A = \dfrac{\theta}{360} \times \pi r^2$

$= \dfrac{145}{360} \times \pi \times 4^2$

$= 20.2458 \ldots$

$\approx 120.25 \text{ cm}^2$

Calculate the area of this shape correct to two decimal places.

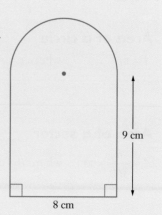

9 cm

8 cm

Solution

Separate the diagram into two sections, a semicircle and a rectangle.

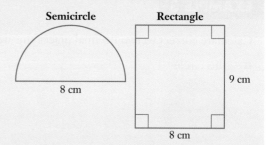

Semicircle

Rectangle

8 cm

9 cm

8 cm

Calculate the area of the two shapes and add them together.

Radius of semicircle $= \frac{1}{2} \times 8$

$= 4$ cm

Area of semicircle $= \frac{1}{2} \times \pi r^2$

$= \frac{1}{2} \times \pi \times 4^2$

$= 25.1327 \ldots$

Area of rectangle $= 8 \times 9$

$= 72$ cm^2

Total area $= 72 + 25.1327 \ldots$

≈ 97.13 cm^2

Exercise 9.02 Areas of circular shapes

Example 3

1 Calculate the area of each shape correct to two decimal places.

a
12 cm

b
25 cm

c
3.9 cm

d
62 cm

e

f

g

h

2 Susan and Ian are putting arch windows in the front room of their house. The diagram shows one of the windows with its dimensions.

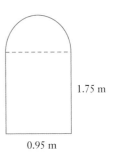

a Find the area of one arch window correct to two decimal places.

b 10% of the area is *not* glass. Find the area of glass used for one window.

Example 4

1.75 m

0.95 m

3 Calculate the area of each shaded region correct to one decimal place.

a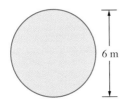
6 m

b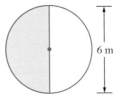
6 m

c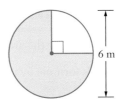
6 m

d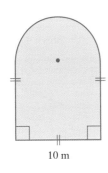
10 m

e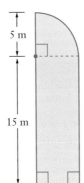
5 m
15 m

f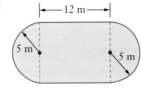
12 m
5 m   5 m

4 This diagram shows a netball court. The shaded area represents the part of the court where the 'wing attack' can play. Calculate correct to one decimal place:

a the area of one of the goal semicircles

b the size of the area where the wing attack can play.

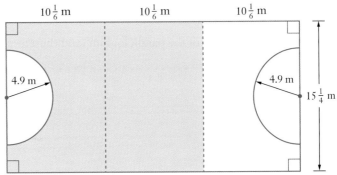

$10\frac{1}{6}$ m $10\frac{1}{6}$ m $10\frac{1}{6}$ m

4.9 m   4.9 m $15\frac{1}{4}$ m

5 Calculate the area of the annulus, the shaded area between the two circles, correct to the nearest square centimetre.

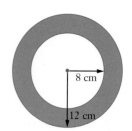
8 cm
12 cm

6 This circle has a radius of 36 cm². Find its radius, r cm, correct to two decimal places.

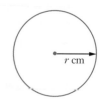
r cm

PRACTICAL ACTIVITY

AREA OF A TRAPEZIUM FORMULA

The formula for the area of a trapezium is $A = \frac{1}{2}h(a + b)$. We can prove this formula by cutting the trapezium into two pieces.

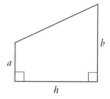
b
a
h
Area = $\frac{1}{2}h(a + b)$

1 Draw this trapezium on a sheet of paper. Draw a vertical dotted line parallel to sides a and b halfway between the sides.

2 Cut out the trapezium, then cut along the dotted line to make two smaller trapeziums. Join side a to side b to make one long parallelogram.

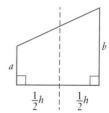
b
a
$\frac{1}{2}h$ $\frac{1}{2}h$

3 Why is the length of this long parallelogram $a + b$?

4 Why is the height of this long parallelogram $\frac{1}{2}h$?

5 How do you find the area of a parallelogram?

6 Show that the area of the parallelogram (and therefore trapezium) is $A = \frac{1}{2}h(a + b)$.

ISBN 9780170413503

9.03 The trapezoidal rule

The area of a trapezium can be used to approximate the area of many irregular shapes, such as the block of land shown.

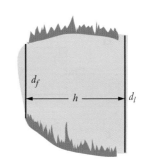

Trapezoidal rule

Approximate the shape as a trapezium by taking two vertical measurements, d_f, and d_l, and the width between them, h.

Then, the formula for the area of a trapezium $A = \dfrac{1}{2} h(a + b)$

becomes $A \approx \dfrac{h}{2}(d_f + d_l)$. This is called the **trapezoidal rule**.

 'Trapezoid' is another way of saying 'trapezium'.

The trapezoidal rule

$$A \approx \frac{h}{2}(d_f + d_l)$$

where h = distance between successive measurements

d_f = first measurement

d_l = last measurement

or, in words:

$$\text{Area} \approx \frac{\text{width of strip}}{2}(\text{first measurement} + \text{last measurement})$$

EXAMPLE 5

Use the trapezoidal rule to approximate the area of this field correct to the nearest square metre.

9.5 m   16.8 m

7.3 m

Solution

In the trapezoidal rule formula, $d_f = 9.5$, $d_l = 16.8$ and $h = 7.3$.

$$A \approx \frac{h}{2}(d_f + d_l)$$

$$= \frac{7.31}{2} \times (9.5 + 16.8)$$

$$= 95.995$$

$$\approx 96 \text{ m}^2$$

Write the answer. The area of the field is approximately 96 m².

---

1 Use the trapezoidal rule to approximate each shaded area.

a

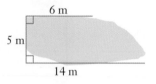

6 m

5 m

14 m

b

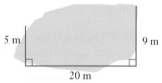

5 m

9 m

20 m

c

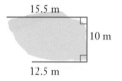

15.5 m

10 m

12.5 m

d

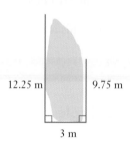

12.25 m   9.75 m

3 m

2 The diagram below shows a salt pan in a desert. The trapezium in the diagram is drawn to a scale of 1:1000. Accurately measure the trapezium, then calculate its area to approximate the area of the salt pan. (Assume that the lengths of the dimensions of the salt pan are in km.)

3 The area of each lake is 60 m². Find the value of x.

a

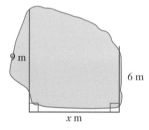

9 m

6 m

x m

b

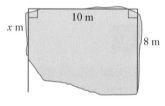

10 m

x m   8 m

Nets of solids

Surface area
of a prism

9.04 Surface areas of prisms

A **prism** is a solid shape that has identical ends. Some examples are shown here: a rectangular prism, a triangular prism and a cube. All side faces are rectangles.

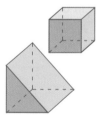

When we make boxes for packaging, we need to know the area of the faces of the box. This is called **surface area** because it is the sum of the areas of all the surfaces of the box. The easiest way to calculate the surface area of an object is to draw its **net**.
We calculate the area of each shape on the net individually and then add them together.

EXAMPLE 6

Find the surface area of each prism.

a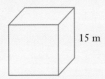
15 m

b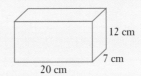
12 cm
7 cm
20 cm

Solution

a   A cube has 6 identical faces, all squares.   Total surface area $= 6 \times (15 \times 15)$

This is the net. $= 1350 \text{ cm}^2$

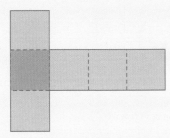

The surface area is 6 times the area of each square.

b   This is the net of a rectangular prism. Notice that the rectangles are in matching pairs.

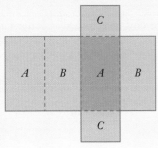

The ends (C) are the same rectangles.   Area of ends $= 2 \times (7 \times 12)$
$= 168 \text{ cm}^2$

The top and bottom (A) are the same.   Area of top/bottom $= 2 \times (20 \times 7)$
$= 280 \text{ cm}^2$

The front and back (B) are the same.   Area of front/back $= 2 \times (20 \times 12)$
$= 480 \text{ cm}^2$

Add the areas together.   Total surface area $= 168 + 280 + 480$
$= 928 \text{ cm}^2$

EXAMPLE 7

Find the surface area of this triangular prism.

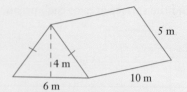

Solution

This is the net of a rectangular prism. There are three rectangles, and two triangles that are the same.

Find the area of the base rectangle.

$$\text{Area of base} = 6 \times 10$$
$$= 60 \text{ m}^2$$

Find the area of the two side rectangles.

$$\text{Area of side faces} = 2 \times (5 \times 10)$$
$$= 100 \text{ m}^2$$

Find the area of the two triangles.

$$\text{Area of two end faces} = 2 \times \left(\frac{1}{2} \times 6 \times 4 \right)$$
$$= 24 \text{ m}^2$$

Add the areas together.

$$\text{Total surface area} = 60 + 100 + 24$$
$$= 184 \text{ m}^2$$

Exercise 9.04 Surface areas of prisms

Example 6

1 Find the surface area of each cube.

a

70 cm

b

10 mm

ISBN 9780170413503

2 Charlotte has a plastic storage cube that is 35 cm long with no top. Calculate its external surface area.

3 This metal cube sculpture has a side length of 450 cm. The surface is to be covered with a weather-resistant finish. How many square metres need to be covered?

4 Find the surface area of each prism.

a

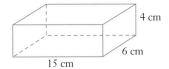

4 cm

6 cm

15 cm

b

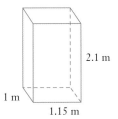

2.1 m

1 m

1.15 m

5 James' new TV was packed in this box. Find the surface area of the box.

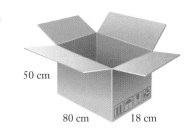

6 Alexa and Jai built a new living room.

  a Calculate the area of the walls and ceiling to be painted. Subtract 6 m² for the windows and door.

  b If paint costs \$23 for a 4-litre can that covers 10 m², calculate how much it would cost to paint the room with two coats.

  c Calculate the area of the floor to be covered with wood parquet.

  d If wood parquet costs \$15.70 per square metre, how much would it cost to cover the floor? Answer to the nearest 10 dollars.

50 cm

80 cm   18 cm

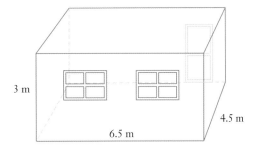

3 m

4.5 m

6.5 m

Example
7

7 Tom's company makes camping equipment, including the tent shown below. What amount of material is required to make the tent (including the floor)?

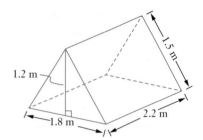

8 The diagram shows a chocolate box in the shape of a triangular prism. The ends of the box are equilateral triangles.

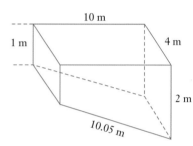

   a   Calculate the surface area of the box.

   b   The manufacturer allows an extra 10% of the surface area for tabs and wastage. How much cardboard is required for one box?

9 Harish and Rachna have a backyard pool shaped as shown. It needs to be repainted. There are tiles around the top edge to a depth of 40 cm. Calculate the area to be painted.

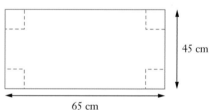

> Remember to convert 40 cm to m.

10 A metal tray is made by taking a rectangular piece of metal and cutting squares from the corners. The edges are then bent upwards and welded at the corners to form the tray.

   a   This piece of metal has dimensions 45 cm by 65 cm. Squares of side length 10 cm are cut out of each corner. What are the dimensions of the tray that can be formed?

   b   The bottom of the tray is to be lined with material. What area of material is required?

   c   The sides and the outside are to be enamelled. What area is to be enamelled?

9.05 Surface areas of cylinders and spheres

A page of
solid shapes

Nets of solids

Surface area
of a cylinder

If we flatten out a cylinder, we get this net of two circles and a rectangle.

Note: The length of the rectangle is the circumference of the circle ($2\pi r$).

Surface area of a cylinder = area of two circles + area of rectangle

$$= 2 \times \pi r^2 + 2\pi r \times h$$

$$= 2\pi r^2 + 2\pi rh$$

There is a simple formula for the surface area of a sphere.

Surface area of a sphere $= 4 \times \pi \times (\text{radius})^2$

$$= 4\pi r^2$$

> This formula is like 4 times the area of a circle.

Surface area of a cylinder = area of 2 circles + area of rectangle

$$SA = 2\pi r^2 + 2\pi rh$$

Surface area of a sphere $= 4 \times \pi \times (\text{radius})^2$

$$SA = 4\pi r^2$$

EXAMPLE 8

Find the surface area of each solid correct to two decimal places.

a

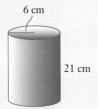

6 cm

21 cm

b

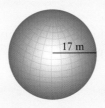

17 m

Solution

a $r = 6$ and $h = 21$

$$SA = 2\pi r^2 + 2\pi rh$$

$$= 2 \times \pi \times 6^2 + 2 \times \pi \times 6 \times 21$$

$$= 1017.8760\ldots$$

$$\approx 1017.88 \text{ cm}^2$$

b $r = 17$

$$SA = 4 \times \pi \times r^2$$

$$= 4 \times \pi \times 17^2$$

$$= 3631.6811\ldots$$

$$\approx 3631.68 \text{ cm}^2$$

Exercise 9.05 Surface areas of cylinders and spheres

Example 8

1 Find the surface area of each solid correct to two decimal places.

a

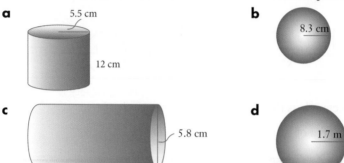

5.5 cm

12 cm

b

8.3 cm

c

5.8 cm

34 cm

d

1.7 m

2 A can of pineapple pieces has height 11 cm and radius 5 cm. The label is wrapped around the curved surface.

Calculate the area of the label correct to two decimal places.

5 cm

11 cm

3 Calculate, correct to one decimal place, the area of recycled plastic used to make this food container. Note that it has an open top.

4 A tennis ball has a diameter of 7 cm.

a Calculate the surface area of the tennis ball correct to the nearest cm².

b If the ball is made of two identical pieces, what is the area of each piece?

3 cm

18 cm

5 How much plastic would be needed to make this water tank? Answer correct to the nearest square metre.

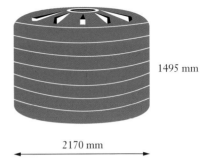

1495 mm

2170 mm

6 The Earth is approximately a sphere with a radius of 6400 km.

   a Calculate the surface area of the Earth correct to the nearest 10 square km.

   b Approximately 70% of the Earth's surface is water. What area of the Earth's surface is covered with water?

   c What area of the Earth's surface is land?

7 A hydroponics shed is made in the shape of a half-cylinder. Find the area of sheet metal needed for one shed, correct to the nearest square metre. Do not include the floor.

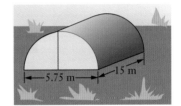

5.75 m   15 m

8 A disco mirror ball is 33 cm in diameter. It is covered with small mirror tiles that have an area of 1.44 cm^2 each.

   a Find the surface area of the sphere correct to two decimal places.

   b How many tiles are used to cover the mirror ball?

9 Elise bought a tube for sending posters through the mail. It is a cardboard cylinder with plastic 'plugs' at each end. The tube is 1.1 m long and has a radius of 2.5 cm.

   a Calculate the amount of cardboard used in creating the tube correct to the nearest square centimetre.

   b Calculate the total area of the plastic ends correct to two decimal places.

1.1 m

2.5 cm

Shutterstock.com/Damon Allen DAVISON

CHRISTMAS BAUBLE BOXES

Irena has been asked to design a new cardboard box for a set of four spherical Christmas baubles with a diameter of 6.5 cm. She has presented two designs:

An open square box

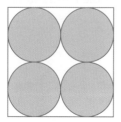

An open rectangular box

An open box has no top.

1 Calculate the area of cardboard required for the open square box. Allow for an extra centimetre on each dimension (length, width, height) to have space for interior packaging. Add 10% extra area for overlapping sections to glue box together.

2 Calculate the area of cardboard required for the open rectangular box. Allow an extra centimetre on each dimension for interior packaging space. Add 10% extra area for overlapping sections.

3 Is there another possible design to hold the four baubles? If so, provide a sketch and calculate the area of cardboard required for it. Allow an extra centimetre on each dimension for interior packaging space. Add 10% extra area for overlapping sections.

4 Which design would you recommend? Justify your decision.

9.06 Surface areas of composite solids

A page of solid shapes

To find the surface area of a composite solid:

- draw a diagram showing all of its faces
- write the dimensions of each face
- calculate the area of each face
- add the areas together.

Exercise 9.06 Surface areas of composite solids

1 Manuel is building a greenhouse as shown in the diagram. Calculate the surface area of his greenhouse, not including the floor.

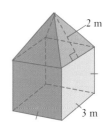

2 This metal water tank supplies water to steam train engines. It has the shape of a square prism, where each side face is a rectangle 3.2 m wide by 1.4 m high. Find the area of metal in the water tank including the base and the lid.

3 The local athletics club has a portable podium for medal presentations.
Every surface, including the base, needs repainting. Calculate the surface area to be repainted.

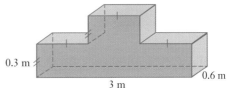

4 Lisa is making a jewellery box shaped as shown. She is going to cover it with special material.

   a Find the surface area of the jewellery box. Give your answer correct to the nearest square centimetre.

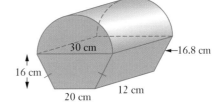

   b Lisa allows an extra 10% of material for the edges and wastage. How much material should Lisa buy? Give your answer correct to the nearest square centimetre.

5 Bruno needs to resurface the interior of his pool. (All measurements are in metres.)

   a Calculate the area of the face *ABCD*, which has the shape of a trapezium.

   b Calculate the total surface area of the five faces Bruno needs to resurface, correct to two decimal places.

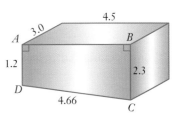

6 Ginny has made a set of wooden bookends in an L-shape as shown. She is going to stain the bookends before she uses them. Find the surface area of the bookends to be stained. Remember there are two bookends in a set.

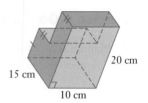

15 cm

10 cm

20 cm

9.07 Volumes of prisms

Solids with identical ends

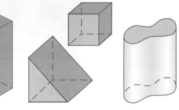

Solids that don't have identical ends

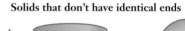

A solid with identical ends, flat faces and a uniform cross-section is called a **prism**. The end or cross-section is also called the **base** of the prism.

A cylinder is not a prism because it has a curved face.

The volume of a prism or solid with identical ends = area of base × height.

$$V = Ah$$

where A is the area of the base and h is the height.

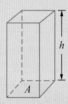
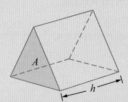
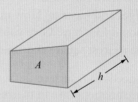

For a **rectangular prism**, the base is a rectangle, so the formula becomes

$$V = l \times w \times h$$
$$= lwh$$

EXAMPLE 9

Find the volume of each solid.

a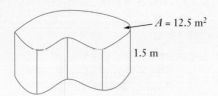

b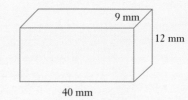

c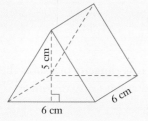

d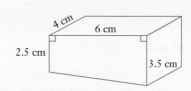

Solution

a $A = 12.5, h = 1.5$

> This solid is not a prism because some faces are not flat, but it does have identical ends.

$V = Ah$

$= 12.5 \times 1.5$

$= 18.75 \text{ m}^2$

b   For a rectangular prism, $V = lwh$ with $l = 40, w = 12, h = 9.$

$V = lwh$

$= 40 \times 12 \times 9$

$= 4320 \text{ mm}^3$

c   For a triangular prism, $V = Ah$ where $A = \dfrac{1}{2}bh.$

$A = \dfrac{1}{2}bh$

$= \dfrac{1}{2} \times 6 \times 5$

$= 15 \text{ cm}^2$

$V = Ah$

$= 15 \times 6$

$= 90 \text{ cm}^3$

d For a trapezoidal prism, $V = Ah$ where $A = \frac{1}{2}h(a+b)$.

$A = \frac{1}{2}h(a+b)$.

$$A = \frac{1}{2}h(a+b)$$
$$= \frac{1}{2} \times 6 \times (2.5 + 3.5)$$
$$= 18 \text{ cm}^2$$
$$V = Ah$$
$$= 18 \times 4$$
$$= 72 \text{ cm}^3$$

Volume and capacity

Volume and capacity units

Volume measures the amount of space inside a container, whereas **capacity** measures the amount of liquid or gas a container will hold. In the metric system, it's easy to convert between the units of volume and capacity.

1 cm³ holds 1 mL   1 m³ holds 1000 L or 1 kL

1 mL

1 cm³

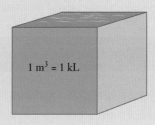

1 m³ = 1 kL

EXAMPLE 10

Convert:

a 5 cm³ to mL

b 3.4 m³ to litres.

Solution

a 1 cm³ holds 1 mL. The number of cm³ and mL are always the same.

5 cm³ holds 5 mL

b 1 m³ holds 1 kL = 1000 L.

$3.4 \text{ m}^3 = 3.4 \text{ kL}$
$$= 3.4 \times 1000 \text{ L}$$
$$= 3400 \text{ L}$$

Exercise 9.07 Volumes of prisms

1 Find the volume of each solid.

a

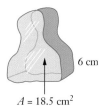

6 cm

$A = 18.5\ \text{cm}^2$

b

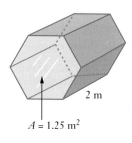

2 m

$A = 1.25\ \text{m}^2$

c

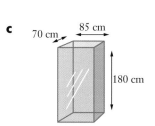

70 cm   85 cm

180 cm

d

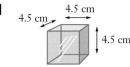

4.5 cm   4.5 cm

4.5 cm

2 Concrete blocks can be used to build walls.

  a One block measures 45 cm by 17 cm by 21 cm. Calculate its volume.

  b Kenny made a wall using 80 concrete blocks. Calculate the volume of the wall.

3 a The volume of a large container is 5000 cm³. How many millilitres does the container hold?

  b How many litres will a container with a volume of 5000 cm³ hold?

4 How many litres can a 2 m³ container hold?

5 How many litres of water does a water truck with a volume of 0.4 m³ hold?

6 Liam is pouring 1500 cm³ of liquid chlorine into a swimming pool. Express this quantity in litres.

7 What is the volume of a carton that holds 1 L of milk?

8 What is the volume in cubic centimetres of a 1.25 litre soft drink bottle?

9 Janet is building a swimming pool that is approximately the shape of a trapezium. The pool will be 1.8 m deep.

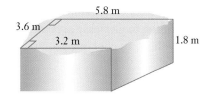

5.8 m

3.6 m

3.2 m

1.8 m

  a Use the trapezoidal rule $A \approx \dfrac{h}{2}(d_f + d_l)$ to approximate the area of the top of the water.

  b Calculate the volume of the pool in cubic metres.

  c Calculate the capacity of Janet's pool in litres. Answer to the nearest 10 000 L.

10 These storage cubes have a side length of 35 cm.

    a Calculate the volume of one cube.

    b Inga has a stack of 11 cubes in her bedroom to store books. What volume of books can she store?

    c Inga's books have an average volume of 1425 cm³ each. Approximately how many books can she store?

11 Find the volume of each prism.

    a
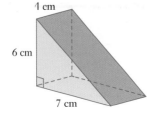

    b
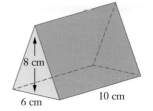

12 Binns Valley Council is building a fishpond in the park.

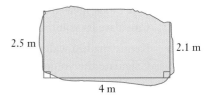

    a Use the trapezoidal rule to approximate the area of the pond.

    b The pond will be 65 cm deep. Calculate the volume of the pond.

    c How many litres will the pond hold?

13 Michael is laying pavers for a new rectangular outdoor area, 4.2 m by 2.7 m. Each paver is 300 mm by 300 mm and 50 mm thick.

> Remember to have all your measurements in the SAME units!

    a How many pavers will he need?

    b Calculate the volume of one paver.

    c Calculate the volume of the total number of pavers required for this job. Write your answer in scientific notation.

    d Pavers come in boxes of 10. How many boxes will Michael need to purchase?

    e Each box costs $56.70. How much will the pavers cost?

14 This concrete ramp was built at Hawk's Garden beach to give easy access to the beach. Calculate the volume of concrete required for this ramp.

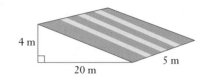

MAINTAINING THE STADIUM

Central Coast Stadium is the home of the Mariners soccer team. In this investigation, you will help the Mariners budget to keep the stadium in the best possible condition.

Fairfax Syndication/Tim Clayton

If you have access to the appropriate information, you can complete this activity for your favourite team's home ground.

Information

- The grass playing area is a rectangle 84 m by 137 m.

- The playing area needs to be watered every 10 days to a depth of 25 mm.

- Water costs $2.26 per kilolitre.

- When new turf is required, it comes in rolls 0.5 m wide by 6 m long at a cost of $7.50 per roll.

- Fertiliser is applied in spring, summer and autumn at a rate of 10 kg per 200 m^2.

- The fertiliser costs $32 per 10 kg bag or $780 per tonne in a bulk delivery.

- The cutting blade for the mower is 120 cm wide.

What you have to do

Prepare a report for the stadium's management committee to help them prepare their budget for next year. Include the cost summary table in your report. The answer to the following questions will provide information you can include in your report.

Calculations

1 The dimensions of the playing area are correct to the nearest metre.

   a What are the limits of accuracy of these measurements?

   b Within what range of values does the exact area fall?

2 What is the area of the rectangular playing field?

3 **a** Express 25 mm in metres.

   b Determine the volume of water required to cover the playing area to a depth of 25 mm.

   c How many times will the playing area require water, assuming watering is only required for the 200 days from mid October to the end of April?

   d How much water will be used?

   e How much will the water cost?

4 The ground staff applies the fertiliser three times per year.

   a How much fertiliser does the ground staff use per year?

   b How much cheaper is it to buy the fertiliser in bulk than in 10 kg bags?

5 On average, 5% of the grass needs to be replaced each year.

   a Calculate the number of rolls of turf required.

   b Calculate the total cost of the turf rolls.

6 A groundskeeper cuts the grass before every match or event. He mows the grass in strips.

   a How far does he travel as he cuts the grass?

   b He mows at a speed of 4 m/s. How long does it take him to cut the grass?

137 m

Complete the cost summary table below, then prepare your report.

Cost summary

Item	Cost	Comments
Water		
Fertiliser		
New turf		

9.08 Volumes of cylinders and spheres

We can also use the formula $V = Ah$ to find the volume of a cylinder, because a cylinder has identical ends. For a cylinder, the base is a circle, with area $A = \pi r^2$.

$$V = Ah = \pi r^2 \times h = \pi r^2 h$$

Volume of a cylinder

$$V = \pi r^2 h$$

where r is the radius of the circular base and h is the height.

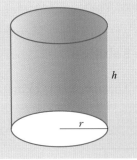

EXAMPLE 11

Find the volume of this cylinder correct to two decimal places.

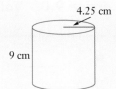

Solution

$r = 4.25$ and $h = 9$

$$V = \pi r^2 h$$
$$= \pi \times 4.25^2 \times 9$$
$$= 510.7051 \ldots$$
$$\approx 510.71 \text{ cm}^2$$

There is a special formula for the volume of a sphere.

Volume of a sphere

$$\text{Volume} = \frac{4}{3} \times \pi \times \text{radius}^3$$

$$V = \frac{4}{3}\pi r^3$$

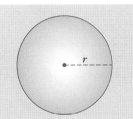

Measurement formulas chart

Formula matching game

A page of solid shapes

Sweet areas and volumes

Volumes of solids

Water tanks

Volumes of prisms and cylinders

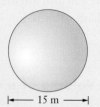

EXAMPLE 12

Find the volume of this sphere correct to the nearest cubic metre.

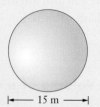

|← 15 m →|

Solution

The diameter is 15 m.

$r = \dfrac{1}{2} \times 15 = 7.5$

$V = \dfrac{4}{3}\pi r^3$

$= \dfrac{4}{3} \times \pi \times 7.5^3$

$= 1767.1458 \ldots$

$\approx 1767 \text{ m}^3$

Exercise 9.08 Volumes of cylinders and spheres

1 Find the volume of each cylinder correct to two decimal places.

a

30 mm   45 mm

b

60 cm

14 cm

c

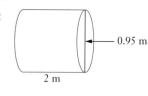

0.95 m

2 m

2 A cylindrical tube of potato chips is 25 cm long with a radius of 3 cm. What is the volume of the container correct to the nearest cubic centimetre?

3 Calculate the volume of each sphere correct to one decimal place.

a

9 cm

b

5.4 mm

c

24 cm

d

13 mm

4 A small cylindrical water tank has a diameter of 720 mm and a height of 970 mm.

   a Calculate the volume of the water tank in cubic metres correct to two decimal places.

> Hint: Change the measurements to metres before you start.

   b What is the capacity of the water tank in litres? Answer correct to the nearest 10 litres.

> Remember: $1 \text{ m}^3 = 1000 \text{ L}$

5 The radius of the Earth is 6400 km. Calculate its volume in cubic kilometres. Answer in scientific notation correct to two significant figures.

6 A soft drink can has a radius of 3.25 cm and a height of 13 cm.

   a Calculate its volume correct to the nearest cubic centimetre.

   b 24 cans are packaged in a box in the shape of a rectangular prism with dimensions 26 cm by 20 cm by 26 cm. What is the volume of the box?

   c How much space is left in the box after the 24 cans are placed in the box? Answer correct to the nearest cubic centimeter.

   d What percentage of the box's volume is the space calculated in part **c**, correct to two decimal places?

7 A hollow cylinder is shown.

   a Find the volume of a cylinder with radius 16 cm correct to two decimal places.

   b Find the volume of the 'hole' with radius 7 cm correct to two decimal places.

   c Calculate the volume of the hollow cylinder by subtracting your answer to part **b** from your answer to part **a**.

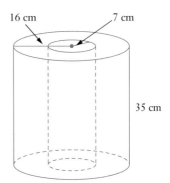

8 Find correct to one decimal place the volume of this hemisphere ('half sphere').

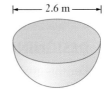

9 Kim used one big breath to blow up a round balloon to a diameter of 23.5 cm.

   a Calculate the volume of the balloon correct to the nearest cubic centimetre.

   b Estimate the capacity of Kim's lungs in litres correct to one decimal place.

Shutterstock.com/gvictoria

10 Harry runs a farm. He uses half-cylinder water troughs for the animals.

880 mm
610 mm

a Calculate the volume of one trough in cubic metres correct to two decimal places.

b How much water is needed to fill one water trough?

c There are 25 water troughs on the farm. How much water does Harry need to fill all the troughs?

d Harry uses a large cylindrical tank on a trailer to fill the water troughs. Its diameter is 1.56 m and its height is 1.21 m. Calculate the volume of the tank correct to the nearest litre.

e Can this tank hold enough water to fill all of the troughs? If not, how many trips will Harry have to make?

11 Oil refineries store gas in large spherical containers with a radius of 5 m.

a Calculate the volume of one of these containers correct to one decimal place.

b Calculate the capacity of a gas container to the nearest 100 L.

12 Two spheres of radius 5 cm fit precisely into a cylinder.

a Calculate correct to one decimal place the total volume of the two spheres.

b What is the radius and height of the cylinder?

c Calculate correct to one decimal place the volume of the cylinder.

PRACTICAL ACTIVITY

DESIGNING A FISHPOND

For this group activity, you will need a large thick crayon, tape measures and a calculator.

1 In the playground (or on the oval), draw an irregular shape for a new fishpond. Record what you have drawn.

2 By taking the appropriate measurements, use the trapezoidal rule to approximate the area of the pond.

3 Your fishpond is going to be 30 cm deep. Calculate the volume of your fishpond.

4 When you are building a fishpond, you need to remove an extra 15% of dirt when you dig the hole. How much dirt will you have to remove for your pond?

5 How much water will you need to fill your pond? Remember: $1 \text{ m}^3 = 1000$ L.

6 Use the Internet to research how many fish you can keep in a pond the size you are making.

BACKWARDS CROSSWORD

Copy the crossword below and fit the words into it. The number of letters in each word is shown in brackets.

net [3]

prism [5]

surface [7]

composite [9]

trapezoidal [11]

cubic [5]

sphere [6]

capacity [8]

dimension [9]

litre [5]

volume [6]

cylinder [8]

rectangular [11]

SOLUTION TO THE CHAPTER PROBLEM

Problem

On a farm, Stuart has to supply and install a septic tank with a capacity of 5000 to 6000 L. The diagram shows the top view of a tank that he is considering. The tank will be 1.7 m deep. Will this tank be suitable for the job?

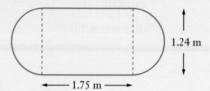

1.24 m

1.75 m

Solution

Area of the top of the tank = 2 × area of the semicircles + area of the rectangle

$$= 2 \times \frac{1}{2} \times \pi \times (0.62)^2 + 1.75 \times 1.24$$

$$= 3.3776 \ldots \text{ m}^2$$

Volume of the tank = area of the top × depth

$$= 3.3776 \ldots \times 1.7$$

$$= 5.7419 \ldots \text{ m}^3$$

$$= 5.7419 \ldots \times 1000 \text{ L} \quad \longleftarrow \boxed{1 \text{ m}^3 = 1 \text{ kL}}$$

$$= 5741.9 \ldots \text{ L}$$

$$\approx 5742 \text{ L}$$

5742 L is between 5000 L and 6000 L. The tank is suitable for the job.

ISBN 9780170413503

- Which parts of this chapter were new to you?

- Suggest two ways you could apply what you've learned in this chapter to your life outside school.

- Identify a career or job that involves surface area or volume calculations.

- Are there any parts of this chapter causing you difficulties or that you don't understand fully? If yes, ask your teacher to help you.

Copy and complete this mind map of the topic, adding detail to its branches and using pictures, symbols and colour where needed. Ask your teacher to check your work.

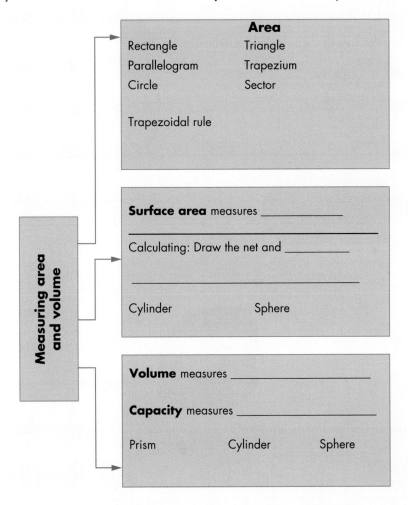

Area

Rectangle	Triangle
Parallelogram	Trapezium
Circle	Sector

Trapezoidal rule

Surface area measures _____

_____

Calculating: Draw the net and _____

_____

Cylinder   Sphere

Volume measures _____

Capacity measures _____

Prism   Cylinder   Sphere

Measuring area and volume

Exercise 9.01

1 Find the area of each shape.

a
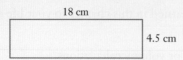
18 cm

4.5 cm

b
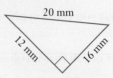
20 mm

12 mm

16 mm

c
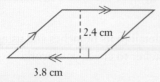
2.4 cm

3.8 cm

d
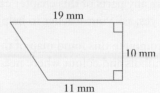
19 mm

10 mm

11 mm

Exercise 9.02

2 Calculate the area of each shaded region correct to one decimal place.

a

16.3 mm

b
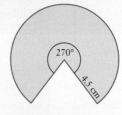
270°

4.5 cm

c
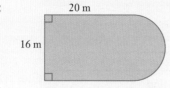
20 m

16 m

Exercise 9.03

3 Use the trapezoidal rule, $A \approx \dfrac{h}{2}(d_f + d_l)$, to approximate the area of this lake.

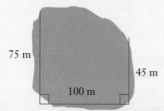
75 m

45 m

100 m

Exercise 9.04

4 Find the surface area of each prism.

a
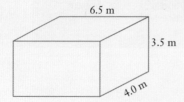
6.5 m

3.5 m

4.0 m

b
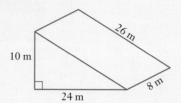
26 m

10 m

24 m

8 m

5 Find the surface area of each solid correct to the nearest square unit.

Exercise
9.05

a

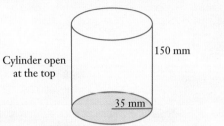

Cylinder open at the top

150 mm

35 mm

b

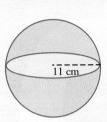

11 cm

6 Find the surface area of this solid.

Exercise
9.06

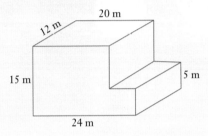

20 m

12 m

15 m

5 m

24 m

7 Find the volume of each prism.

Exercise
9.07

a

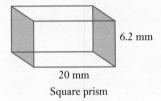

6.2 mm

20 mm

Square prism

b

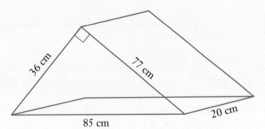

36 cm

77 cm

85 cm

20 cm

8 Find the volume of each solid correct to one decimal place.

Exercise
9.08

a

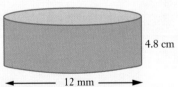

4.8 cm

12 mm

b

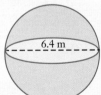

6.4 m

10.

MANAGING A HOME

Chapter problem

Brittany receives her electricity bill quarterly. Each quarter, her bill is approximately $370.

a　How much does Brittany pay for electricity annually?

b　Brittany is paid fortnightly. How much should she budget each fortnight to pay her electricity bill? Round up to the nearest $10.

CHAPTER OUTLINE

WHAT WILL WE DO IN THIS CHAPTER?

- Identify and calculate the costs of renting a home
- Read and interpret a household water bill
- Analyse and calculate water use and costs in the home
- Identify strategies for conserving water in the home
- Read and interpret a household electricity bill
- Analyse and calculate electricity use and costs in the home
- Calculate and compare the costs of running electrical appliances in the home
- Identify strategies for saving energy in the home

HOW ARE WE EVER GOING TO USE THIS?

- Understanding household bills and accounts
- Saving water and electricity in our daily lives in order to reduce costs and help the environment
- Purchasing appliances for a home
- Preparing a household budget to cover annual bills

10.01 Renting a home

If you are renting an apartment or a house, then you are called a **tenant** who pays rent regularly to a **landlord**, the owner of the property. Landlords are required to keep their property safe and tenants should take care of the property.

Tenants pay a **rental bond** at the start, a deposit equivalent to 4 weeks rent. The bond covers any rent you may owe when you leave the property or any damage you cause to the property.

EXAMPLE 1

Samoa is going to rent a two-bedroom flat in Yass for $240 per week. Before he can move in, he has to pay a rental bond and 2 weeks rent in advance.

a What is the bond if it is equivalent to 4 weeks rent?

b Calculate the total amount Samoa pays before he moves in.

Solution

a 4 weeks rent

$$\text{Rental bond} = 4 \times \$240$$
$$= \$960$$

b Bond + 2 weeks rent

$$\text{Total amount} = \$960 + 2 \times \$240$$
$$= \$1440$$

Exercise 10.01 Renting a home

1 Olivia is renting a unit for $265 per week. How much is the rent:
   a per fortnight?   **b** annually?

2 Jose pays $864 rent per fortnight.
   a How much is Jose's rent per week?
   b Calculate Jose's annual rent.

3 King's annual rent is $26 520.
   a What is his weekly rent?

> Weekly rent = Annual rent ÷ 52
> Monthly rent = Annual rent ÷ 12

   b King pays his rent monthly. How much does he pay each month?

Example 1

4 Calculate the bond for each of the following properties if it is equal to 4 weeks rent.
   a A studio apartment with a weekly rent of $195.
   b A three-bedroom townhouse with a weekly rent of $278.
   c A small house with fortnightly rent of $620.
   d A garden villa with an annual rent of $15 080.

---

5 Before Surinder can move into a unit he has to pay 2 weeks rent in advance and a bond equivalent to 4 weeks rent. The weekly rent for the unit is $230. How much does Surinder have to pay before he moves in?

6 Samantha paid a bond of $880 for the unit she is renting which is equivalent to 4 weeks rent.

    a How much is Samantha's weekly rent?

    b Samantha owed 1 week's rent when she left the unit. She also left the unit in a very dirty state. The agent deducted the rent owing and $185 for cleaning from Samantha's bond. How much of the bond was refunded to Samantha?

7 Sally, Aisha and Megan rent a three-bedroom unit in Katoomba for $475 per week. This table records their initial costs.

Item	Cost
2 weeks rent in advance	**i**
Bond equivalent to 4 weeks rent	**ii**
Lease fee	$20
Internet connection	$180
Electricity connection	$65
Moving costs	$210

    a From the table, calculate the missing values **i** and **ii**.

    b Calculate the total amount the women will have to pay before they can move in.

    c The women are going to share all costs equally. How much will be each woman's share of the costs?

    d The table below shows the women's weekly expenses. Calculate the total weekly expenses.

Item	Weekly cost
Rent	$475
Food and cleaning items	$225
Electricity	$60
Internet/Phone	$30

    e How much is Megan's share of the weekly expenses?

    f Megan's weekly take-home pay is $720. How much of her pay will be left after she pays her share of the weekly costs?

    g Do you think Megan can afford to share this house with Sally and Aisha? Give a reason for your answer.

10.02 Reading a water bill

Your home has a **water meter** that measures the amount of water used in your home. A water bill is sent to your home every quarter (3 months) listing the cost of the water service and sewerage (which is fixed), and the cost of the water used in your home (which is variable). The rate at which water usage is charged is called the **tariff**, which is measured in **dollars per kilolitre ($/kL)**.

Exercise 10.02 Reading a water bill

Questions **1** to **14** refer to the household water bill next page.

1 Which organisation issued this water bill?

2 What is the date of this notice?

3 When does this bill need to be paid?

4 What is the total amount payable on this bill?

5 What is Mr Bill's account number?

6 Did he pay his previous bill? (*Hint*: Check if there is any money in arrears [owing].)

7 What was the consumption period for this bill?

8 How many days are in the consumption period?

9 What was the total consumption of water over this period in kilolitres?

10 What was the average daily consumption of water in litres?

11 The bar chart shows the average daily water usage for the property for this billing quarter as well as for the four previous quarters. Briefly describe the pattern of consumption. Which quarters have the highest usage and which have the lowest usage?

12 Give one reason why consumption in the January quarter may be higher than for the other periods.

13 List the options available to pay this account.

14 What penalty applies for overdue payments?

   ISBN 9780170413503

Greater Middleton Council (GMC)

Civic Centre,
215 Main Street, Middleton NSW 2901
Ph: (02) 4811 7392 **Fax:** (02) 4811 0000
Email: council@middleton.nsw.gov.au

Opening Hours
8.30am to 5.00pm Monday to Friday

1.1658   019

Mr Walter Bill,
17 Reservoir Rd,
Middleton NSW 2901

Consumption Notice

Account Number
30789999
Issue Date
10/02/2018
Due Date
09/03/2018
Arrears Due Immediately

General Manager

ACCOUNT SUMMARY

Meter No GMC3188	Consumption Period 26/10/2017–24/01/2018	Consumption in kilolitres 43	Daily Ave in litres 478	Amount $71.81

Total (GST does not apply) $71.81

See reverse for details

The daily interest accrues on overdue charges at a rate of 11% per annum.

PAYMENT SUMMARY

Arrears	$0.00
Total this account	$71.81
Amount due	**$71.81**
Deduct payments since 06/02/2018	

Your average daily water use for this account is: 478

See reverse for details

TOTAL DUE: $71.81

- -

Remittance Advice

Greater Middleton Council (GMC)

Account Name:	Mr Walter Bill
Account Number:	30789999
Total Due:	$71.81
Due Date:	09/03/2018

Greater Middleton Council (GMC)
Civic Centre,
215 Main Street, Middleton NSW 2901

☐ Please tick box if a receipt is required

Present this remittance advice at our office
Civic Centre, 215 Main Street,
Middleton NSW 2901

✉ Post this remittance advice with your cheque

Credit card payments can be made at
Civic Centre or via the Postbillpay service.

POST billpay   Billpay Code: 9159
Ref: 30789999 57

*71 159 30789999 57

Pay in person at any Post Office
phone 13 18 or go to
postbillpay.com.au

B PAY   Biller Code: 485276
Ref: 30789999

Call your bank, credit union or building society
to make this payment from your cheque or
savings account. More info www.bpay.com.au

ISBN 9780170413503

Questions **15** to **19** refer to page 2 of the water bill, shown below.

ACCOUNT DETAILS

Meter No	From	To	Days	Current	Last	Usage
GMC3188	26/10/2017	24/01/2018	90	1183	1140	43

Charge	Tariff Rate	Consumption	Amount
Water Usage 20 mm meter	Refer below	43.00	$71.81

NOTE: Any consumption* shown above has had an Apportionment Factor applied as specified by the Best Practice Guidelines issued by the NSW State Government.

ADDITIONAL INFORMATION

Greater Middleton Commercial Water Tariffs		
Meter Size	**Water Tariff 1** $1.67 per kL	**Water Tariff 2** $2.25 per kL
20 mm	0 – 0380 kL per day	over 0.80 kL per day
25 mm	0 – 1.30 kL per day	over 1.30 kL per day
32 mm	0 – 2.00 kL per day	over 2.00 kL per day
40 mm	0 – 3.30 kL per day	over 3.30 kL per day
50 mm	0 – 5.15 kL per day	over 5.15 kL per day
65 mm	0 – 9.15 kL per day	over 9.15 kL per day
75/80 mm	0 – 13.15 kL per day	over 13.15 kL per day
100 mm	0 – 20.55 kL per day	over 20.55 kL per day
150 mm	0 – 46.23 kL per day	over 46.23 kL per day

Liquid Trade Waste	
Compliant	**Non Compliant**
$2.20 per kL	$19.85 per kL
Commercial Sewer	
Middleton	Barang
$2.55 per kL	$2.55 per kL
20 mm minimum $73.75	20 mm minimum $51.25
25 mm minimum $23.25	
Residential & Farmland Water	
Water Tariff 1	**Water Tariff 2**
0 – 0.080 kL per day	Over 0.80 kL per day
$1.67 per kL	$2.25 per kL

15 What is the size of the water meter in millimetres?

16 Is the average daily water usage less than or greater than 0.8 kL?

17 Is this household on Water Tariff 1 or Water Tariff 2?

18 How much is the water tariff for this household?

19 Copy and complete the following table to calculate the water usage costs for households with different water meters and water consumption figures. The first one has been completed for you.

Meter size (mm)	Consumption (kilolitres)	No. of days in billing period	Average daily use (kL per day)	Tariff ($ per kL)	Cost of supply ($)
25	150	90	150 ÷ 90 = 1.67	$2.25 (from table)	150 × 2.25 = $337.50
100	980	90			
50	450	90			
20	86	90			
65	540	90			
32	165	90			

20 Jesse uses his dishwasher four times per week. The dishwasher uses 25 litres of water each time.

   a How much water does Jesse's dishwasher use in a year?

   b At $1.67 per kilolitre, what is the annual cost of using this dishwasher?

21 Heidi uses a lawn sprinkler that sprays 1200 litres of water per hour.

   a How much water does it spray per minute?

   b She had the sprinkler on for 45 minutes. How much water did it spray?

   c If Heidi waters her lawn for 45 minutes twice a week for 6 months of the year, how much water will she use?

   d At $1.67 per kilolitre, what is the annual cost to Heidi for watering her lawn?

INVESTIGATION

YOUR WATER BILL

Read and interpret the water bill for your household, or for someone you know.
Answer Questions **1** to **14** in Exercise **10.02** for your water bill.

DID YOU KNOW?

Water restrictions

During times of drought, local councils often apply **water restrictions**. This may mean you can't use garden hoses or sprinklers and you can't wash your car. Sometimes, councils will ask residents to take short showers of 3 minutes or less. Councils also encourage residents to install rainwater tanks by offering financial help. They also promote the use of dual-flush toilets and showerheads that use less water.

Water usage
in your home

Water usage
and costs

10.03 Household water usage

Here is a listing of typical rates of water usage in the home.

- Washing your hands/face   5 L
- Brushing your teeth: tap running   5 L
- Brushing your teeth: tap not running   1 L
- Cooking and making coffee/tea   8 L per day
- Flushing the toilet   9 L to 13 L
- Flushing the toilet: half-flush   4.5 L to 6 L
- Household tap   18 L per minute
- Washing the dishes: hand   18 L
- Washing the dishes: dishwasher   25 L per cycle
- Bath   85 L to 150 L
- Shower: 8 minutes   80 L to 120 L
- Washing machine: front loading   120 L per cycle
- Washing machine: top loading   180 L per cycle
- Washing the car (with hose)   100 L to 300 L
- Garden sprinkler   1 kL to 1.5 kL per hour
- Garden hose   1.8 kL per hour

Use these figures to answer the questions in Exercise **10.03**.

Exercise 10.03 Household water usage

Adriana and Luis have bought a new house and want to calculate how much water they will use in one year. They will work out approximately how much it will cost annually and hence how much to budget fortnightly.

1 **Bathrooms (including toilets)** use up to 40% of the water consumed inside the house.

a For an 8 minute shower, how much water is consumed per minute? Use the higher rate listed above in your calculation.

b Luis has two 6 minute showers per day. How much water does he use?

c Luis flushes the toilet (full flush) on average six times a day. How much water does he use? Assume the lower rate in your calculation.

d Luis washes his face and hands three times a day and brushes his teeth twice a day, with the tap running. How much water does he use?

e What is the total amount of water Luis uses in the bathroom in a day?

f How much water will this amount to in a year? Answer in kilolitres.

g Adriana has a similar water usage pattern. How much water will Adriana and Luis use in the bathroom in one year?

2 In the **kitchen**:

a Adriana and Luis use the standard amount of water for cooking and coffee/tea making and the dishwasher is used once per day. How much water do they use per day in the kitchen?

b How much is this per year? Answer in kilolitres.

3 In the **laundry**:

a The couple do three loads of washing per week in a top-loading washing machine. How much water do they use for washing each week?

b Calculate, correct to one decimal place, the average amount of water they use in the laundry per day.

c How many kilolitres of water do they use in the laundry each year?

4 **Outside the house** is where the majority of water consumption occurs in most households.

a Each week, Adriana and Luis water the garden with a hose for a total of 2 hours, use the garden sprinkler for 1 hour and wash the car once. How many litres of water do they use outside the house per week? Use the higher rates in your calculation.

b How much water is this, on average, per day? Answer correct to one decimal place.

c How many kilolitres of water do they use outside the house each year?

5 a Summarise your results for this exercise by copying and completing the following table.

Water consumption	Litres/day	Kilolitres/year
Bathroom		
Kitchen		
Laundry		
Outside		
Total		

b Adriana and Luis are charged a water tariff of $2.55 per kilolitre. How much in total will they pay for water for one year?

c How much will they pay for water per fortnight?

10.04 Conserving water

Adriana and Luis believe they should use water more wisely to save money and be more environmentally conscious. They decide to implement some water-saving measures in the home.

Exercise 10.04 Conserving water

1 In the **bathroom**:

   a Luis showers twice a day for 6 minutes each time and uses 15 L of water per minute. How much water could he save per day by reducing each shower by one minute?

   b A *water-efficient showerhead* uses only 9 L per minute. How much water would Luis save each day with 5 minute showers and a water-efficient showerhead?

   c Luis flushes the toilet six times per day at 9 L per flush. A *dual-flush toilet* uses an average of 6 L per flush. How much water will Luis save each day by installing a dual-flush toilet?

   d Luis washes his face and hands three times a day and brushes his teeth twice a day, with the tap running. Each time, he uses 5 L of water. An *aerator on the tap* will reduce the water flow by 50%. How much water will Luis save each day with an aerator on the tap?

2 In the **kitchen**:

Adriana uses the dishwasher once per day using 25 L of water. If she only *uses the dishwasher once every 2 days*, how much water will she save per day?

3 In the **laundry**:

Adriana and Luis do three loads of washing per week with their top-loading washing machine. The machine uses 180 L of water per load. Adriana and Luis decide to buy a *front-loading washing machine* which uses 120 L per load.

   a How much water will they save per week by changing the washing machine?

   b How much is this per day, correct to one decimal place?

4 Outside the house:

Adriana and Luis use 5400 L of water outside each week watering the garden and washing the car. They decide to use *recycled grey water* for one-third of the outside activities.

a How much water will they save per week?

b What is this, on average, per day? Answer correct to one decimal place.

5 a Complete the following table to calculate the total amount of water saved by Adriana and Luis.

b If water is charged at $2.55 per kilolitre, calculate how much Adriana and Luis will save in one year?

Water savings	Litres/day	Kilolitres/year
Bathroom		
Kitchen		
Laundry		
Outside		
Total		

6 Compile a list of all of the water-saving strategies mentioned in this exercise.

10.05 Reading an electricity bill

Your home has an **electricity meter** that measures the amount of electricity used in your home. An electricity bill is sent to your home every quarter showing the cost of using the service (which is fixed) and the cost of the electricity used in your home (which is variable).

Electricity usage is measured in **kilowatt-hours (kWh)**. The rate at which electricity usage is charged is called the **tariff**, which is measured in **cents per kilowatt-hour (c/kWh)**.

Domestic electricity refers to electricity that is used during the day.

Off-peak electricity refers to electricity that is used during the late evening and early morning, such as between 11 p.m. and 7 a.m., when the demand from households and businesses is much lower. Off-peak electricity is charged at a cheaper rate than domestic electricity, and is often used to heat water in homes.

Exercise 10.05 Reading an electricity bill

Questions **1** to **8** refer to this household electricity bill.

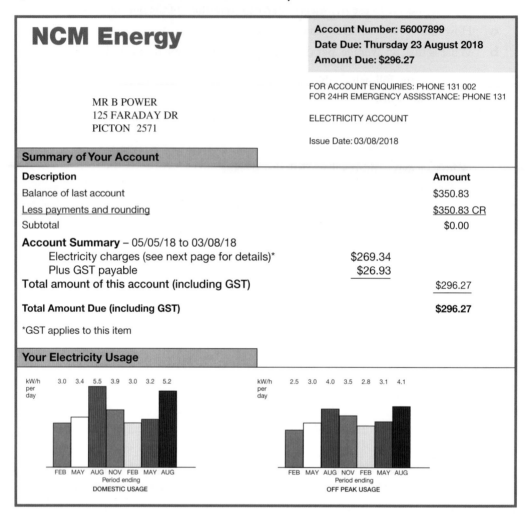

1 How much was the previous bill?

2 Was this amount paid?

3 What are the electricity charges (excluding GST) for the current quarter?

4 What is the GST on this amount?

5 What is the total cost of electricity including GST?

6 What percentage of the total cost is the GST? Answer correct to two decimal places.

7 Did domestic usage increase or decrease between May and August? By how much?

8 For this quarter, use the graph to find (in kWh) the average daily consumption of:

    a   domestic electricity   **b**   off-peak electricity.

Questions **9** to **14** refer to this table on page 2 of the electricity bill.

Itemised Details					
Meter information for Period 05/05/18 to 03/08/18 – 90 days					
Meter No.	**Const**	**This Read**	**Previous Read**	**Days**	**Usage (kWh)**
1234530	1	5533	4776	90	757
OFF PEAK – Total Usage					757
1234532	1	17384	16462	90	922
DOMESTIC – Total Usage					922

Your Account Calculations

Pricing Option	GST	Usage Applies	Rate	Amount
OFF PEAK				
Off Peak 1	Yes	757.000	$0.063100	$47.77
Off PeakAccess Chg $0.037300	Yes	90 days		$3.36
DOMESTIC				
Domestic	Yes	922.000	$0.189300	$174.53
System Access Chg $0.485300	Yes	90 days		$43.68
Electricity subject to GST				**$269.34**

9 How many days does this bill cover?

10 For this quarter, find (in kWh) the total usage of:

    a   off-peak electricity   **b**   domestic electricity.

11 What is the rate charged (in $/kWh) for using:

    a   off-peak electricity   **b**   domestic electricity?

12 For this quarter, find the amount charged for using:

    a   off-peak electricity   **b**   domestic electricity.

13 Which type of electricity is used more often but costs more?

14 What is the off-peak access charge for this quarter?

INVESTIGATION

YOUR ELECTRICITY BILL

Read and interpret the electricity bill for your household, or for someone you know.
Answer the questions in Exercise **10.05** for your electricity bill.

10.06 Household electricity usage

Household energy costs

Power problems

Energy consumption

A **watt (W)** is a unit of **power**. A **kilowatt (kW)** is 1000 watts.

A **kilowatt-hour (kWh)** is a unit of **electricity usage** equal to the amount of electrical energy used by a 1000 W load drawing power for 1 hour.

We can calculate the running costs of different appliances if we know the power rating or wattage of the appliance, our usage time and the electricity tariff for our area.

EXAMPLE 2

Lina has a 200 W television that she uses for 6 hours per day. The domestic tariff for electricity is 27 cents per kilowatt hour, written as $0.27/kWh or 27 c/kWh.

a How many kWh of electricity will Lina's TV use each day?

b How much will the electricity cost?

Solution

a Change 200 W to kW, then multiply by hours used for kWh.

$$200 \text{ W} = 200 \div 1000$$
$$= 0.2 \text{ kW}$$
$$\text{Electricity usage} = 0.2 \text{ kW} \times 6 \text{ h}$$
$$= 1.2 \text{ kWh}$$

b $\dfrac{\text{cents}}{\text{kWh}}$: to find cents, multiply by the rate.

$$\text{Electricity cost} = 1.2 \text{ kWh} \times \$0.27$$
$$= \$0.324 \text{ (or 32.4c)}$$

Exercise 10.06 Household electricity usage

1 Copy and complete this table to calculate typical daily and monthly costs of some home appliances. Assume the electricity costs 27 cents per kWh and that there are 30 days in a month.

Example 2

	Appliance	Power rating (W)	Hours used per day	kWh per day	Daily cost (cents)	Monthly cost ($)
a	Fridge (600 L)	800	24			
b	Clothes dryer	2400	1.5			
c	Washing machine	900	0.5			
d	Bathroom/fan/heater/light	1100	1			
e	Incandescent light globe	100	6			
f	Food processor	380	0.5			
g	Electric kettle	1500	1			
h	Stove hotplate (max. setting)	1500	0.25			
i	Dishwasher	1900	1			
j	Toaster	650	0.5			
k	Vacuum cleaner	950	0.25			
l	Stereo system	40	5			
m	Hairdryer	1400	0.5			
n	TV (medium to large)	550	6			
o	Iron	950	0.75			

2 Which of the appliances from Question 1 are in your household? How many of each appliance?

3 Use the information from Question 1 to calculate the approximate cost of electricity for running these appliances in your household for one year.

4 How close is your calculation to your household's actual electricity bill? Give reasons why your calculation might be different to that of your bill.

INVESTIGATION

THE ENERGY RATING SYSTEM

Some household appliances now display **energy rating labels** to help people choose energy-efficient appliances. These labels contain a star rating to show how efficient the appliance is – the more stars, the more efficient the energy use. The label also lists an estimate of the energy the appliance will use over a year.

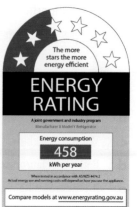

Use the Internet to research the answers to these questions.

1 What information does the energy rating label on electrical appliances provide?

2 What does it mean for an appliance to have more stars than another?

3 Apart from the number of stars, what other important information is printed on the label?

4 There are eight appliances that are required to carry energy rating labels. List them.

5 The two key features of the label are:

  a the star rating band: how many stars can this have and what does the coloured section indicate?

  b the energy consumption box: what does the number in this box indicate?

6 For all star-rated appliances other than air conditioners, the energy consumption figure is given in what units and for what period?

7 How is the information given for air conditioners?

8 Explain how the energy consumption figure can be used to calculate the annual electricity cost of the appliance.

9 Calculate the annual cost of running a refrigerator rated at 750 kWh/year if the electricity tariff is 27 cents/kWh.

COMPARING ENERGY RATINGS

The government website Energy Rating allows us to compare the star and energy ratings of different brands and models of appliances.

1 Visit the website and select **Consumers** and **Search the Registration database**.

2 Search for all two-door refrigerator/freezers with a capacity of 400–500 litres.

3 Find the most efficient and the least efficient refrigerator/freezer (they may be on different pages). Copy and complete this table, including calculating the annual running cost based on a tariff of 27 cents per kWh.

	Brand	Model	Star rating	Energy consumption (kWh/year)	Annual cost ($/year)
Most efficient					
Least efficient					

4 Calculate the annual cost difference between the two refrigerator/freezers you found.

5 Repeat Questions **1** to **4** for 90–110 cm plasma TVs.

6 Repeat Questions **1** to **4** for 90–110 cm LCD TVs.

7 How do LCD and plasma TVs compare in costs for:

    a the most efficient TVs?   **b** the least efficient TVs?

8 Repeat Questions **1** to **4** for an 8 kg load, top-loading washing machine.

9 Is the cost of running an appliance the only factor in deciding what appliance to buy? List other considerations. What is the most important factor in your opinion? Justify your answer.

10.07 Conserving electricity

Let's examine the electricity usage and costs for a typical home, and look at ways we can reduce the amount of electricity we use. We will base our calculations on a **quarter**; that is, either one-quarter of a year if we have yearly figures or 90 days if we have daily figures.

Electrical appliances

EXAMPLE 3

Calculate the quarterly electricity usage and cost for each of the following appliances. Assume that the electricity tariff is 27 cents per kWh.

a A dishwasher with a rating of 220 kWh per year.

b A 200 W television used for 5 hours per day.

Shutterstock.com/P.S._2

Solution

a Divide yearly use by 4 for a quarter.

$\dfrac{\text{cents}}{\text{kWh}}$: to find cents, multiply by the rate.

Quarterly usage = 220 kWh ÷ 4

= 55 kWh

Quarterly cost = 55 × $0.27

= $14.85

b Change 200 W to kW.

5 hours per day, 90 days in a quarter.

200 W = 200 ÷ 1000

= 0.2 kW

Quarterly usage = 0.2 kW × 5 × 90

= 90 kWh

Quarterly cost = 90 × $0.27

= $24.30

Exercise 10.07 Conserving electricity

In this exercise, calculate all percentages correct to one decimal place and assume that the electricity tariff is 27 cents per kWh.

1 Copy and complete this table to estimate the quarterly electricity consumption and cost for a typical family. The shaded boxes do not require any values.

	Appliance	kWh per year	Wattage	Usage hours per day	Quarterly usage (kWh)	Quarterly cost @ 27c/kWh
a	Fridge	900				
b	Hot water system	1100				
c	Space heating		2400	4		
d	Lighting		1875	4		
e	Washing machine	250				
f	Clothes dryer	150				
g	Oven		1800	2		
h	Dishwasher	200				
i	Microwave		1350	$\frac{1}{2}$		
j	Toaster		600	$\frac{1}{2}$		
k	Stove hotplate		1200	2		

	Appliance	kWh per year	Wattage	Usage hours per day	Quarterly usage (kWh)	Quarterly cost @ 27c/kWh
l	Other kitchen		500	2		
m	Iron		800	1		
n	Hairdryer		1200	$\frac{3}{4}$		
o	TV		220	6		
p	Other		500	2		
q	Total					

2 The hot water system in Question **1** has instantaneous heating, but a hot water system that operates during off-peak hours is a cheaper option. The electricity suppliers offer a substantially lower tariff for off-peak use.

a Assuming an off-peak tariff of 19.8 cents per kWh, copy and complete this table for an off-peak hot water service.

Appliance	kWh per year	Quarterly usage (kWh)	Quarterly cost @ 19.8c/kWh
Off-peak hot water system	1100		

b How much money does this save per quarter?

c What percentage of the total from Question **1** is this?

$$\text{Percentage} = \frac{\text{Saving}}{\text{Total cost from Question 1}} \times 100\%$$

3 Households can save a substantial amount on lighting. The table in Question **1** assumed a household with 25 incandescent light globes each being 75 watts and switched on for 4 hours a day.

a Lights should be switched off in rooms where they are not needed.

  i Calculate how much money you would save if you reduced the average daily use of the lights from 4 hours to 3 hours.

  ii What percentage of the total from Question **1** is this?

b Compact fluorescent light globes (CFLs) are 80% more efficient than incandescent globes. This means that a 15 watt CFL globe will do the same job as a 75 watt incandescent globe.

  i Suppose all 25 globes from Question **1** were replaced by 15 watt CFL globes. Copy and complete this table.

Appliance	Wattage	Usage hours per day	Quarterly usage (kWh)	Quarterly cost @ 27c/kWh
Lighting		4		

  ii How much money would this save?

  iii What percentage of the total from Question **1** is this?

4 Heating costs can be reduced by using reverse-cycle air conditioners such as 5- and 6-star rating air conditioners which save 40% to 60% in power. They can also be used for cooling in the summer.

a Copy and complete this table for the cost of a 900 W reverse-cycle air conditioner.

Appliance	Wattage	Usage hours per day	Quarterly usage (kWh)	Quarterly cost @ 27c/kWh
Reverse-cycle air conditioner	900	4		

b How much does this save per quarter?

c What percentage of the total from Question **1** is this?

5 Many appliances use electricity even when they are not being used. This is called 'standby' energy, common in computers, printers, TVs, video players, games consoles and microwave ovens that are not switched off completely. The standby wattage is very low, but because these appliances are on all the time, they consume a considerable amount of power in a quarter of a year.

a Copy and complete this table to calculate the quarterly usage and cost of these appliances in standby mode.

	Appliance	Wattage	Usage hours per day	Quarterly usage (kWh)	Quarterly cost @ 27c/kWh
i	Clock radio	4	24		
ii	Computer monitor	5	24		
iii	Cordless phone	3	24		
iv	Microwave oven	4	24		
v	Computer	2	24		
vi	Printer	8	24		
vii	Stereo	10	24		
viii	Television	10	24		
ix	DVD player	8	24		
	Total				

b What is the total quarterly cost if all of these items are in standby mode?

c What percentage is this of the total quarterly bill calculated in Question **1**?

d If we switched off all of these appliances completely for 12 hours each day instead of leaving them on standby, how much would this save?

6 Compile a list of all of the electricity-saving strategies mentioned in this exercise.

ISBN 9780170413503

PRACTICAL ACTIVITY: HOUSEHOLD BUDGETS

Running and maintaining a home is an expensive activity. We need to budget carefully to make sure we have the money to pay our bills on time in order to avoid costly interest charges.

In this activity, we are going to prepare a budget for Adriana and Luis to manage their annual expenses.

1 This table lists the main expenses for Adriana and Luis' house. Copy this table and complete the missing values. The first and last rows have been completed as examples.

Bill	Description	Amount	Bill frequency	Annual cost calculation	Annual cost
Water	Water supply and usage	$325	Quarterly	$325 \times 4 = 1300$	$1300
Electricity	Electricity supply and usage	$970	Quarterly		
Insurance	Building and contents insurance	$151	Monthly		
Council rates	Local council charges	$305	Quarterly		
Mobile phone, phone and Internet	All three bundled together	$100	Monthly		
Repair and replacement fund	Savings to cover unexpected expenses	$500	Monthly	$500 \times 12 = 6000$	$6000
				Total annual cost	

2 Calculate the total annual cost of Adriana and Luis' household bills.

3 Increase the total for Adriana and Luis' annual household bills by 10% to allow for future price rises.

4 Both Adriana and Luis are paid fortnightly. Divide the answer to Question **3** by 26 to calculate the amount they should budget each fortnight to cover household expenses.

5 What other expenses would Adriana and Luis have to budget for?

DEFINITION MATCH

Copy each word in the first column and write its correct definition from the second column.

Annually	Four times per year
Bond	Someone who rents a property
Budget	Twelve times per year
Kilolitre	Owner of properties for rent
Kilowatt-hours	Rate at which water and electricity are charged
Landlord	Money paid in case of damage to a property
Monthly	Unit for measuring water usage
Off-peak electricity	Per year
Quarterly	Units of electricity
Tariff	Electricity used overnight
Tenant	Unit for measuring electricity usage
Watt/kilowatt	An outline of how our income will be spent

SOLUTION TO THE CHAPTER PROBLEM

Problem

Brittany receives her electricity bill quarterly. Each quarter, her bill is approximately $370.

a How much does Brittany pay for electricity annually?

b Brittany is paid fortnightly. How much should she budget each fortnight to pay her electricity bill? Round up to the nearest $10.

Solution

a Brittany's annual electricity bill $= \$370 \times 4$ Four quarters in a year

$$= \$1480$$

b Brittany's fortnightly budget for electricity $= \$1480 \div 26$ 26 fortnights in a year

$$= \$56.92307 \ldots$$

$$\approx \$60 \qquad \text{Rounding up}$$

- Give examples of how the information about water use in this chapter might change your habits?

- Give examples of how the information about electricity use in this chapter might change your habits?

- Is there any part of the calculations you didn't understand? If so, ask your teacher for help.

Copy and complete this mind map of the topic, adding detail to its branches and using pictures, symbols and colour where needed. Ask your teacher to check your work.

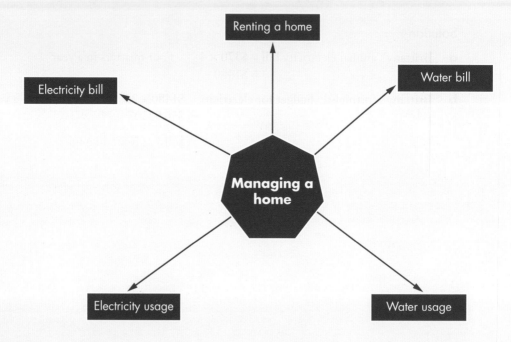

1 Jack pays $235 per week to rent a room in a terrace house in inner Sydney. Calculate how much he pays in rent:

a annually   **b** monthly   **c** fortnightly.

Exercise
10.01

2 Before Jack moved into the house he had to contribute 2 weeks rent in advance and a bond equivalent to 4 weeks rent. Calculate how much Jack paid.

Exercise
10.01

3 This is Aaron's water bill for 6 months.

Exercise
10.02

New Century Water

|||||||||||||||||||||||||||
Mr Aaron Wong
60 Glenning Rd
GLENNING VALLEY NSW 2261

021
1002534

Account Number	Due Date
0240661	12-Dec-18
Total Amount Due	**Date of Issue**
$277.25	11-Nov-18

Water Usage and Charges

Water Usage

Meter No.	Previous Reading Date	Reading	Current Reading Date	Reading	Usage In Kilolitres	Net Usage In Kilolitres
2023243	10-Mar-18	5330	21-Oct-18	5474	144	144

Total Usage   **144 kL**

Charges   *Water usage up to June 30 2018 @ $1.87 per kL

              *Water usage from July 1 2018 @ $1.98 per kL

Description	Meter No.	Billed Quantity (kL)	Tariff	Amount
Water Usage	2023243	71.68	1.87	$134.05
Water Usage	2023243	72.32	1.98	$143.20

Total Current Charges   **$277.25**

Your Average Daily Water Usage in Litres

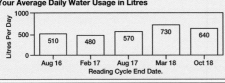

Account Summary	
Current Charges	$277.25
TOTAL DUE	**$277.25**

a On which date did the council issue the account?

b By what date does Aaron have to pay the account?

c How much does Aaron owe on the account?

d On 1 July, the price of water increased. By how much per kL did the price increase?

Use the information on pages 297–8 to answer Questions **4** to **7**.

4 Each day, Daniel has a 5 minute shower (use the rate per minute calculated in Question **1a** on page 298), flushes the toilet four times (use the higher rate), does the cooking and washes up by hand. Find his total water usage for these activities.

5 Daniel does four loads of washing each week using a top-loading washing machine.

a How many litres does he use for washing each week?

b Calculate correct to one decimal place the usage per day.

6 Daniel decides to buy a dishwasher. Instead of washing up each day he uses the dishwasher three times a week. How many litres of water does he save in a week?

7 Daniel is considering buying a front-loading washing machine. If he does so, he will need to do six loads of washing per week.

a How many litres of water would he now use for washing each week?

b Will he save water by purchasing a front-loading washing machine?

8 This is Aaron's electricity bill for 6 months.

Absolute Power

ḷıḷḷıḷḷıḷḷıḷḷıḷḷıḷḷıḷḷı

BN10457 - 0001 - 12230

Mr Aaron Wong
60 Glenning Rd
GLENNING VALLEY NSW 2261

Due Date
5 September 2018

Amount Payable
$203.91

Quarterly Electricity Account

Total amount payble of your last Electricity bill	247.65
Payments - Thankyou	–247.65 cr
Electricity (17/05/2018 – 15/08/2018)	185.37
Subtotal of charges before GST	185.37
Total GST payable 10%	18.54
Total charges including GST	203.91
Total Amount Payable	**$203.91**

a How much is Aaron's current electricity bill?

b How much GST is included in the bill?

c What would the bill be if the government made electricity bills GST free?

d Aaron has been trying to reduce the amount of electricity he uses. How much less expensive is his current bill than his previous one?

9 Daniel owns a 550 watt TV which he has on for 5 hours each day. The domestic tariff is 27c/kWh.

a How many kWh of electricity will Daniel's TV use each day?

b How much will the electricity cost?

10 Daniel has 13 light globes in his home. He has six 100 W globes and seven 60 W globes. On average, he has the lights on for 7 hours per day.

a How many kWh of electricity will Daniel use for lighting each day?

b How much will the electricity cost?

11 Daniel sells his large TV and buys a smaller 200 W TV.

a Assuming the same usage as in Question **9**, how much does Daniel save per day by having a smaller TV?

b How much would he save per year?

12 a How much would Daniel save on lighting if he reduced the average time he has the lights on to 4 hours per day?

b How much would he save per year?

Practice set 2

Section A: Multiple-choice questions

For each question, select the correct answer A, B, C or D.

Exercise
8.01

1 The favourite party food of a group of 3-year-old children was recorded. What is the only statistical measure that can be found for this data?

  A mean   **B** median   **C** mode   **D** range

Exercise
6.01

2 What is the probability that a student randomly chosen has a birthday in a month beginning with the letter J?

  A $\dfrac{1}{3}$ **B** $\dfrac{1}{4}$ **C** $\dfrac{1}{6}$ **D** $\dfrac{1}{12}$

Exercise
10.03

3 For the month of April, Julie's family used 395 L of water per day. How much water did they use for the entire month?

  A 11.06 kL   **B** 11.85 kL   **C** 12.245 kL   **D** 15.8 kL

Exercise
6.03

4 Which of the following is the complementary event to 'winning a race'?

  A coming last   **B** coming second

  C coming second or third   **D** not winning the race

Exercise
9.08

5 Which calculation could be used to find the volume of this cylinder?

  A $\pi \times 3 \times 1$

  B $\pi \times 3^2 \times 1$

  C $\pi \times 1.5 \times 1$

  D $\pi \times 1.5^2 \times 1$

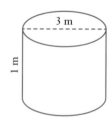

Exercise
7.02

6 A tub of margarine weighs 500 g. This mass is greater than:

  A 60 000 mg   **B** 0.8 kg   **C** 2.5 kg   **D** 0.01 tonnes

Exercise
7.07

7 This figure is made up of a rectangle and an equilateral triangle. Find its perimeter.

  A 22 cm

  B 44 cm

  C 54 cm

  D 64 cm

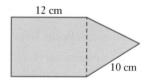

NCM 11. Mathematics Standard (Pathway 1)

ISBN 9780170413503

8 Calculate the median for this set of data.

Exercise 8.01

$$10 \quad 3 \quad 5 \quad 8 \quad 4 \quad 2 \quad 10$$

A 5 **B** 6 **C** 8 **D** 10

9 Linda has two $5 notes, four $10 notes and three $20 notes in her wallet. She takes out a note without looking. What is the probability that it is a $20 note?

Exercise 6.01

A $\dfrac{60}{110}$ **B** $\dfrac{20}{35}$ **C** $\dfrac{3}{35}$ **D** $\dfrac{1}{3}$

10 Justin watches a 30 W TV for 2 hours and 50 minutes. How much energy is used by the TV?

Exercise 10.06

A 0.085 kWh **B** 85 kWh **C** 0.075 kWh **D** 75 kWh

11 A window is 2 metres long and 90 cm wide. Find the area of the window.

Exercise 9.01

A 0.18 m² **B** 1.8 m² **C** 18 m² **D** 180 m²

12 The internal dimensions of a refrigerator are 150 cm (height), 60 cm (width) and 40 cm (depth). What is the capacity of the refrigerator in litres?

Exercise 9.07

A 250 L **B** 300 L **C** 360 L **D** 430 L

Section B: Short-answer questions

Exercise 8.01

1 Find the area of each shape.

a

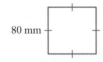

80 mm

b

9 cm

11 cm

Exercise 8.01

2 The front yard of a house is rectangular and measures 20 m by 8 m. The backyard is also rectangular and measures 35 m by 7.5 m. Calculate the cost of covering both yards with turf which costs $18.60/m^2.

Exercise 7.05

3 Find the length of the unknown side in each triangle. Answer correct to one decimal place.

a

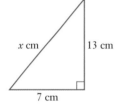

x cm   13 cm

7 cm

b

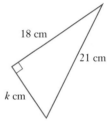

18 cm

21 cm

k cm

Exercise 10.01

4 Annaliese is going to rent a room in a share house. Her rent will be $245 per week. Before she can move in she must pay 2 weeks rent in advance and a bond of 4 weeks rent. She must also pay $85 for Internet connection. Calculate the total she must pay before she can move in.

Exercise 8.07

5 The ages of children at a play centre are shown in this frequency table. Find correct to two decimal places:

a the mean

b the standard deviation.

Age (years)	Frequency
5	3
6	4
7	5
8	2
9	2

6 For the above data set, find:

   a   the mode   **b**   the range   **c**   the median.

Exercise
8.06

7 For each sphere, calculate correct to two decimal places:

   i   its volume   **ii**   its surface area.

Exercise
9.05

   a   **b**

Exercise
9.08

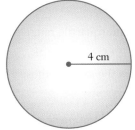

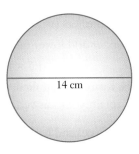

8 Write each number in scientific notation.

   a   423 000   **b**   0.000 34   **c**   24

Exercise
7.05

9 Write each number in normal decimal form.

   a 8.1×10^{4} **b** 2.99×10^{-3} **c** 5×10^{2}

Exercise
7.05

10 Kayleigh receives her water bill for a quarter. It shows the following information.

- Current meter reading: 439 kL
- Previous meter reading: 365 kL
- Tariff: $2.80/kL
- Number of days: 91

   a   How many kilolitres of water did Kayleigh use this quarter?

   b   Calculate the cost of the water Kayleigh used.

   c   Find Kayleigh's average water use per day correct to the nearest litre.

Exercise
10.02

Exercise
8.04

11 This data set shows the number of goals scored in each game of the season by a local soccer team.

| 2 | 1 | 0 | 0 | 7 | 2 | 3 | 2 | 4 |
| 3 | 2 | 9 | 1 | 2 | 1 | 2 | 4 | 3 |

a Identify any outliers (without using the formula).

b Find the upper and lower quartiles and the interquartile range for this data.

c Calculate: **i** $Q_1 - 1.5 \times \text{IQR}$ **ii** $Q_3 + 1.5 \times \text{IQR}$.

d When the formula is applied, are the outliers the same as those identified in part **a**? Explain your answer.

Exercise
6.05

12 The weather this weekend has an equal chance of being fine or raining.

a Copy and complete this tree diagram to show the possible outcomes for the weather on Saturday and Sunday.

Saturday   Sunday   Outcomes

fine

rain

b What is the probability that it will be fine on both days?

Exercise
9.07

13 Calculate the volume of each prism.

a

15 cm
3 cm 3 cm

b

4 m
4 m
4 m

c

6 mm
8 mm
10 mm
11 mm

Exercise
9.04

14 Find the surface area of each prism in Question **13** above.

15 Anish rolled a die 90 times and recorded the results shown below.

Number	1	2	3	4	5	6
Frequency	12	15	9	16	13	25

Exercise 6.06

    a What is the theoretical probability of rolling a 6? Answer correct to three decimal places.

    b What is the relative frequency of rolling a 6? Answer correct to three decimal places.

    c Explain why Anish's relative frequency and the theoretical probabilities aren't the same.

    d Based on this data, if the die was rolled 200 times, how many times would you expect the number 3 to come up?

16 Copy and complete this table.

Exercise 7.03

Measurement	Absolute error	Lower limit of accuracy	Upper limit of accuracy
55 cm correct to the nearest cm			
75 kg correct to the nearest 5 kg			
280 km correct to the nearest 10 km			

17 This frequency table gives a summary of a judge's scores for student speeches in an English assessment task.

Exercise 8.05

    a Copy this table and complete the cumulative frequency column.

    b Construct a cumulative frequency histogram and polygon for the data.

    c Use the cumulative frequency polygon to find an estimate for the median of this data.

Judge's score	Frequency	Cumulative frequency
2	2	
3	1	
4	3	
5	3	
6	0	
7	6	
8	7	
9	4	
10	1	

Exercise
9.02

18 Calculate the shaded area of each shape. Answer correct to two decimal places.

a

120°

15 mm

b

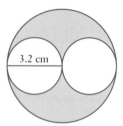

3.2 cm

Exercise
7.08

19 Find the perimeter of the sector in Question **18a** correct to two decimal places.

20 A die is rolled twice.

Exercise
6.07

a Copy and complete this probability tree for the chances of rolling numbers divisible by 3 (i.e. a 3 or a 6).

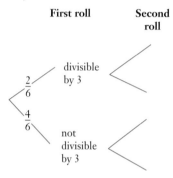

First roll   Second roll

$\frac{2}{6}$ divisible by 3

$\frac{4}{6}$ not divisible by 3

b What is the probability of rolling two numbers divisible by 3?

c What is the probability of rolling one number divisible by 3 and one number not divisible by 3?

Exercise
10.05

21 Jackson receives his electricity bill for a quarter. It shows the following information.

This reading: 18 075 kWh

Previous reading: 17 384 kWh

Domestic tariff: 23.8c/kWh

Number of days: 90

a How many kilowatt hours of electricity did Jackson use this quarter?

b Calculate the cost of the electricity Jackson used.

c Find Jackson's average use per day for this quarter correct to the nearest kilowatt hour.

22 Kane and Anne have kept details of their household expenses from last year.

Exercise
10.08

Electricity: $3874   Council and water rates: $2613

Insurance: $1297   Internet/phones: $1200

a Calculate the total annual cost of Kane and Anne's household bills.

b In the coming year, they expect increases in all of these expenses. Increase the total for Kane and Anne's annual household bills by 10% to allow for future price rises. Round up to the nearest dollar.

c Both Kane and Anne are paid monthly. Use your answer to part **b** to calculate the amount they should budget each month to cover household expenses. Round up to the nearest dollar.

d What other items might Kane and Anne have to budget for?

11.

GRAPHING LINES

Chapter problem

At 5 a.m. when Dwayne was jogging along a country lane, he discovered a car that had crashed into a tree. The male driver was unconscious in the car. Dwayne called an ambulance and the police. This is the information that the police were able to obtain.

- The driver had been drinking at a party the previous night.
- He left the party at 1 a.m.
- Blood samples taken from the driver showed the following blood alcohol content (BAC).

Time	6 a.m.	7 a.m.	8 a.m.	9 a.m.
BAC	0.043	0.031	0.019	0.007

Graph this table of values, find the gradient of the line and use the gradient to estimate the driver's BAC at 1 a.m.

CHAPTER OUTLINE

WHAT WILL WE DO IN THIS CHAPTER?

- Graph linear functions such as $y = 2x + 9$
- Find the gradient and y-intercept of a line
- Use linear functions to model practical situations
- Interpret the meaning of the gradient and y-intercept in a linear model
- Solve and graph problems involving direct variation, $y = kx$
- Use conversion graphs

HOW ARE WE EVER GOING TO USE THIS?

- When we want to represent relationships graphically
- When we model real-life situations with algebra and a graph
- When converting from one measurement into another; for example, converting currencies for overseas travel

Graphing
linear
functions

Graphing
linear
equations

11.01 Linear relationships

We can graph algebraic rules on a number plane. In this chapter, we will examine algebraic rules the graphs of which are straight lines. These rules are called **linear functions** or linear equations.

To graph a linear function:

- construct a table of values for the rule

The word 'linear' means 'of a line'.

- plot the points from the table of values on a number plane

- rule a straight line through the points.

We can choose any numbers we want for x, but we need to make sure the points will fit on our graph and be easy to calculate. It is easiest to choose whole numbers close to 0.

EXAMPLE 1

Graph the linear function $y = x + 2$.

Solution

Draw a table and choose some x values.

x	-2	-1	0	1	2	3
y						

Calculate the y values by substituting the x values in the rule.

$y = -2 + 2 = 0$ $y = 1 + 2 = 3$

$y = -1 + 2 = 1$ $y = 2 + 2 = 4$

$y = 0 + 2 = 2$ $y = 3 + 2 = 5$

Complete the table.

x	-2	-1	0	1	2	3
y	0	1	2	3	4	5

Write the points from this table.

$(-2, 0), (-1, 1), (0, 2), (1, 3), (2, 4), (3,5)$

Draw a set of axes and plot the points. Rule a straight line through the points, place arrows at each end and label the line with its equation.

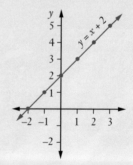

Exercise 11.01 Linear relationships

1 Draw the graph of each linear function.

Example 1

a $y = x - 4$　　　　　**b** $y = 2x$　　　　　**c** $y = x + 3$

d $y = \dfrac{x}{2}$　　　　　**e** $y = -3x$　　　　　**f** $y = 2x - 4$

g $y = -x + 2$　　　　　**h** $y = 3x$　　　　　**i** $y = \dfrac{x}{4}$

11.02 Gradient and *y*-intercept

Notice that we can always write the equation of a linear function in the form $y = mx + c$, where *m* and *c* are numbers.

$y = x - 4$　　　　$y = 2x$　　　　$y = x + 3$　　　　$y = \dfrac{x}{2}$　　　　$y = -3x$

Drawing gradients

$y = 2x - 4$　　　　$y = -x + 2$　　　　$y = 3x$　　　　$y = \dfrac{x}{4}$

A page of number planes

> The equation of a line has the form $y = mx + c$ where *m* is the **gradient** of the line and *c* is the **y-intercept**.
>
> The **gradient** measures how steeply the line goes up or down.
>
> The **y-intercept** is the value where the line crosses the *y*-axis.

Gradient and y-intercept

A **positive gradient** means the line goes up from left to right.

A **negative gradient** means the line goes down from left to right.

y = mx + c

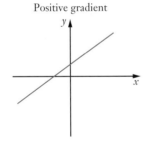
Positive gradient

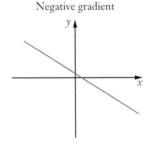
Negative gradient

Finding the equation of a line

Linear functions code puzzle

EXAMPLE 2

a What is the gradient of this line?

b What is the value of the *y*-intercept?

c What is the equation of the line?

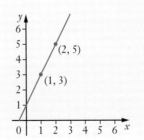

Gradient and y-intercept of a line

　　　　11. Graphing lines

Solution

a Draw a right-angled triangle off the two points using the line as the hypotenuse.

The line goes up 2 units for every 1 unit across to the right.

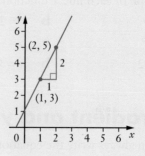

$$m = \frac{2}{1} = 2$$

The gradient is 2.

b The *y*-intercept is the value where the line crosses the vertical axis.

The *y*-intercept is 1.

c The equation of a straight line is $y = mx + c$. From above, $m = 2$, $c = 1$.

The equation of the line is $y = 2x + 1$.

EXAMPLE 3

For each linear function, state:

   i the gradient   **ii** the *y*-intercept.

a $y = 4x - 1$ **b** $y = -2x$ **c** $y = 8 - 3x$

Solution

When the equation is in the form $y = mx + c$, the gradient is the number in front of the *x* (*m*) and the *y*-intercept is the **constant** (*c*).

a $y = 4x - 1$

  i gradient $= 4$

  ii *y*-intercept $= -1$

b $y = -2x$

  i gradient $= -2$

  ii *y*-intercept $= 0$

c $y = 8 - 3x = -3x + 8$

  i gradient $= -3$

  ii *y*-intercept $= 8$

ISBN 9780170413503

Exercise 11.02 Gradient and *y*-intercept

1 For each line, find its gradient, *y*-intercept and equation.

a

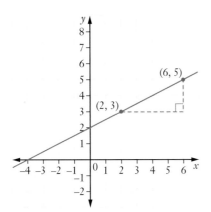

b

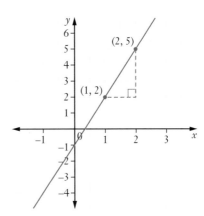

c

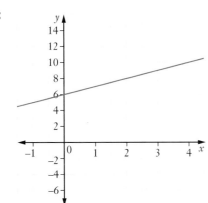

d

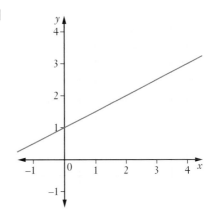

2 Find the equation of each line. Remember that the gradient is negative when the line goes down to the right.

a

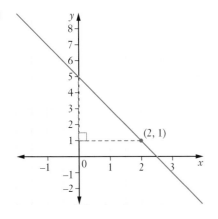

b

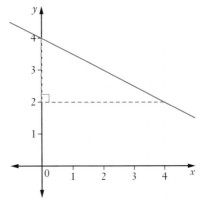

c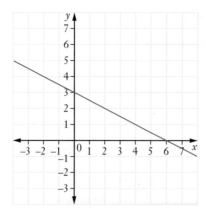

d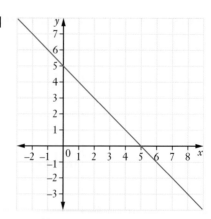

3 What is the equation of this line?
Select the correct answer **A**, **B**, **C** or **D**.

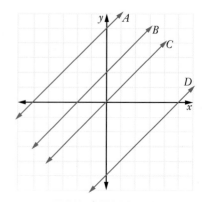

A $y = -\dfrac{1}{2}x + 8$

B $y = \dfrac{1}{2}x + 8$

C $y = -2x + 8$

D $y = 2x + 8$

4 For each linear function, state:

i the gradient   **ii** the *y*-intercept.

> Remember: $\dfrac{x}{4} = \dfrac{1}{4}x$

a $y = 3x + 2$ **b** $y = -2x + 3$ **c** $y = 7 - x$

d $y = 4x$ **e** $y = \dfrac{x}{4} - 2$

5 Look at your graphs for Exercise 11.01 on page 329. For each graph, state whether the gradient is positive or negative and write the gradient from the equation.

6 Match each equation to its graph by considering the *y*-intercept.

a $y = x$ **b** $y = x + 2$

c $y = x - 5$ **d** $y = x + 4$

7 Match each equation to its graph by considering the gradient.

a $y = x$ **b** $y = 3x$

c $y = -x$ **d** $y = 6x$

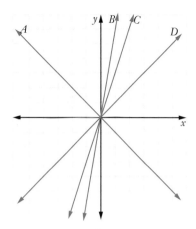

GRAPH THESE LINES

Use a graphics program/app or website to enter appropriate linear functions to graph each set of lines shown. Use the same scale on both the *x*- and *y*-axes.

Graphing
lines:
graphics
calculator

a

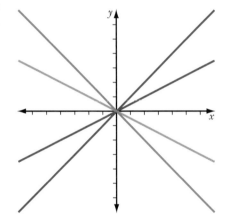

b

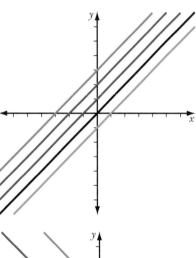

c

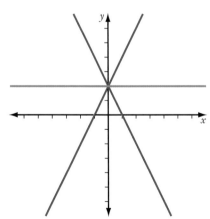

d

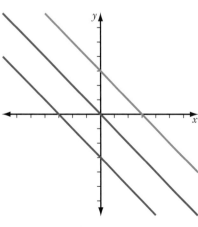

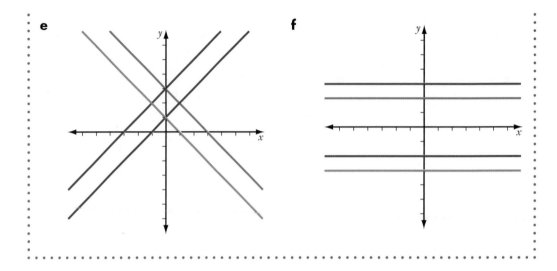

e

f

Linear
modelling

11.03 Linear modelling

Linear functions can be used to model many real-world situations. This is called
linear modelling.

EXAMPLE 4

The concentration of a drug in a person's body decreases
as time passes, according to a linear function. This is
represented by the graph shown. Note that the variables
x and y have been replaced by t and C.

> μg means micrograms or
> one-millionth of a gram.

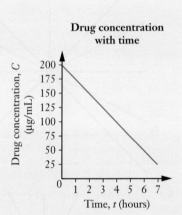

**Drug concentration
with time**

a   What is the gradient of the line?

b   What does the gradient represent?

c   Find the **vertical intercept** of the line.

d   What does the vertical intercept represent?

e   What is the equation of the line?

f   When will there be no drug remaining in the body?

Solution

a Draw a triangle on the line to calculate the gradient.

The line goes down 125 units for every 5 units across. The gradient will be negative because the line goes down rather than up.

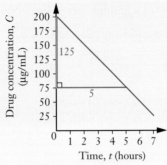

Drug concentration with time

$$m = -\frac{125}{5}$$

$$= -25$$

b The gradient represents the decrease in drug concentration per hour. It goes down 25 µg/mL every hour.

c The vertical intercept is the general name for the y-intercept, where the line crosses the vertical axis.

The vertical intercept is 200.

d The vertical intercept is the drug concentration when $t = 0$.

The vertical intercept represents the initial concentration of the drug: 200 µg/mL.

e The equation of a straight line is $y = mx + c$, but, in this case, it's $C = mt + c$. Here, $m = -25$ and $c = 200$.

The equation is $C = -25t + 200$.

f There is no drug remaining when $C = 0$. We can either extend the line to see where it crosses the horizontal axis, or substitute $C = 0$ into the equation and solve the equation.

When $C = 0$,

$$0 = -25t + 200$$

$$25t = 200$$

$$\frac{25t}{25} = \frac{200}{25}$$

$$t = 8.$$

There will be no drug in the body after 8 hours.

Nabil's taxi charges $3.10 flagfall and $1.85 per kilometre.

> 'Flagfall' is the initial charge, before any kilometres are travelled.

a Construct a graph to show the amount Nabil charges, $C, when passengers travel n km.

b Find an equation relating C to n.

Solution

a Construct a table of values, then draw a graph with n on the horizontal axis.

Distance, n km	0	5	10	15	20
Charge, $C	3.1	12.35	21.6	30.85	40.1

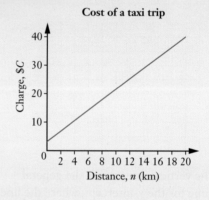

Cost of a taxi trip

b The equation of a straight line is $y = mx + c$, but, in this case, it's $C = mn + c$.

The vertical intercept, c, is the flagfall, 3.1.

The gradient, m, is the charge per km, 1.85.

The equation is $C = 1.85n + 3.1$.

Exercise 11.03 Linear modelling

Example 4

1 The number of times a cricket chirps in a minute is related to the temperature. The relationship is shown on the graph. The variable T represents the temperature in °C and n represents the number of chirps.

a What is the graph's vertical intercept?

b What does the vertical intercept represent in this situation?

c Calculate the gradient of the line.

d In this context, what does the gradient represent?

e Write the equation of the line.

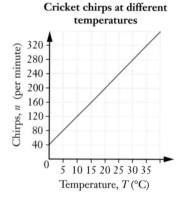

Cricket chirps at different temperatures

f   How many times per minute does a cricket chirp when it is 32°C?

g   At what temperature do crickets chirp 160 times per minute?

h   Using the formula, calculate how many times per minute a cricket chirps at 100°C. Is this realistic? What are the limitations of this model?

2   At sea level, water boils at 100°C. At different altitudes, the boiling temperature changes.

Boiling point of water at different altitudes

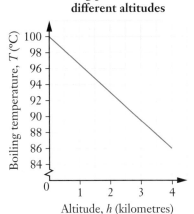

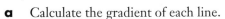
'Altitude' is another word for 'height'.

This graph shows the relationship between altitude and the temperature at which water boils.

a   Water boils at 100°C at sea level. What altitude is sea level?

b   Calculate the gradient of the line.

c   What physical quantity does the gradient represent?

d   Find the equation of the line.

e   Use the equation of the line to determine the boiling point of water at an altitude of 2.8 km.

f   Using the formula, calculate the boiling point of water at an altitude of 8 km.

Is this realistic? What are the limitations of this model? Hint: think about space.

3   After school, three students went home by walking, jogging or riding a bike. Each of them travelled at a constant speed and took one hour to reach home. The graph shows the distances they travelled.

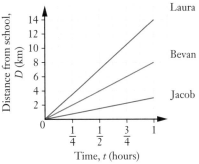

a   Calculate the gradient of each line.

b   In this context, what does the gradient represent?

c   Determine the equation of each line.

d   Which one of the students, Laura, Bevan or Jacob, travelled at the slowest speed? Give a reason for your answer.

e   Match Laura, Bevan and Jacob to their method of travelling home; that is, walking, jogging or riding a bike.

4 When babies are born, the medical staff measures the baby's length and head circumference. The graph shows the length and head circumference of normal baby boys.

Baby boys' length and head circumference

Head circumference, h (cm) vs Length, l (cm)

Points shown: (55, 37.6) and (45, 31.3)

a Find the linear function that relates l and h.

b A newborn baby boy is 51 cm long and his head circumference is 35.1 cm. Do you think the mother should be worried about her new son? Give a reason for your answer.

5 Anastasia operates a printing business. When she quotes a price for printing tickets, she charges an initial fee to cover the cost of design and an additional fee based on the number of tickets required. The price she charges for printing tickets is shown on the graph.

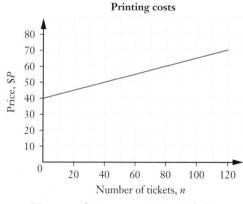

Printing costs

Price, $P vs Number of tickets, n

a How much is Anastasia's initial charge for the design?

b What is the gradient of the line?

c What physical quantity does the gradient represent?

d Write an equation that relates the price, P dollars, that Anastasia charges to print n tickets.

e How much does she charge to design and print 500 tickets?

f When Anastasia designs and prints elaborate wedding invitations, she charges an initial fee of $110 and 50c per invitation. Write a formula to determine the price in D dollars for designing and printing n elaborate wedding invitations.

6 Hawkes Landscaping Company supplies garden soil. The charge is $70 plus $30 per tonne to deliver up to 25 tonnes of soil.

a Copy and complete this table of values.

Number of tonnes, n	1	2	3	4	5	6
Cost of soil, $C						

The graph shows Cost in $, C on the vertical axis (scaled 0 to 275 in intervals of 25) and Number of tonnes, n on the horizontal axis (0 to 6).

b Copy the grid and graph the table of values.

c What is the vertical intercept of the line?

d What does this value represent?

e What is the gradient of the line?

f What does the gradient represent?

g Explain how you could calculate the cost of having 20 tonnes of soil delivered.

h Write a sentence to explain why the formula $C = 30n + 70$ can be used to calculate the cost C dollars of having a load of n tones of soil delivered.

i The company is planning to change to a delivery fee of $80 and a charge of $27 per tonne. Write a formula relating the cost C dollars of having n tonnes of soil delivered with the new price.

j Why do you think the company limits this pricing system to deliveries of up to 25 tonnes?

7 Kerri makes hot chips as part of her job in a fast-food shop. Each day, the fresh oil costs $48 and each container of chips costs her 90 cents to make.

a Copy and complete this table.

Containers of chips, n	0	10	20	30	40	50	60
Cost, $C							

b Write a sentence to explain how you could calculate the cost of producing 95 containers of chips.

c Write a formula relating total cost, $C, of producing n containers of chips.

d Use your formula to calculate:

  i the cost of producing 126 containers of chips

  ii the number of containers of chips produced when the cost was $180.30.

e Copy the grid and graph the line representing the cost of producing n containers of chips in one day.

f What is the vertical intercept and what physical quantity does it represent?

g Calculate the gradient and write a sentence to explain what physical quantity it represents.

h Kerri sells the containers of chips for $2.50 each. Explain why the formula $C = 2.5n$ represents the number of dollars Kerri receives from selling n containers of chips.

i Graph the line $C = 2.5n$ on the same grid as the graph of Kerri's costs.

j What does the point where the lines intersect represent?

k What is the smallest number of containers of chips that Kerri has to sell each day to make a profit?

l What are the limits to this model? Could Kerri make 1000 containers of chips per day?

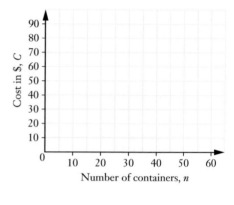

8 Alan runs a coffee stall at sporting events. The graph shows how much it costs Alan to set up the stall and make mugs of coffee.

a How much does it cost Alan to set up the stall?

b How much does it cost Alan to make each mug of coffee?

c If n is the number of mugs sold and C are Alan's costs in dollars, what is the equation of the line on the graph?

d What are the limits to this model? At a venue, could Alan make 1000 mugs of coffee?

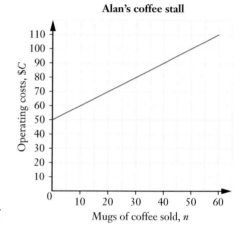

Alan's coffee stall

11.04 Direct linear variation

One common example of linear modelling is direct variation.

When two quantities x and y vary together; i.e. when x increases, y increases; and when x decreases, y decreases, this can be modelled using a linear function. This is called **direct linear variation** (or **direct proportion**).

This relationship between two quantities can be written in the form $y = kx$, where k is a number called the **constant of variation** (or **constant of proportionality**).

ISBN 9780170413503

EXAMPLE 6

A car travels at an average speed of 90 km/h. The distance it travels and the time it takes are related by direct linear variation. This is shown in the table below.

Time (*t* hours)	0	1	2	3	4	5
Distance (*d* km)	0	90	180	270	360	450

a   Graph the relationship between time and distance.

b   What is the gradient of the graph?

c   Find the constant of variation.

d   Find the equation for *d* in terms of *t*.

e   Complete this sentence: For every _____ , the car travels _____ km.

Solution

a   Time goes on the horizontal axis and distance goes on the vertical axis.

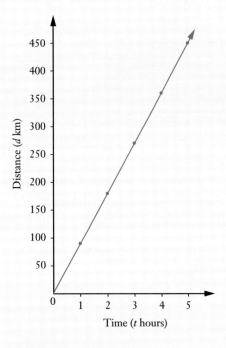

b   For every 1 unit across, the graph goes up 90 units.   The gradient is 90.

c	The variation equation is in the form $y = kx$, or, in this case, $d = kt$. To find k, substitute one of the points into the equation.	Substitute (3, 270): $d = kt$ $270 = k \times 3$ $3k = 270$ $k = 90$

The gradient of the line is equal to the constant of variation.

d		$d = kt$ $d = 90t$

e		For every hour, the car travels 90 km.

Variation can be described in words, given as an equation or shown graphically as we saw in the previous example.

Exercise 11.04 Direct linear variation

Example 6

1 The cost of filling your petrol tank is directly related to the number of litres it holds. This is shown in the table below:

Petrol amount (*L* litres)	0	10	20	30	40	50
Cost ($*C*)	0	14	28	42	56	70

a Graph the relationship between petrol amount and cost.

b What is the gradient of the graph?

c Find the constant of variation.

d What does the constant of variation represent in this context?

e Find the equation for C in terms of L.

f Calculate the cost of buying 75 litres of petrol.

2 The weight of an object on the Moon (*M* kg) is $\frac{1}{6}$ the weight of the same object on Earth (*E* kg).

a Sam weighs 66 kg on Earth. How much does he weigh on the Moon?

b Construct a table of values for this relationship for weights on Earth from 0 kg to 96 kg.

c Graph the relationship between E (horizontal axis) and M (vertical axis).

d What is the gradient of the graph?

e Find the constant of variation.

f Write the equation for M in terms of E.

g Calculate the weight on Earth of a machine weighing 44 kg on the Moon.

3 Angus is a store manager for Newsentry Speedy Pizzas. He is paid according to the formula $W = 21.2h$, where W = wages in dollars and h = hours worked.

   a How much is Angus paid for 7 hours work?

   b What is the constant of variation in this variation equation?

   c What does the constant of variation represent in this context?

   d How much is Angus paid in a week where he works 37 hours?

   e How many hours does Angus need to work to be paid $360.40?

4 Chantelle is organising a raffle for a mental health charity. Prizes have been donated by local businesses. The graph shows the relationship between the number of tickets sold (n) and the amount raised (A).

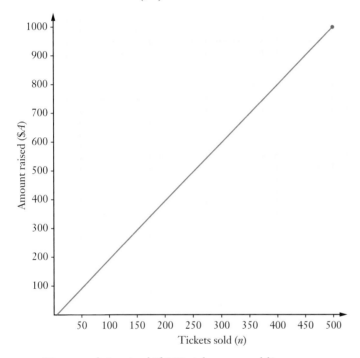

   a How much is raised if 200 tickets are sold?

   b How many tickets have to be sold to raise $900?

   c What is the gradient of this graph?

   d What is the meaning of the gradient in this context?

   e Write the variation equation for A in terms of n.

   f How much will they raise if they sell 1375 tickets?

   g The charity needs office supplies to the value of $3200. How many tickets need to be sold to raise this amount?

5 A small car has a fuel consumption of 8 L/100 km. This means the car uses 8 L for every 100 km travelled.

a Copy and complete this table relating the distance travelled and the amount of petrol used.

Distance (*d* km)	100	200	300	400	500
Petrol used (*P* L)					

b Graph the relationship between *d* (horizontal axis) and *P* (vertical axis).

c What is the gradient of the graph?

d Find the constant of variation.

e What does the constant of variation represent in this context?

f Find the equation for *P* in terms of *d*.

g How many litres are used to travel 250 km?

h How realistic is this model? What other factors may affect how much petrol a car uses?

6 When a car's speedometer shows a different speed to the car's GPS, in most cases, the GPS is right and the speedometer wrong. When Gavin was driving on the freeway he recorded the following table of values.

Speedometer, *x* km/h	0	48	80
GPS, *y* km/h	0	51	85

a Construct a graph to show the information in the table.

b Draw a line through the data.

c What is the gradient of the line?

d Determine the equation of the line.

e Use the equation to estimate Gavin's car's true speed (the GPS speed) when the car's speedometer is showing 100 km/h.

f Gavin's speedometer showed he was travelling at 39 km/h through a 40 km/h school zone. Use the equation to determine whether Gavin was speeding.

g What advice should Gavin give anyone who is driving his car?

11.05 Conversion graphs

A conversion graph is a line graph used to convert between different units, such as between metric and imperial units, or between currencies in foreign currency exchange. A conversion graph is an example of direct linear variation.

Currency
conversion
graph

EXAMPLE 7

Karen discovered some recipes in an old cookbook, but the measurements are in pounds. She found a table showing some conversions from pounds to grams.

Pounds	0	1	1.5	2
Grams	0	454	680	907

a Construct a conversion graph that Karen can use to convert pounds to grams.

b Karen needs $\frac{3}{4}$ pound of meat to make pizzas. How many grams of meat is this?

Solution

a Draw a set of axes and label the horizontal axis 'Pounds' and the vertical axis 'Grams'. Graph the straight line given by the table of values.

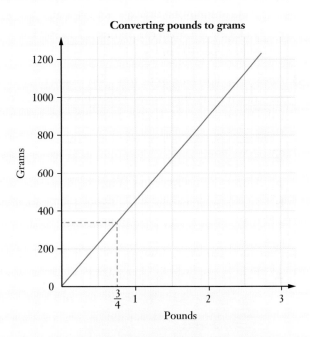

Converting pounds to grams

b Find $\frac{3}{4}$ or 0.75 pounds on the horizontal axis. Read up to the line, then across to the vertical axis.

Karen needs approximately 340 g of meat for the pizzas.

Exercise 11.05 Conversion graphs

1 Ariel has old dressmaking patterns that show material required in yards, but she needs to convert to metres.

a Use this table of values to construct a conversion graph between yards and metres.

Yards	0	1	3	5	10
Metres	0	0.9	2.7	4.5	9.1

b Use your graph to estimate the number of metres of material Ariel needs when the pattern states $2\frac{3}{4}$ yards.

2 Luke's hobby is restoring old British cars. Old car manuals give the cars' petrol consumption in gallons per 100 miles. This table shows some equivalent values for gallons per 100 miles and litres per 100 km.

Gallons per 100 miles	0	5	10	20
Litres per 100 km	0	14.12	28.25	56.5

a Construct a conversion graph between gallons per 100 miles and litres per 100 km.

b One powerful old sports car uses 7 gallons of fuel per 100 miles. What is the car's fuel consumption in litres per 100 km?

c Luke owns a small car that uses 7 L/100 km. Use the graph to express this in gallons per 100 miles.

3 Adele is backpacking in Europe. She doesn't have much money, so she needs to know how much everything costs. This table shows the conversion between Australian dollars (AUD) and euros (EUR), the common currency in Europe.

Australian dollars	0	10	20	50
Euros	0	6.63	13.26	33.15

a Construct a conversion graph for Australian dollars to euros.

b Approximately how many euros should Adele receive for 30 AUD?

c Adele's lunch cost 9 EUR. How much is this in AUD?

4 One Australian dollar buys $0.76 US dollars (USD).

a Copy and complete the table of values.

Australian dollars (AUD)	0	10	100	200
US dollars (USD)				

b Construct a conversion graph between AUD and USD.

c Tahira ordered headphones online for $25 USD. How much did the headphones cost in AUD? Round to the nearest cent.

d Rafi is going to change $400 AUD into USD before he flies to Hawaii. How many USD will he receive?

5 This conversion graph converts between miles per hour (*m*) and kilometres per hour (*k*).

a What speed in km/h is equivalent to 30 miles/h?

b What is the gradient of the conversion graph?

c Write an equation relating *m* and *k*.

d The speed limit on a British motorway is 100 miles per hour. What is this speed in km/h?

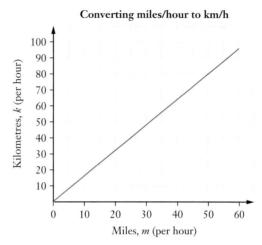

Converting miles/hour to km/h

Kilometres, *k* (per hour)

Miles, *m* (per hour)

KEYWORD ACTIVITY

WORD SCRAMBLE

Unscramble the letters in each term, then write the meaning of each term in your own words.

SAIX	RTDGENAI	AAIIONTRV
NVCOSNOIRE	TINPETCER	-YPTTCEERNI
ERICDT	EARNIL	TSAOTNCN
QTAEIOUN	LLEDNIOMG	ABTEL FO VAULSE

SOLUTION TO THE CHAPTER PROBLEM

Problem

At 5 a.m. when Dwayne was jogging along a country lane, he discovered a car that had crashed into a tree. The male driver was unconscious in the car. Dwayne called an ambulance and the police. This is the information that the police were able to obtain.

- The driver had been drinking at a party the previous night.

- He left the party at 1 a.m.

- Blood samples taken from the driver showed the following blood alcohol content (BAC).

Time	6 a.m.	7 a.m.	8 a.m.	9 a.m.
BAC	0.043	0.031	0.019	0.007

Graph this table of values, find the gradient of the line and use the gradient to estimate the driver's BAC at 1 a.m.

Solution

Gradient of line $= \dfrac{0.007 - 0.043}{3}$

$= -0.012$

This means the driver's BAC was decreasing at 0.012 per hour.
His BAC at 6 a.m. was 5×0.012 lower than that at 1 a.m.
Working backwards from 6 a.m.:

Driver's BAC at 1 a.m. $= 0.043 + 5 \times 0.012$
$= 0.103$.

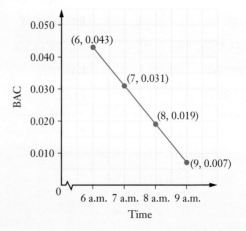

- Give an example of a context that could be modelled using a linear function.
- Explain what the gradient and *y*-intercept would mean for your example.
- Explain direct linear variation in your own words.
- Is there any part of the topic you didn't understand? If so, ask your teacher for help.

Copy and complete this mind map of the topic, adding detail to its branches and using pictures, symbols and colour where needed. Ask your teacher to check your work.

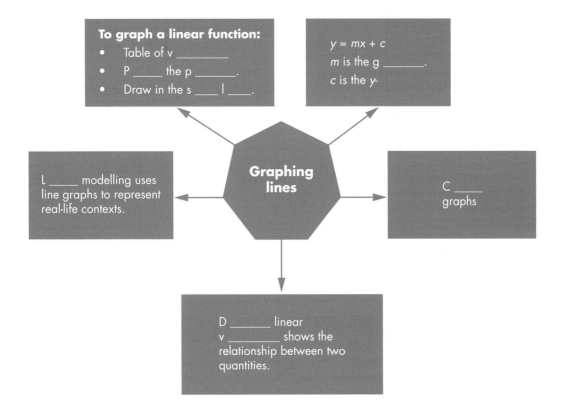

To graph a linear function:
- Table of v _____
- P _____ the p _____.
- Draw in the s ____ l ____.

$y = mx + c$
m is the g _____.
c is the y-

L _____ modelling uses line graphs to represent real-life contexts.

Graphing lines

C _____
graphs

D _____ linear
v _____ shows the relationship between two quantities.

Exercise
11.01

1 Graph each linear function.

   a $y = 2x - 2$ **b** $y = -x + 3$

Exercise
11.02

2 For each of the following graphs, find:

   i the gradient   **ii** the y-intercept   **iii** the equation of the line.

a

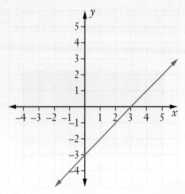

b

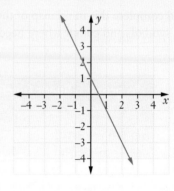

Exercise
11.02

3 For each equation, state:

   i the gradient   **ii** the y-intercept.

     a $y = 4x - 3$ **b** $y = -x + 2$ **c** $y = \dfrac{x}{3}$

Exercise
11.03

4 Goran packs glass items in boxes. He has to assemble the box before he can pack it. The graph shows the time, T minutes, it takes him to assemble one box and pack n glass items in it.

   a How long does it take Goran to assemble a box before he starts to pack it?

   b Calculate the gradient of the line.

   c What physical amount does the gradient represent?

   d Calculate the time it takes Goran to assemble a box and pack 15 items in it.

   e Find the equation of the graph.

NCM 11. Mathematics Standard (Pathway 1)   ISBN 9780170413503

5 Liza sells mugs of coffee for $4 each.

Exercise
11.03

   a Complete this table of values.

Mugs of coffee sold, n	0	1	2	3	4
Sales, $C					

   b Write an equation relating C and n, the dollars Liza receives.

   c What is the vertical intercept?

   d What does this value represent?

   e What is the gradient of the line?

   f What does the gradient represent?

6 The monthly cost of renting retail space is directly related to the area of the floor space. A space of 60 m² rents for $1800 per month.

Exercise
11.04

   a What is the monthly cost to rent a space of 120 m²?

   b Construct a table of values for this relationship from 60 m² to 3000 m².

   c Graph the relationship between floor space (S) and monthly rent (R).

   d What is the gradient of the graph?

   e Find the constant of variation.

   f What does the constant of variation represent in this context?

   g Find the equation for R in terms of S.

   h Calculate the monthly cost of renting a retail space of 1850 square metres.

7 This graph can be used to convert Australian dollars (AUD) to euros (€).

Exercise
11.05

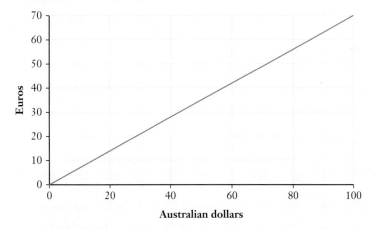

   a Convert $15 to euros.

   b Convert €50 to Australian dollars.

   c A meal in Paris costs €35. What is this in Australian dollars?

   d Gustav is in Australia. He has €25 left on his debit card. Is this enough to pay for a $25 meal at the local club?

   e Calculate how many euros you would get for $220.

12.

INTERESTING FIGURES

Chapter problem

Isabella received a phone call from an investment advisor she doesn't know telling her about a good investment opportunity that is virtually risk-free. The advisor said that the offer will be fully subscribed quickly, and that he will phone her again tomorrow morning to start getting the investment organised. Isabella would invest $10 000 at an interest rate of 15% p.a. compounding annually for the next 4 years.

"Annually" means yearly

a Calculate the amount of interest the advisor claims $10 000 will earn in the next 4 years.

b Why should Isabella be very cautious about investing in this opportunity?

CHAPTER OUTLINE

WHAT WILL WE DO IN THIS CHAPTER?

- Increase and decrease an amount by a percentage
- Make percentage calculations involving profit and loss
- Calculate simple interest and read simple interest graphs
- Use the straight-line method of calculating depreciation
- Calculate compound interest and read compound interest graphs

HOW ARE WE EVER GOING TO USE THIS?

- When determining whether an investment is safe
- When choosing a safe investment that will give us reasonable return
- When making calculations to check that we have received the correct investment returns
- To avoid falling for a financial scam

12.01 Percentage increase and decrease

Percentage increase means to make a quantity bigger by a given percentage; for example, in pay increases or adding GST to prices.

Percentage decrease means to make a quantity smaller by a given percentage; for example, in discounts and sale prices.

EXAMPLE 1

a   Increase $500 by 12%.

b   Decrease $500 by 7%

There are two ways to do this. Choose the method you like!

Solution

a   Find 12% of $500 $12\% \times \$500 = \$60$

     Do $\dfrac{12}{100} \times 500$ or 0.12×500

     Add this amount to $500 $\$500 + \$60 = \$560$

     OR New amount is 100% plus 12% $100\% + 12\% = 112\%$

     Find 112% of $500 $112\% \times \$500 = \$560$

     Do $\dfrac{112}{100} \times 500$ or 1.12×500

b   Find 7% of $500 $7\% \times \$500 = \$35$

     Do $\dfrac{7}{100} \times 500$ or 0.07×500

     Subtract this amount from $500 $\$500 - \$35 = \$465$

     OR New amount is 100% minus 7% $100\% - 7\% = 93\%$

     Find 93% of $500 $93\% \times \$500 = \$465$

     Do $\dfrac{93}{100} \times 500$ or 0.93×500

a   All Government employees receive a 2.5% pay increase. Georgina is on a salary of $82 000 per year. What is her new salary?

b   During an end-of-financial-year sale, computer games are discounted by 25%. Find the cost of a game regularly priced at $115.

Solution

a   Find 2.5% of $82 000 $2.5\% \times 82\,000 = \$2050$

Add this amount to $82 000 $\$82\,000 + \$2050 = \$84\,050$

Answer the question.   Georgina's new salary is $84 050.

OR New amount is 100% plus 2.5% $100\% + 2.5\% = 102.5\%$

Find 102.5% of $82 000 $102.5\% \times \$82\,000 = \$84\,050$

b   Find 25% of $115 $25\% \times 115 = \$28.75$

Subtract this amount from $115 $\$115 - \$28.75 = \$86.25$

Answer the question.   The computer game will cost $86.25.

OR New amount is 100% minus 25% $100\% - 25\% = 75\%$

Find 75% of $115 $75\% \times \$115 = \$86.25$

Exercise 12.01 Percentage increase and decrease

1 a   Increase 95 kg by 60%   b   Increase $2500 by 6%.

c   Increase 150 m by 5%.   d   Increase 10 L by 33%.

e   Decrease $150 by 57%.   f   Decrease 2000 L by 28%.

g   Decrease 110 kg by 4.5%.   h   Decrease 840 students by 15%.

Example 1

Example 2

2 Jenna earns $32.75 per hour. Her manager gives her a pay increase of 8% because she produces quality work and to cover inflation. What is her new wage rate?

3 In the January sales, all whitegoods are discounted by 15%. Liong buys a new washing machine usually priced at $799. Calculate the discount price he pays for the washing machine.

4 Jamiela runs a fashion business. She calculates the selling price of each item of clothing by increasing the cost price by 95%. Find the selling price of a jacket that costs Jamiela $120.

5 Non-essential items in Australia attract a GST (goods and services tax) of 10%. Find the GST-inclusive price of a car costing $18 900.

6 Last year, the Australian road toll was 1290 deaths. This year, the road toll decreased by 10.3%. Calculate the road toll for this year, rounded to the nearest whole number.

7 Renee and her sister Elise have singing lessons. Renee pays full fees of $450 per term, but Elise gets a 5% family discount.
  a Calculate the discount for Elise.
  b Calculate the total amount of fees paid for the two sisters.

8 Ken and Amanda buy a house for $64 000 and sell it 10 years later at a profit of 147%. What is the selling price of the house?

9 The Luxury Hotel charges an additional 1.5% when the customer pays with a credit card. Find the total cost of 3 nights accommodation at $185 per night that is charged to a credit card. Round your answer to the nearest cent.

10 An electricity company offers a 2% discount when customers pay the account on time.
  a Calculate the discount if a bill of $1978.70 is paid on time. Round your answer to the nearest cent.
  b How much does the customer pay?

11 Katie works at the local supermarket. She is given a staff discount of 3.5% on all purchases. Find how much she pays for purchases totalling $178.25.

12 For tax purposes, the value of a computer is depreciated (decreased) at a rate of 20% per year. Sunil buys a computer priced at $1290. Calculate its depreciated value after one year.

13 Filomena buys TVs at a cost price of $295. She adds 75% of the cost price to get her selling price, then she adds 10% GST to calculate the final selling price.
  a Find the price of a TV after Filomena has added her 75% markup.
  b Find the price of the TV after she adds GST. Round your answer to the nearest cent.
  c Is it unlikely that Filomena will charge her customers exactly that price. Why?

14 Harry's Hardware is having an end-of-financial-year sale. All items in the store are discounted by 18%.

   a Lauren is a painter. She buys ten 4 L tins of paint regularly priced at $67 per tin.

      i Calculate the cost of the paint at its regular price.

      ii Calculate the cost of the paint after the 18% discount.

   b Lauren is also entitled to a trade discount of 10%. Calculate the cost of the paint to Lauren.

12.02 Profit and loss

All successful businesses make a **profit**. Businesses that make a **loss** quickly close down. Most businesses express the profit or loss they make as a percentage of their costs.

Percentage profit and loss

$$\text{Percentage profit} = \frac{\text{profit}}{\text{cost price}} \times 100\%$$

$$\text{Percentage loss} = \frac{\text{loss}}{\text{cost price}} \times 100\%$$

The **cost price** is the price at which a business buys an item from a supplier.

The **selling price** is the price at which a business sells an item to a customer.

EXAMPLE 3

Jane buys a necklace for her shop for $36. She sells it for $47.

   a How much profit does Jane make when she sells the necklace?

   b Calculate the percentage profit correct to one decimal place.

Shutterstock.com/mikolajn

Solution

a Profit = selling price − cost price

                                     Profit = $47 − $36

                                           = $11

b Write the profit over the cost price and multiply by 100%.

$$\text{Percentage profit} = \frac{11}{36} \times 100\%$$

$$= 30.5555\ldots\%$$

$$\approx 30.6\%$$

EXAMPLE 4

During a sale, a bed is sold for $599. The bed cost the store $750.

a Calculate the loss for the store on the bed.

b Calculate the percentage loss correct to one decimal place.

Solution

a Loss = cost price − selling price

$$Loss = \$750 - \$599$$
$$= \$151$$

b Write the loss over the cost price and multiply by 100%.

$$Percentage\ loss = \frac{151}{750} \times 100\%$$
$$= 20.1333 \dots \%$$
$$\approx 20.1\%$$

Exercise 12.02 Profit and loss

1 Calculate, correct to one decimal place, the percentage profit for each sale.

a cost price $35, selling price $45

b cost price $625, selling price $810

c cost price $149.75, selling price $274.25

d cost price $2.60, selling price $4.05

2 Calculate, correct to one decimal place, the percentage loss for each sale.

a cost price $516, selling price $420

b cost price $2175, selling price $1300

c cost price $12 500, selling price $11 000

d cost price $5.20, selling price $4.40

3 Carrie pays $40 for a small cupboard at a garage sale and paints it. She sells it for $75 to a furniture dealer. Calculate her percentage profit.

4 Jason pays $165 for a ticket to the football grand final. When his team doesn't make the grand final, he sells the ticket for $90. Calculate his percentage loss.

5 Voula buys an old bicycle for $20.
She spends $115 replacing some parts
and painting it. She sells it for $150.

   a Has she made a profit or a loss?
How much is the profit or loss?

   b Calculate the percentage profit/loss.

6 Keira buys a used car for $2500 and
spends $1750 replacing some parts.
At the end of the year she sells it
for $3750.

   a Does she make a profit or a loss?
How much is the profit or loss?

   b Calculate the percentage profit or loss.

   c Comment on whether you think Keira would be happy with this outcome.

7 To cover costs, Stefan needs to make 65% profit on the items in his leisure store.
What is the minimum price he can charge for hiking boots that cost him $165?

8 Mardi buys 2000 shares valued at $23 400. She pays a stockbroker's fee of 2.5% of the
value of the shares. She later sells the shares for $10.55 per share.

   a Calculate the total amount Mardi paid for the shares.

   b How much money does she get for her shares when she sells them?

   c Has she made a profit or a loss on the shares?

   d Calculate, correct to the nearest whole number, the percentage profit or loss on
the shares.

9 Gianni buys seven boxes of bananas for $45 each. Each box contains 10 kg of bananas.
He sells four of the boxes at $5.10 per kg. He reduces the price to $4 per kg to clear the
remaining boxes before the fruit goes off.

   a How much do the bananas cost Gianni?

   b How much does he earn from the bananas?

   c Does Gianni make a profit or a loss?

   d Calculate his percentage profit or loss. Answer correct to one decimal place.

10 A shoe shop bought pairs of shoes for $48 each and priced them at $90 per pair. During
the end-of-season sale, the manager reduced the price by $20. Calculate the percentage
profit the shop makes when it sells the shoes in the sale. Answer correct to one
decimal place.

12.03 Simple interest

When we invest money with **simple interest** or **flat rate interest**, the amount of **interest** earned is the same every year. The interest is a percentage of the **principal**, the original amount invested.

The **simple interest formula** is:

$$I = Prn$$

where P = principal (what we invest)

r = the rate of interest per time period, as a decimal

n = number of time periods.

EXAMPLE 5

Mike invested $2000 at 3.25% p.a. simple interest for 3 years.

a How much interest will he earn?

p.a. = per annum = per year

b How much will be in his account at the end of 3 years?

Solution

a In the simple interest formula $I = Prn$, $I = Prn$

$P = 2000$, $r = 3.25\% = 0.0325$ and $n = 3$. $= 2000 \times 0.0325 \times 3$

$= 195$

Write the answer.   Mike will earn $195 in simple interest.

b Add the interest to the principal.   Account total $= \$2000 + \$195$

$= \$2195$

Write the answer.   Mike will have $2195 in his account.

EXAMPLE 6

Ruby invested $5600 at 3.9% p.a. for 18 months. How much interest did she earn?

Solution

The interest rate is per year but the time is in *months*. We must change 18 months to years by dividing by 12.

$$\text{Time} = \frac{18}{12} \text{ years}$$

$$P = 5600, r = 0.039, n = \frac{18}{12}$$

$$I = 5600 \times 0.039 \times \frac{18}{12}$$
$$= 327.6$$

Write the answer.

Ruby earned $327.60 in interest.

EXAMPLE 7

What percentage interest per month is equivalent to 5.16% p.a.?

Solution

Divide by 12 to change it to a monthly interest rate.

$5.16\% \div 12 = 0.43\%$

Write the answer.

5.16% p.a. is 0.43% per month.

Exercise 12.03 Simple interest

1 Calculate the simple interest on each principal.

  a $700 at 6.2% p.a. for 3 years   **b** $520 at 4.2% p.a. for 5 years

  c $3400 at 4.85% p.a. for 2.5 years   **d** $750 at 3.1% p.a. for 1.5 years

Example 5

2 What is the simple interest on each investment?

  a $940 at 3.25% p.a. for 18 months   **b** $12 560 at 7.05% p.a. for 6 months

  c $1560 at 4.5% p.a. for 3 months   **d** $380 at 5.25% p.a. for 1 month

Example 6

3 What percentage interest per month is equivalent to 4.68% p.a.?

Example 7

4 The Advantage Bank offers investors 8.4% p.a. simple interest. Express this rate of interest as a:

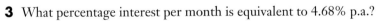

  a   monthly rate   **b**   weekly rate

  c   six-monthly rate   **d**   daily rate

  e   fortnightly rate   **f**   quarterly rate (three-monthly rate).

Divide by 52 for the weekly rate, divide by 2 for the six-monthly rate, and divide by 4 for the quarterly rate.

5 Nicole borrowed $2400 from a finance company for 2 years at 18% p.a. interest.

   a How much interest did she have to pay?

   b How much, including interest, did Nicole have to repay the finance company?

6 Matthew owed $365 on his credit card. The credit card company charged him one month's interest at 22% p.a.

   a How much interest was he charged?

   b Calculate the total amount he had to repay the credit card company.

7 Rohan won $5000 as part of his prize for being the 'Apprentice of the Year'. He invested it for one month at 4.5% p.a. interest. How much interest did he earn?

8 The Great Aussie Credit Union pays different interest rates for large investments.

Balance	Interest rate
$1–$4999	3.00% p.a.
$5000–$9999	3.50% p.a.
$10 000–$19 999	4.00% p.a.
$20 000–$49 999	4.75% p.a.
$50 000 and over	5.50% p.a.

   a Kate invested the $38 000 she inherited from her great-aunt with the credit union for 2 years. How much interest will she earn?

   b Yuri won $54 000 on lotto. He invested his win with the credit union for 18 months. How much interest will he earn?

   c Felicia saved $15 050 from her part-time job at the hardware store. How much interest will she earn if she invests it with the credit union for 3 months?

9 Sunny borrowed $265 000 from a finance company to set up a plumbing supplies business. He borrowed the money for 8 months and was charged 16.5% p.a. interest. How much did he have to repay the finance company, including interest?

10 During the global financial crisis, bank interest rates for savings in the UK fell to 0.5% p.a. Keith in London invested $30 000 for 2 years at 0.05% p.a. simple interest.

   a How much interest did the investment earn?

   b At the time, Keith met a man at a party who told him of an alternative, non-bank investment that was paying 8% p.a.

     i How much more interest would Keith earn in the non-bank investment?

     ii Why do you think Keith ignored the non-bank investment opportunity?

Some companies and advisors can't be trusted. Before we make an investment we should check the investment company carefully. Any investment that offers significantly higher returns could be a scam. Remember the investor's rule: 'If it sounds too good to be true, it probably is.'

ISBN 9780170413503

IS MY INVESTMENT SAFE?

Visit the MoneySmart website to learn how to determine whether or not an investment is safe.

MoneySmart

What you have to do

1 On the MoneySmart website, search 'Scams' to see the list of unlicensed companies you should not deal with.

2 Should you invest with the company Thomas Moore Global? Explain.

3 List four different companies, each in a different country, with which you shouldn't invest.

4 Randomly choose eight letters from the alphabet; for example, B, F, G, K, N, P, Q and Z.

5 List all the companies that begin with each of your eight letters, along with their country of origin.

6 Copy and complete this frequency table using the companies you have listed.

Region of origin	Tally	Frequency
Africa		
Asia		
Australia		
Europe		
Middle East, including United Arab Emirates		
South America		
USA		
Total		

7 What conclusion can you draw about the locations of these companies?

8 Write a paragraph describing your findings. Include a pie chart to display your data visually.

12.04 Simple interest graphs

Straight-line graphs can be used to illustrate the return (interest) obtained on investments.

Exercise 12.04 Simple interest graphs

1 This graph shows the return (interest) earned on a $100 investment at 5% p.a., 8% p.a. and 10% p.a. simple interest.

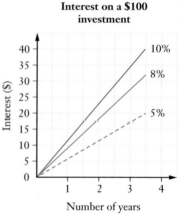

Interest on a $100 investment

    a How much interest will be earned on a $100 investment at 8% p.a. for 3 years?

    b How much more interest would be earned on a $100 investment at 10% p.a. compared with 8% p.a. for 3 years?

    c Use your answer to part **a** to determine the interest on a $400 investment at 8% p.a. for 3 years.

    d Kane is going to draw a graph to show the return on $100 at 7% p.a. simple interest. Predict the position of the 7% p.a. graph compared with the 5% p.a., 8% p.a. and 10% p.a. graphs.

    e Write a sentence to explain how the gradient of the line is related to interest rates.

2 When Tyson constructed these three graphs he forgot to label them. One line shows the simple interest return on a $1 investment at 6% p.a., another at 4% p.a. and the third at 5% p.a.

Interest on a $1 investment

    a Which interest rates should be written in positions *X*, *Y* and *Z*?

    b Use the graph to determine the return on a $7300 simple interest investment at 6% p.a. for 5 years.

    c Tyson said: 'If you double the length of the time of a simple interest investment and keep the interest rate the same, you double the return.' Is Tyson's statement correct?

    d How many times larger is the return if you double the rate of simple interest and triple the length of the investment?

3 The graph shows the return on a simple interest investment of $2000 at 6% p.a. Use the simple interest formula to determine the values of A, B and C.

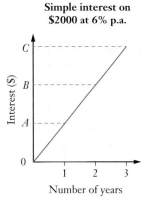

Simple interest on $2000 at 6% p.a.

4 This graph shows the simple interest return on Melissa's investment of $1000.

a How much interest did Melissa earn after 3 years?

b Calculate the rate of simple interest for Melissa's investment.

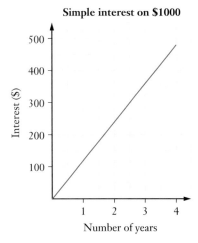

Simple interest on $1000

5 Zishan is going to graph the return on a $500 simple interest investment at 8% p.a. for n years.

a In the formula $I = Prn$, what values should she substitute for P and r?

b Explain why the formula for her graph will be $I = 40n$.

c Copy and complete the table of values for $I = 40n$.

n	0	1	2	3	4	5	6
I							

Simple interest on $500 at 8% p.a.

d Copy the grid and show the graph of the return on the $500 investment for the first 6 years.

6 The graph shows the total value of a $100 simple interest investment at 15% p.a. and 20% p.a. for 5 years.

a How much is the investment worth after 4 years at 15% p.a.?

b How much simple interest is earned in 3 years at 20% p.a.?

c How much more interest is earned in 5 years at 20% p.a. than at 15% p.a.?

d Both of the graphs start at 100 on the vertical axis. What is the significance of the 100?

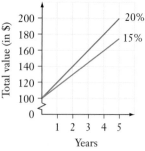

Total value of $100 simple interest investment

12.05 Straight-line depreciation

Straight-line depreciation formula practice

Items such as motor vehicles, computers, office equipment and tools depreciate in value with age. This means they lose value or are worth less than their original purchase price. If you own any of these items and you use them in your work, you can claim the **depreciation** (loss of value) as an **allowable tax deduction**.

In **straight-line depreciation**, we decrease the value of the item by the same amount each year. If we graph the value of the item as it gets older, the graph is a **straight line**. The value of the item at any time is called its **salvage value**.

EXAMPLE 8

Carol bought a filing cabinet for her office for $420. She is depreciating it at a rate of $42 per year. Calculate the salvage value of the filing cabinet after 3 years.

Solution

The filing cabinet depreciates $42 each year. We have to subtract 3 lots of $42 from the new price.

Salvage value $= \$420 - 3 \times \$42$

$= \$294$

Straight-line depreciation

The straight-line depreciation formula

With **straight-line depreciation**, the value of an item decreases by the same dollar amount each year. The formula for straight-line depreciation is:

$$S = V_0 - Dn$$

where: S is the salvage (current) value

V_0 is the original value

D is the dollar amount of depreciation per period

n is the number of periods.

EXAMPLE 9

Roberto's new office computer cost him $3200 and will depreciate 22% of its original value each year. Use the straight-line method of depreciation to calculate:

a   the amount it will depreciate each year

b   the amount Roberto can claim as a tax deduction each year

c   the value of the computer after 3 years

d   the age of the computer when its salvage value is $1792.

Solution

a   We must calculate 22% of $3200. $0.22 \times \$3200 = \$704$

Write the answer.   The computer depreciates $704 each year.

b   Roberto can claim a tax deduction of $704 each year for his computer.

c   Subtract 3 lots of $704 from the original value.

$$\text{Salvage value} = \$3200 - 3 \times \$704$$
$$= \$1088$$

Write the answer.   The value of the computer after 3 years is $1088.

d   In the formula $S = V_o - Dn$, $S = \$1792$, $V_o = \$3200$ and $D = \$704$.

$$\$1792 = \$3200 - \$704 \times n$$
$$\$704n + \$1792 = \$3200$$
$$\$704n = \$1408$$
$$n = 2$$

We can set up an equation and solve it to calculate n.

Write the answer.   The computer will be 2 years old when its salvage value is $1792.

Exercise 12.05 Straight-line depreciation

Example
8

1 Find the salvage value of an item when:

 a original value = \$5000, depreciation = \$650 per year and number of years = 4

 b number of years = 7, original value = \$9600 and depreciation = \$970 per year.

Example
9

2 Use the formula $S = V_0 - Dn$ to find the value of:

 a V_0 when $S = 3640$, $n = 3$ and $D = 920$

 b n when $V_0 = 2900$, $S = 590$ and $D = 330$

 c D when $S = 94$, $V_0 = 550$ and $n = 8$.

3 A fibreglass spa at Mittagong Fitness Centre costs \$125 000. Its annual depreciation rate is 5% of this original cost.

 a By how much does the value drop each year?

 b What is the salvage value of the spa when it is 8 years old?

 c After how many years will the spa be valued at \$50 000?

 d When the spa has a salvage value of approximately \$12 000, it will be considered unsafe to use and be scrapped. At what age will this happen?

4 Brady bought a tractor for \$743 500 and, every year, he claims 15% of this amount on his tax using straight-line depreciation.

 a How much can he claim each year on his tractor?

 b Calculate the salvage value of the tractor when it is 4 years old.

 c After how many years will the salvage value of the tractor be zero?

5 Two years ago, the ACE Accounting Company purchased a computer for \$4500.

 a With a straight-line depreciation rate of 20%, how much can the company claim as a taxation deduction each year?

 b What is the salvage value of the computer after 2 years?

 c The company plans to sell the computer when its salvage value reaches \$1800. After how many years will the salvage value be \$1800?

6 A university library photocopier has a working life of 5 years, so its salvage value after 5 years is zero. What is its annual rate of depreciation?

7 A tea trolley used at the Better Baking Company was originally valued at \$200 and is expected to last for 20 years.

 a Explain why the company can claim \$10 per year as depreciation on the trolley.

 b What percentage rate of depreciation is \$10 on an item with an original value of \$200?

 c Calculate the salvage value of the trolley after 15 years.

8 The cash register purchased by P&S Stationery cost \$2400. The straight-line method of depreciation allows an annual depreciation of \$240.

a Copy and complete the table.

Age in years, n	Yearly depreciation, D	Salvage value, S
0	\$0	\$2400
1	\$240	\$2160
2	\$240	
3		
4		
5		

b Construct a graph showing the salvage value of the cash register for $n = 0$ to 10.

c After how many years is the salvage value of the cash register \$480?

d Why is this method of depreciation called 'straight-line depreciation'?

12.06 Compound interest

When you invest your money in **compound interest**, you get 'interest on your interest'. After interest is added to the investment, the next time interest is calculated, it will be based on this larger amount. The amount of interest grows or is **compounded** as the size of the investment grows.

Simple and compound interest

EXAMPLE 10

Simple vs compound interest: spreadsheet

Scott invested \$2000 at 5% p.a. interest compounded annually.

a Calculate the value of his investment at the end of 3 years.

b Calculate the total interest earned.

Solution

a We need to calculate the interest for the first year and add it to the principal before we calculate the following year's interest.

$I = Prn$ where $r = 5\% = 0.05$, $n = 1$ (year) and P changes each year, but its initial value is 2000.

	Interest	Balance
End of the 1st year	$I = \$2000 \times 0.05 \times 1$ $= \$100$	$\$2000 + \$100 = \$2100$
End of the 2nd year	$I = \$2100 \times 0.05 \times 1$ $= \$105$	$\$2100 + \$105 = \$2205$
End of the 3rd year	$I = \$2205 \times 0.05 \times 1$ $= \$110.25$	$\$2205 + \$110.25 = \$2315.25$

At the end of 3 years, Scott's investment is worth \$2315.25.

Alternatively, calculate the value of 5% interest added to the principal by multiplying by 1.05. For example, $2000 × 1.05 = $2100, giving you the answers in the Balance column straight away.

b Total interest = final balance –
original principal

Total interest = $2315.25 – $2000

= $315.25

OR add the three interest amounts in
the middle column in the table above.

Total interest = $100 + $105 + $110.25

= $315.25

Write your answer.

The total interest earned over the 3 years is
$315.25.

Exercise 12.06 Compound interest

Example
10

1 Lidija invested $8000 at 6% p.a. interest compounded annually for 3 years. Copy and complete this table to calculate the value of her investment at the end of 3 years.

	Interest	Balance
End of the 1st year	$I = Prn$ $= \$8000 \times 0.06 \times 1$ $= \$\_\_\_\_$	$\$8000 + \$\_\_\_\_ = \$\_\_\_\_$
End of the 2nd year	$I = Prn$ $= \$\_\_\_\_ \times 0.\_\_\_\_ \times 1$ $= \$\_\_\_\_$	$\$\_\_\_\_ + \$\_\_\_\_ = \$\_\_\_\_$
End of the 3rd year	$I = Prn$ $= \$\_\_\_\_ \times 0.\_\_\_\_ \times 1$ $= \$\_\_\_\_$	$\$\_\_\_\_ + \$\_\_\_\_ = \$\_\_\_\_$

2 Jayden invested $12 000 for 3 years at 4% p.a. compounding yearly.

a How much interest did he earn in the first year?

b How much was in his account at the end of the first year?

c How much was in his account at the end of the second year?

d Calculate the value of his investment at the end of 3 years.

e How much interest will Jayden earn during his 3-year investment?

3 Nazneen invested $4000 at 3.2% p.a. for 2 years compounding annually.

a Calculate the total value of her investment at the end of the 2 years.

b How much interest will she earn during the 2 years?

c How much less interest would Nazneen have earned if the interest had been simple interest rather than compounding interest?

4 To buy a car, Suresh has saved $14 000 from his after-school job. He is going to invest the money for 3 months until he has his P-plates. Suresh's investment is going to pay 0.7% per month, interest compounded monthly.

    a Copy and complete the following table.

	Interest	Balance
End of the 1st month	$I = Prn$ $= \$14\,000 \times 0.007 \times 1$ $= \$\_\_\_\_$	$\$14\,000 + \$\_\_\_\_ = \$\_\_\_\_$
End of the 2nd month	$I = Prn$ $= \$\_\_\_\_ \times 0.\_\_\_\_ \times 1$ $= \$\_\_\_\_$	$\$\_\_\_\_ + \$\_\_\_\_ = \$\_\_\_\_$
End of the 3rd month	$I = Prn$ $= \$\_\_\_\_ \times 0.\_\_\_\_ \times 1$ $= \$\_\_\_\_$	$\$\_\_\_ + \$\_\_\_\_ = \$\_\_\_\_$

    b How much interest will Suresh make during the 3-month investment?

    c Use repeated multiplication by 1.007 to check your answers to parts **a** and **b**.

5 Kirrilly invested some money in monthly compounding interest. The table shows her interest and balance calculations.

	Interest	Balance
End of the 1st month	$I = Prn$ $= \$9600 \times 0.008 \times 1$ $= \$76.80$	$\$9600 + \$76.80 = \$9676.80$
End of the 2nd month	$I = Prn$ $= \$9676.80 \times 0.008 \times 1$ $= \$77.41$	$\$9676.80 + \$77.41 = \$9754.21$

    a How much money did Kirrilly invest?

    b What was the monthly rate of compound interest?

    c Convert the monthly rate to an annual rate of compound interest.

    d How much interest did Kirrilly's investment earn?

12.07 Compound interest graphs

Spreadsheets are useful for calculating and graphing compound interest. Ask your teacher to download the 'Compound interest' spreadsheet from NelsonNet for this exercise.

Compound
interest

EXAMPLE 11

Max invested $5000 at 3.5% p.a. annually compounding interest for 20 years.

Use the compound interest spreadsheet to graph the value of his investment and determine the total amount of interest he will earn.

Solution

• Enter 5000 in cell E3 and 3.15 in cell E4.

• Don't type a dollar sign or a percentage sign and don't alter any other values.

• Watch as the spreadsheet does all the work for you!

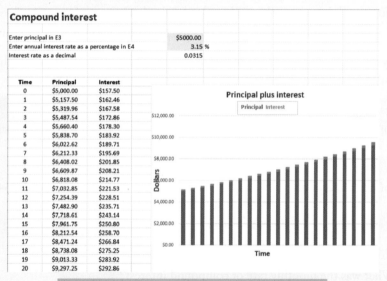

The compound interest graph shows that the investment is growing as a curve, which is faster than a straight line.

At the end of 20 years, Max will have $9297.25.

Interest = $9297.25 − $5000

= $4297.25

Exercise 12.07 Compound interest graphs

Use the spreadsheet to answer each of the following questions.

1 Claire invested $7500 at 2.9% p.a. annually compounding interest.

    a Find the value of her investment in 10 years. Answer to the nearest dollar.

    b How much interest will she earn in the first 10 years?

    c Will the interest for the 10 years from year 11 to year 20 be the same as for the first 10 years? Why or why not?

2 Determine the interest, correct to the nearest dollar, that Jamie will earn on an investment of $8000 at 3% p.a. annually compounding for 6 years.

3 Enter a principal of $12 500 and an interest rate of 5% into the spreadsheet.

    a Describe how the spreadsheet determines the value in cell C9.

    b What does the spreadsheet do to determine the value in cell B10?

    c Describe two different ways you could calculate the total interest this investment earns in 20 years.

4 Rachel invested $6000 at 4% p.a. annually compounding interest.

    a Copy and complete this table of values. Express each value correct to the nearest dollar.

Number of years	0	1	2	3	4	5	6
Interest earned during the year	0	$240					
Value of the investment at the end of the year	6000	$6420					

    b Sketch a graph to show the value of Rachel's investment from the initial investment to the end of 6 years.

    c Sketch a graph of the *interest* Rachel's investment earned during the 6 years.

5 This graph shows the total value on $1 compound investment at 3% p.a., 6% p.a. and 10% p.a.

    a Describe how the shape of a compound interest graph is different to the shape of a simple interest graph.

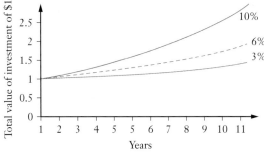

    b Approximately how long does it take for a $1 investment to double its value at 10% p.a. compound interest?

6 Nick and Adam invested the same amount of money at 9% p.a., but Nick's investment was 9% p.a. compounded annually while Adam's was 9% p.a. simple interest. The graphs show the value of their investments after 5 years.

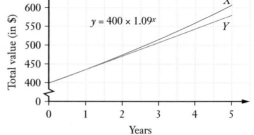

a How much was Nick and Adam's original investment?

b Which graph shows Adam's investment and which shows Nick's investment?

c After 5 years, which of the two investments has the greater value and by how much?

7 Compound interest graphs are *not* straight lines. Why not? Explain.

KEYWORD ACTIVITY

CALCULATION MATCH

Match the six statements in column A to the numerical expressions in Column B.

Column A		**Column B**	
1	Simple interest earned when $400 is invested at 5% p.a. for 2 years	**A**	$(400 \times 1.05) \times 1.05$
2	The value of a $400 investment after 2 years at 5% p.a. annually compounding interest	**B**	$400 + 0.05 \times 400$
3	The salvage value of an item valued at $400 when new, and now it is 2 years old, using straight-line depreciation at 5% p.a.	**C**	$400 - 105$
4	The price of a $400 item after a 5% price increase	**D**	$400 \times 0.05 \times 2$
5	The loss when an item costing $400 is sold for $105	**E**	$400 - 0.05 \times 400$
6	The price of a $400 item with a 5% discount	**F**	$400 - (400 \times 0.05) \times 2$

SOLUTION TO THE CHAPTER PROBLEM

Problem

Isabella received a phone call from an investment advisor she doesn't know telling her about a good investment opportunity that is virtually risk-free. The advisor said that the offer will be fully subscribed quickly, and that he will phone her again tomorrow morning to start getting the investment organised. Isabella would invest $10 000 at an interest rate of 15% p.a. compounding annually for the next 4 years.

a Calculate the amount of interest the advisor claims $10 000 will earn in the next 4 years.

b Why should Isabella be very cautious about investing in this opportunity?

Solution

a Calculate the interest for each year.

	Interest	Balance
End of the 1st year	$I = \$10\ 000 \times 0.15 \times 1$ $= \$1500$	$\$10\ 000 + \$1500 = \$11\ 500$
End of the 2nd year	$I = \$11\ 500 \times 0.15 \times 1$ $= \$1725$	$\$11\ 500 + \$1725 = \$13\ 225$
End of the 3rd year	$I = \$13\ 225 \times 0.15 \times 1$ $= \$1983.75$	$\$13\ 225 + \$1983.75 = \$15\ 208.75$
End of the 4th year	$I = \$15\ 208.75 \times 0.15 \times 1$ $= \$2281.3125$	$\$15\ 208.75 + \$2281.3125 = \$17\ 490.0625$ $\approx \$17\ 490.06$

$$\text{Interest} = \$17\ 490.06 - \$10\ 000$$
$$= \$7490.06$$

b The investment offer is probably a scam. Scam indicators include:

- cold (unexpected) phone call from someone she didn't know
- push from the 'advisor' to act quickly
- very high and unrealistic interest rate compared to the market rate
- the opportunity is implied to be 'risk-free'.

- What parts of this chapter did you remember from previous years?
- List any parts of this chapter that you have seen in everyday life outside school.
- What part of the chapter do you think you might use in your job when you leave school?
- Are there any parts of this chapter you're not sure about? If yes, ask your teacher to explain it to you.

Copy and complete this mind map of the topic, adding detail to its branches and using pictures, symbols and colour where needed. Ask your teacher to check your work.

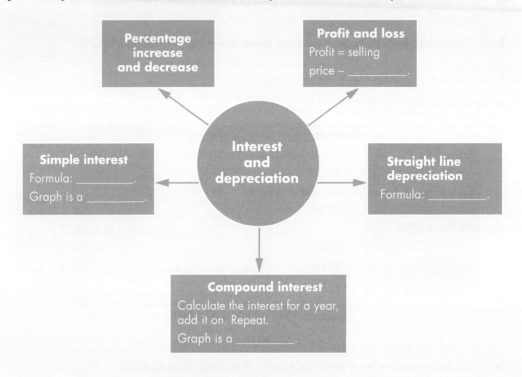

Percentage increase and decrease

Profit and loss
Profit = selling
price − _____.

Simple interest
Formula: _____.
Graph is a _____.

Interest and depreciation

Straight line depreciation
Formula: _____.

Compound interest
Calculate the interest for a year, add it on. Repeat.
Graph is a _____.

12. TEST YOURSELF

1 a Increase $450 by 12%.

  b Decrease $312 by 5%.

2 As a result of inflation, wages increased by 3%. Before the increase, Josephine earned $17.90 per hour. What was her hourly rate after the wage rise?

3 Gerry paid $1280 for an old car. He spent $2420 renovating the car and then he sold it for $6950.

  a How much profit did Gerry make?

  b Calculate his percentage profit on total costs.

4 Donna bought a rural block of land near a creek for $450 000. Shortly after, the council changed the flood line in the locality and the value of the land fell. Donna was able to sell the land for only $125 000. Calculate her percentage loss correct to the nearest per cent.

5 Calculate the simple interest on each investment.

  a $560 invested for 3 years at 2.75% p.a.

  b $4500 invested at 4% p.a. for 18 months

6 Jake has $2800 in an online savings account that pays 3% p.a. simple interest.

  a Complete the following table of values.

Year	0	1	2	3	4	5	6
Amount in the account at the end of the year	$2800						

  b Construct a graph showing the amount in Jake's account at the end of each year for 6 years.

7 Hamish is using the straight-line depreciation method at 15% p.a. to depreciate his office equipment. Calculate the salvage value of his 3-year-old boardroom table that cost $5200 new.

8 Alexis invested $3000 in an account paying annually compounding interest at 2.5% p.a.

  a How much interest did she earn in the first year?

  b How much more interest did she earn in the second year than in the first?

13.

BUYING A CAR

Chapter problem

Sally and Harry are going to drive to Perth to visit Sally's parents. Their SUV (sport utility vehicle) has a fuel consumption of 9.8 litres per 100 kilometres. On some parts of the trip, they will have long distances to travel with few petrol stations and Sally would like to work out how far the car will go on a full tank of petrol. The petrol tank holds 64 litres but, to be safe, she will base her calculations on 90% of this figure. How far can they travel between fuel stops?

CHAPTER OUTLINE

WHAT WILL WE DO IN THIS CHAPTER?

- Consider the basics of buying a car
- Calculate stamp duty, insurance and registration costs
- Compare loans and loan repayments for buying a car
- Compare the different types of car insurance
- Compare fuel consumption and costs, and the amount of fuel used by a car
- Calculate the annual running costs of owning a car
- Compare the costs with a budget

HOW ARE WE EVER GOING TO USE THIS?

- When choosing a car to buy
- When working in a career involving cars such as selling cars, renting cars or fleet management
- When using a company car or your own car as part of your job

13.01 Starting up

Before buying a vehicle, we must learn some of the terminology involved and investigate some issues for consideration.

RACQ

Exercise 13.01 Starting up

1 Terminology

Use the Internet to find the meaning of each term.

a ANCAP   **b** REVS

c RMS   **d** Used Car Safety Ratings

2 Things to consider

Use the RACQ website (the Queensland version of NRMA) to answer Questions **2** to **5**.

a Find **Ten tips for buying a car**. Write these tips and highlight the four that you consider most important.

b The safety of a car is an important consideration in buying a car. Where can you find the safety ratings for new and used cars?

3 New or used car?

Use the information on the website to list the advantages and disadvantages of each option.

4 Where to buy

a List the four possible ways of purchasing a car, together with one advantage and one disadvantage for each.

b Which option would you choose? Why?

5 New car extras

a There is more to the cost of a new car than just its price. List some of the extra costs (also called **on-road costs**) that need to be considered.

b When buying a new car, you are often offered extra accessories. List at least three of these and the advantages and disadvantages of the accessories offered.

6 Essential checks for used cars

Before you buy a used car you should check on the history of the car.

a Search **Vehicle checks** on the Internet and list three places that offer this service.

b List what information is provided with a vehicle history check.

c List three reasons you should have a history check before buying a used car.

7 **Vehicle inspections for used cars**

A vehicle inspection details the current mechanical state of the car.

a Search for **Vehicle inspections**. List three places near you that will inspect a used car and, if possible, note the cost of a vehicle inspection.

b List some of the items that are checked in the inspection.

c Do you think a pre-purchase inspection for a used car is a good idea? Give reasons for your answer.

INVESTIGATION

LET'S BUY A CAR

In this investigation, use the **Car Sales** website or something similar to choose a car.

Part A

Car Sales

You will use the website to find the 4 cars listed below.

1 **Ford Fiesta**

Select **New Cars**

- Complete as many of the search boxes as you wish: choose Ford for **Make** and Fiesta for **Model**. If you know the body type and the price range you want, these can also be chosen. If not, you can leave them as they are and click **Search**.

- All vehicles meeting your criteria will be listed and can be browsed.

- You can find more detailed information on any model by clicking the **View** tab.

2 **A new SUV (sport utility vehicle)**

If you are not sure of the model of car you want, you can browse by body type.

For example, you can choose SUV and enter information about the make, model, price range etc. Then, the SUVs meeting these criteria will be shown and details can be found in the same way as in part **1**.

3 **A new car with a good safety or green rating**

By clicking the **Research** tab and selecting **Research Cars**, you are able to search cars by a number of different categories.

You can select by the ANCAP rating or the Green rating as well as by a number of others. Any of these specialist searches can be narrowed down.

---

4 **A used car within a price range and close to home**

For a used car, click the **Cars for Sale** tab and select **Private seller cars**.

- The search boxes can be completed to the level of detail you wish.

- Important criteria are entering your location (postcode) and how far you are prepared to travel out of your area to locate the car.

- You can also restrict the search to only private sellers or allow it to include dealers.

- Again, details of the cars can be obtained by clicking on the enquiry buttons when they are listed. Often, the information for private enquiry cars is not as good as for dealer cars.

Part B

1 Select a new car of your choice using the **Car Sales** website. Once you have made your selection, copy and complete the grid below to show the following details.

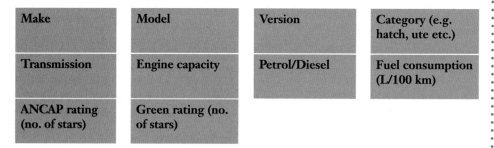

Make	Model	Version	Category (e.g. hatch, ute etc.)
Transmission	Engine capacity	Petrol/Diesel	Fuel consumption (L/100 km)
ANCAP rating (no. of stars)	Green rating (no. of stars)		

2 Choose a used car of your choice, either privately or from a dealer, within 100 km of your postcode. Record as many of the car's details as are available. Also, show its advertised price.

13.02 On-road costs

Stamp duty

In addition to the basic cost of a new car, the buyer has to pay a number of additional costs, called **on-road costs**. The major on-road costs are **vehicle registration, Compulsory Third Party insurance** (CTP insurance) and **stamp duty**.

Vehicle registration is an annual charge paid to the Roads and Maritime Services (RMS).

Compulsory Third Party (CTP) insurance, often called a **green slip**, is also an annual payment to provide cover for people injured as a result of an accident in which the car is involved.

Stamp duty is a tax paid to the state government when ownership of a vehicle changes hands.

Stamp duty is charged on the market value of the vehicle or the price paid, whichever is higher. In NSW, stamp duty is calculated as follows:

- for cars under $45 000, 3% of the value of the vehicle
- for cars over $45 000, $1350 + 5% of the value greater than $45 000.

For a new car, these on-road costs are sometimes included in the purchase price of the car, usually called the 'drive-away price'. If you buy a used car, registration and CTP payments are not required because these are transferred with the car sale.

EXAMPLE 1

Danielle bought a used car for $25 400. How much stamp duty will Danielle have to pay on the transfer of ownership?

Solution

The value of the car is less than $45 000.
Therefore, stamp duty on the transfer is 3%.

$$\text{Stamp duty} = 3\% \times \$25\ 400$$
$$= \$762$$

EXAMPLE 2

Jerry has just bought a new Volkswagen Passat V6 costing $61 400. How much stamp duty did he pay?

Solution

The value of Jerry's car is greater than $45 000.

The stamp duty will be $1350 plus 5% of the value greater than $45 000.

$$\text{Value greater than } \$45\ 000 = \$61\ 400 - \$45\ 000$$
$$= \$16\ 400$$
$$\text{Stamp duty} = \$1350 + 5\% \times \$16\ 400$$
$$= \$2170$$

Exercise 13.02 On-road costs

1 Ethan has purchased a used car with a market value of $13 800. How much stamp duty will he have to pay on the transfer of ownership?

Example
1

2 Milo buys a new car for $29 990. How much stamp duty is payable?

3 Tegan is a florist and purchases a van for $37 000 to deliver flowers. Calculate the stamp duty payable on this purchase.

Example
2

4 How much stamp duty will Cooper have to pay on his new car that is valued at $55 800?

5 Aiden is buying a new crew cab ute priced at $59 990.

    a   Calculate the stamp duty payable.

    b   Registration is $649 and CTP insurance is $417. Calculate the total of the on-road costs.

6 Calculate the stamp duty payable on the purchase of a Ferrari costing $419 990.

7 a   Copy and complete the table below by calculating the stamp duty payable on the sales of the vehicles shown.

Vehicle	Vehicle price	Stamp duty
1	$5000	
2	$20 000	
3	$35 000	
4	$45 000	
5	$55 000	
6	$65 000	
7	$80 000	

Remember that the formula changes for values greater than $45 000.

    b   Use the table to create a line graph showing the stamp duty payable on vehicles of various prices.

        •   Show vehicle price on the horizontal axis from $0 to $80 000.

        •   Show stamp duty on the vertical axis.

        •   If drawn correctly, the graph should consist of two line segments that join.

        •   Complete your graph by giving it a title and labelling both axes.

8 Tanya is buying a used car whose registration is due next month. She will pay $14 000 for the car. The registration is $357 and CTP insurance is $368.

    a   Calculate the stamp duty payable on this car.

    b   How much will Tanya pay in total for the purchase of the car, including the stamp duty, the registration and the CTP insurance?

9 Billy's father has a second car valued at $14 000. However, he is happy to sell it to Billy for $1000.

    a   Calculate the stamp duty payable on $1000.

    b   Calculate the stamp duty payable on $14 000.

    c   The RMS requires that stamp duty be paid on the market value of the car, in this case $14 000. Why do you think this is a legal requirement?

13.03 Car loans

You can either save up to buy a car or take out a loan and pay it off over a number of years. This depends on the price of the car and your income.

EXAMPLE 3

Megan has found the car of her dreams for $32 500. She will take out a personal car loan from a bank for this amount. The interest rate of 9% p.a. and the loan is to be repaid in monthly instalments over 5 years.

a Use an online calculator to find the monthly payment required.

b Calculate the total repayment amount over the life of the loan.

c How much interest will Megan pay on the loan?

Solution

a There are many online loan calculators available on the Internet. We will use the one found on the **MoneySmart** website.

MoneySmart

- Search for Personal loan calculator.
- Select How much will my repayments be?
- Set Amount borrowed to 32 500.
- Set Interest rate to 9.
- Set Repayment frequency to Monthly.
- Set Length of loan to 5 years.
- Set Fees to $10 monthly.
- Click the **Calculate** button.
- The repayments are shown to be $685 per month (rounded to the nearest dollar).

b Calculate the number of monthly repayments over 5 years. Multiply this by the monthly repayment amount.

> Remember, there are 12 months in a year.

Number of monthly repayments = 5×12

$$= 60$$

Total repayments = $60 \times \$685$

$$= \$41\ 100$$

c The interest is how much greater than the sale price of $32 500 are the total repayments.

Interest = $\$41\ 100 - \$32\ 500$

$$= \$8600$$

Exercise 13.03 Car loans

1 Janine borrowed $14 000 to buy a car at 7.8% p.a. interest over 3 years. Her monthly repayment was calculated at $447 per month.

   a How much will Janine repay over the 3 years?

   b How much interest will Janine pay?

2 The interest rates offered by financial institutions vary a great deal. At another bank, Janine would have paid 14.5% p.a. interest and her repayments would have been $492 per month.

   a What is the difference per month between the two loans?

   b How much extra over the 3 years would Janine pay with the second loan?

3 Loans for a greater amount of money are paid off over a longer period of time. Khiem borrowed $50 000 at 7.8% p.a. over 10 years to purchase a van to carry musical equipment for his band. The monthly repayment was $611.

   a How much will Khiem repay over the 10 years?

   b How much interest will he pay?

4 If Khiem can afford to pay more per month, he can pay off the loan more quickly and pay less interest. If he repays the loan over 6 years, his monthly repayment will be $882.

   a For this shorter time period, how much will Khiem repay?

   b How much interest will he pay?

   c How much interest will Khiem save by paying the loan off over 6 years rather than 10 years?

Use an online calculator for Questions 5 to 7.

5 Dominik has taken a loan of $25 000 to buy a used car. He repays the loan in monthly instalments at an interest rate of 9.5% p.a. over 4 years.

   a What will Dominik's monthly repayment be?

   b How much will he pay in total over the 4 years?

   c How much interest will he pay?

6 How does the interest rate affect the overall cost of a car loan? Assume we borrow $20 000 to buy a car with monthly repayments over 5 years. Using interest rates of 7% p.a, 9% p.a. and 11% p.a, copy and complete the table below and describe how the interest rate affects the total amount repaid.

	Interest rate	Monthly repayment	Total amount repaid	Interest
a	7% p.a.			
b	9% p.a.			
c	11% p.a.			

7 Sophia is negotiating a car loan for $18 000 with a finance company that charges an interest rate of 8.7% p.a. They have given her the option of repaying the loan over 4, 5 or 6 years with repayments paid monthly. Sophia has set up the following table to investigate these options.

Loan term (years)	Monthly repayment	Total repaid	Interest
4			
5			
6			

a Copy and complete the table.

b Taking into account the monthly repayment and the interest paid, which of these options do you think Sophia should take? Give a reason for your choice.

INVESTIGATION

THE INTEREST RATE ON A CAR LOAN

Banks, credit unions and finance companies offer car loans or personal loans that can be used to buy vehicles.

1 Search the Internet for 10 different financial institutions that provide such loans.

2 Copy this table.

Institution	Interest rate	Comparison rate	Other conditions

The comparison rate is a rate that takes into account all the additional fees and charges that a financial institution adds to the loan. It is a legal requirement that this rate be shown in all advertising or information about loans, which allows the consumer to compare loans fairly.

3 For each institution, write the name in the first column, the interest rate they offer in the second column and the stated comparison rate in the third column.

4 Note any other special conditions in the last column.

5 Find the average interest rate of the 10 institutions.

6 Find the average comparison rate of the 10 institutions. Is this higher than your answer to Question **5**? Why?

7 Which institution would you choose if you needed a loan? Why?

13. Buying a car   387

13.04 Car insurance

There are three main types of car insurance.

Compulsory Third Party (CTP) insurance is compulsory and covers injuries to any persons involved in an accident in which you are at fault.

Comprehensive insurance is not compulsory but can be purchased in addition to CTP insurance. It covers the cost of repairs or replacement to your car and other property damage whether or not you are at fault.

As an alternative, **Third Party Property insurance** covers the damage you cause to other people's property as the result of an accident, but does not cover damage to your own vehicle.

If you are in an accident and you are *not* at fault, then the other driver is responsible for the costs of repairing your car.

The amount you pay for insurance is called a **premium**.

When buying car insurance, many insurance companies offer discounts such as:

- **no-claim discount** – for every year you *don't* make a claim you receive a percentage discount up to a maximum of 60%

- **multi-policy discount** – if you have more than one insurance policy (home, life, another car) with the same insurer

- **loyalty discount** – if you stay with an insurer for a long period of time.

EXAMPLE 4

Harrison is insuring his new car and has been quoted an annual premium of $1226.50 for comprehensive insurance. He is given a multi-policy discount of 20% as he has a home and contents insurance policy with the same company. How much does he pay for his insurance?

Solution

Calculate the discount and subtract from the price.	Discount = 20% × $1226.50
	= $245.30
	Price paid = $1226.50 − $245.30
	= $981.20
OR use 100% − 20% = 80%.	Price paid = 80% × $1226.50
	= $981.20

Exercise 13.04 Car insurance

1 Mandy's comprehensive car insurance annual premium is $1392. She receives a 10% no-claim discount off this premium. How much does she pay for her insurance?

2 Josh has Third Party Property insurance on his car as the car is old and worth only $3000. When it is time to renew, the premium is $240.50. Josh is given a 30% no-claim bonus as he has had no claims over the last five years. How much will Josh pay?

3 Rachna's renewal premium for her comprehensive car insurance is $1782.50. She is given a multi-policy discount of 10% and a no-claim discount of 40%.

    a Calculate Rachna's premium after the multi-policy discount.

    b The no-claim discount is calculated on the amount after the multi-policy discount has been deducted. Calculate Rachna's premium after this second discount has been applied.

    c Is this equal to a 50% discount? Explain your answer.

In Questions **4** to **6**, we will obtain an online quote for comprehensive and Third Party Property insurance. Most insurance companies offer online quotes. In this case, we will seek one from the NRMA for a Hyundai Accent Active, with the following information:

- Year of manufacture: 2016
- Make: Hyundai
- Model: Accent Active
- Shape: Hatch
- Engine size: 1.4 L
- No finance owing on the car.

4 Go to the **NRMA car insurance** website.

- Click **Get a quote**.
- Choose the location as NSW, the insurance type as **car insurance** and the type of cover as **comprehensive**.
- Complete the information for the Hyundai Accent described above and click **Agree** to the terms of use. Assume the car is for private use and is insured under your name. You have no existing policies with the NRMA. Assume you are aged **18 years**.
- Once you reach the confirmation page, check the details and, if correct, continue.
- Quotes will appear under a number of different policy types.

NRMA car insurance

    a In a table, write the different policy options, together with the quoted annual premium for each.

    b For each option, also write the excess payable in case of a claim. If you make an insurance claim, **excess** is the maximum amount you first have to pay before the insurance company pays the rest.

5 Repeat the process with the same car. This time, assume you are 30 years old and get a new quote.

    a List the policy options and annual premiums in this case.

    b What is the difference between this price and that for an 18-year-old driver?

    c Give a reason for this difference.

6 Choose a different insurance company to the NRMA and get quotes for the same circumstances as in Question **4**.

   a   Is this second quote more, less or about the same as in Question **4**?

   b   Why might different companies have different prices for insurance?

BOCSAR

INVESTIGATION

HOW THEFT AFFECTS INSURANCE COSTS

1 Visit the **NSW Bureau of Crime Statistics and Research (BOCSAR)** website.

2 Search for crime information by clicking on the map of NSW.

3 Choose Offence Type – Motor vehicle theft and Local Government Area – Sydney.

4 Scroll over the map showing motor vehicle theft hotspots.

5 Write two suburbs that are shown as hotspots and two suburbs that are not hotspots.

6 Return to one of the insurance sites you used in Exercise 13.04. Using the same details as you used in Question **4**, obtain a comprehensive insurance quote for each of the suburbs you chose.

7 How much more expensive is it to insure the same car in the hotspot suburbs?

8 Repeat this investigation for different cars and different suburbs. Write a paragraph about the effect of the levels of motor vehicle theft on insurance premiums.

13.05 Fuel consumption and costs

How much
petrol?

The most common fuels for cars are **petrol** and **diesel**, but **LPG** (liquefied petroleum gas) is also used. **Electric** cars have been developed but are not widely used, although this might change in the future.

Graph of
petrol prices

The amount of fuel a vehicle uses depends on many factors, including the size and efficiency of the engine, the weight of the vehicle and whether it is driven on city or country roads.

Fuel
consumption

> The fuel consumption of a vehicle is a rate, given as the number of litres used per 100 kilometres travelled (L/100 km).

For example, the fuel consumption for a Honda Jazz GLi is 5.8 L/100 km, and for the much larger V6 Holden Commodore it is 9.8 L/100 km.

EXAMPLE 5

Emma bought a new Mazda 5-door hatch and wants to check its fuel consumption. She fills the tank and takes a long drive in the country. When Emma returns, she again fills the tank and finds she has used 12 litres of petrol and travelled 160 km. Calculate the car's fuel consumption.

Solution

First, find the fuel consumption in L/km.

$$\text{Fuel consumption} = 12 \text{ L} \div 160 \text{ km}$$
$$= 0.075 \text{ L/km}$$

To travel 100 km, Emma's car will use 100 times this amount.

$$\text{Fuel consumption} = 0.075 \times 100 \text{ L/100 km}$$
$$= 7.5 \text{ L/100 km}$$

$$\text{Fuel consumption} = \frac{\text{fuel used in L}}{\text{distance travelled in km}} \times 100$$

EXAMPLE 6

Sofia's car has a fuel consumption rate of 5.8 L/100 km.

a How many litres of petrol will it consume on a 550 km trip?

b What will be her fuel cost for the trip if petrol costs $1.40/L?

Solution

a Express 5.8 L/100 km as $\dfrac{5.8 \text{ L}}{100 \text{ km}}$.
To find litres of petrol used, multiply by this rate.

$$\text{Fuel consumption} = \frac{5.8 \text{ L}}{100 \text{ km}} = 0.058 \text{ L/km.}$$
$$\text{Petrol required} = 550 \times 0.058$$
$$= 31.9 \text{ L}$$

b Express $1.40/L as $\dfrac{\$1.40}{\text{L}}$.
To find fuel cost for the trip, multiply by this rate.

$$\text{Fuel cost} = \text{petrol used} \times \text{cost per litre}$$
$$= 31.9 \times \$1.40$$
$$= \$44.66$$

EXAMPLE 7

Greg drives a VW Golf using diesel fuel at a rate of 6.1 L/100 km. How far (to the nearest km) can he travel on 30 litres of diesel fuel?

Solution

Express 6.1 L/100 km as $\dfrac{6.1\ \text{L}}{100\ \text{km}}$.
To find how many km, divide by this rate.

$$\text{Fuel consumption} = \dfrac{6.1\ \text{L}}{100\ \text{km}} = 0.061\ \text{L/km}$$

$$\text{Distance} = 30 \div 0.061$$

$$= 491.8032 \ldots.$$

$$\approx 492\ \text{km}$$

Exercise 13.05 Fuel consumption and costs

Example 5

1 Chloe drives 190 km and finds that her car has consumed 16 L of petrol. Calculate its fuel consumption.

2 The fuel consumption for a number of vehicles was tested by measuring the fuel consumed for a particular journey. Calculate the fuel consumption of each vehicle correct to two decimal places.

	Vehicle	Kilometres travelled	Litres of fuel used
a	Holden Barina	420	29.4
b	Hyundai i30 diesel	1230	55.4
c	Holden Commodore V6	879	83.5
d	Ford Territory	334	30.1
e	Mazda CX-9	1075	120.0

3 Is fuel consumption higher when driving in the city or the country? During a week of driving around Sydney, Noah travels 120 km and uses 10.5 L of petrol. The next week, he visits his sister near Canberra travelling 275 km and using 20 L of petrol.

    a Calculate the car's fuel consumption when driving in Sydney.

    b What was the consumption rate for the trip to Canberra?

    c Is the fuel consumption higher in Sydney?

Example 6

4 Sebastian's Lexus 8-cylinder Sport uses premium petrol and has a fuel consumption of 11.1 litres per 100 kilometres.

    a How much petrol will his car use on a 355 km trip?

    b What is the cost of fuel for the journey if premium petrol costs $1.62 per litre.

5 This table shows the fuel consumption of a selection of vehicles. For each vehicle, calculate the amount of fuel it would use to travel 450 km.

	Vehicle	Fuel	Consumption in L/100 km
a	Mitsubishi Triton	Diesel	9.1
b	VW Scirocco	Petrol	8.1
c	Jeep Cherokee	Diesel	7.5
d	VW Polo	Diesel	5.0
e	Mazda 2 Neo	Petrol	6.8

6 Liz lives in Batemans Bay and owns a Toyota RAV4 which has a petrol consumption of 9.1 L/100 km.
  a How much petrol will she use on a 1500-km round trip to Melbourne?
  b What will be Liz's petrol bill for the trip if petrol costs 145 cents per litre?
  c If Liz travels 12 000 km each year, how many litres of petrol will her car use?
  d What is Liz's annual petrol bill for her car?

7 Callum's used Toyota Camry has an economy rating of 7.5 L/100 km. Its petrol tank holds 60 litres of fuel. How far can Callum travel on one tank of petrol?

Example 7

8 The fuel consumption of a number of vehicles is given in the following table. How far will each vehicle travel on the amount of fuel shown? Answer to the nearest kilometre.

	Vehicle	Fuel consumption in L/100 km	Fuel in tank in litres
a	Mazda 6	8.7	64
b	Porsche Cayenne	12.2	100
c	Mitsubishi Lancer	6.9	59
d	Range Rover Sport	14.9	70
e	Ford Focus	6.2	55

9 Antonia is required to do a lot of travel in her job. She drives an Audi A1 Sport car with a fuel consumption of 5.3 L/100 km.
  a Last year, Antonia drove a total of 17 500 km.
    i How many litres of petrol did her car use last year?
    ii At $1.45 per litre, how much would Antonia have paid for petrol last year?
  b Antonia travelled to a nearby town to meet some friends. She left with a full tank. When she returned home, it cost her $19.60 at $1.40 per litre to refill the tank.
    i How many litres of fuel did her car use on the trip?
    ii How far is the town from Antonia's home?

13. Buying a car

10 Jose intends to buy a Hyundai i30 1.6 L hatchback. He has the choice of either a diesel or a petrol model. The diesel model has fuel consumption of 4.7 L/100 km whereas the petrol model's fuel consumption is 6.5 L/100 km.

   a If both models have a 45 L fuel tank:

      i how far will Jose be able to drive on one tank of fuel in the diesel model?

      ii how many kilometres will he get from a full tank in the petrol model?

   b If Jose drives his car, on average, 15 000 km per year:

      i how many litres of fuel will he use in the diesel model?

      ii what will be his annual consumption for the petrol model?

      iii If the price of petrol and diesel fuel are nearly the same at $1.50/L, how much money would he save on his yearly fuel bill if he bought the diesel model?

   c The diesel model priced at $23 050 is more expensive than the petrol model priced at $19 590. How many years will it take for the fuel savings to compensate for this price difference? Answer to the nearest year.

11 Shauna has a Mitsubishi Triton that consumes 12.5 L/100 km. She is thinking of converting the vehicle to run on LPG as it is much cheaper than petrol. However, the consumption rate on LPG will be higher at 16.2 L/100 km.

   a If Shauna travels 16 000 km/year, how much petrol does she use annually?

   b At $1.50/L for petrol, how much does she pay for petrol in a year?

   c If she converted her vehicle, how many litres of LPG would she use in a year?

   d LPG costs only 90 cents/L. How much would Shauna pay per year for LPG?

   e What would she save annually on fuel if she converted?

   f The conversion will cost Shauna $3000. However, the Federal Government provides a rebate of $1250 to those who convert to LPG. So, Shauna will pay only $1750. How long will it take for Shauna to recover her conversion cost?

DID YOU KNOW?

Solar cars

Solar cars are powered by sunlight that has been converted to electricity. The World Solar Challenge is a race across Australia held every two years, covering a distance of 3000 km. Most solar cars are built for races only with space for just one person.

Guinness World Records recognises a land speed record of 88.8 km/h for solar cars. In 2012, the record was held by the University of NSW with the car Sunswift IV.

Getty Images/Look at Sciences/Laurent Douek

FUEL PRICES THEN AND NOW

In this investigation, we will look at how fuel prices have changed over time, both in the city and in the country.

AAA

What you have to do

1 Visit the **Australian Automobile Association** website.

2 Search for the map of Australia showing latest fuel prices.

3 Clicking on the State of NSW gives you a spreadsheet with average monthly prices from 1998 to the present for Sydney metro and NSW cities or towns.

4 a Choose one year and copy the petrol prices for that year (January to December) for Sydney metro and two NSW cities or towns.

b Draw a line graph for each of the three locations. Draw all three on one graph and use a different colour for each.

c Write one or two sentences about the information you have found.

5 a Choose one place and review the petrol prices over the years.

b Calculate the amount the price has changed since January 2003.

c Calculate to one decimal place the percentage change over this time.

> Percentage change = amount of change ÷ original price × 100

d Find the year when the price of petrol was half its current price.

e Write a paragraph about the change in petrol prices over a 10-year period.

13.06 Annual running costs

We can now examine all the costs involved in car ownership. This will help us to see the relative importance of each cost.

Exercise 13.06 Annual running costs

In this exercise, we are going to calculate the approximate annual cost of owning three different cars: the 2016 Hyundai Accent, the 2009 Holden Commodore, and a car of your choice.

	2016 Hyundai Accent	2009 Holden Commodore
Price ($)	16 480	12 525
Financing	Loan	Paid cash
Tare weight (kg)	1570	1684
Odometer reading (km)	95	117 389
Registration ($)	556	556
CTP insurance ($)	599	610
Fuel type	Petrol	Petrol
Fuel consumption (L/100 km)	6.2	11.3
Engine capacity (L)	1.4	3.6

Make 3 copies of the following table, which we will complete for each car during this exercise. For each car, answer the questions below.

	Yearly cost ($)	Weekly cost ($/week)	Percentage of total
Loan interest			
Registration and CTP insurance			
Comprehensive insurance			
Fuel			
Other running costs			
Total cost			

Example 3

MoneySmart

1 Determine the financing cost. Assume the car is financed by a loan at 9% p.a. interest over 5 years.

   a Use the **MoneySmart** online calculator to calculate how much you will pay.

   b The calculator also shows the interest paid over the 5 years. Divide the interest by the number of years to obtain an average financing cost and enter it into the yearly cost column of the table. Note: the financing cost for the Holden is $0 as we have not borrowed any money.

2 Calculate the total for Registration and CTP insurance and enter your result into the first column of the table.

3 Determine a figure for comprehensive insurance. When we looked at insurance, the cost varied as a result of age and other factors. Find a comprehensive insurance quote for each car, as you did in Exercise 13.04 as an 18-year-old. Enter this insurance cost into the table.

4 Assume that, on average, the vehicle travels 15 000 km each year.

   a Calculate the number of litres it will use in a year.

   b Determine the yearly fuel cost, assuming petrol costs $1.45/L. Enter this fuel cost into the table.

Example 6

5 Other running costs, which include tyres, servicing and repairs, are very specific to the individual car and we can only make a broad guess. They tend to vary with the size of the car and the distance travelled. We will use the following formula to estimate them.

    Other running costs = 0.033 × engine capacity (in L) × distance travelled

Use this formula based on the engine capacity of your vehicle and assume a driving distance of 15 000 km to calculate the yearly cost. Enter this value into the table.

6 Total the first column.

7 Complete the second column by converting each yearly cost to a cost per week.   ←—— There are 52 weeks in one year.

8 Complete the third column by calculating the percentage contribution each cost makes to the total cost.

$$\text{Percentage} = \frac{\text{cost of item}}{\text{total cost}} \times 100\%$$

9 **a** List the cost categories in order from most expensive to least expensive.

   b Which costs depend on the number of kilometres driven per year?

   c Which costs are independent of the number of kilometres driven?

   d Is it possible that the order of categories would change if the kilometres per year were greater than or less than 15 000 km?

   e How much should you budget per week to run each car?

10 Repeat what you have done for Questions **1** to **9** for a car of your choice.

11 When you are first working full-time you often don't have a very high wage. Joe, a first-year apprentice, is paid approximately $465 per week.

   a For each of the three cars you have looked at in this exercise, calculate the weekly running cost as a percentage of Joe's weekly income.

   b Which car could Joe afford to buy?

WORD MATCH

Match each term to its correct definition.

Word

1 ANCAP rating
2 comprehensive insurance
3 CTP insurance
4 running costs
5 fuel consumption
6 on-road costs
7 registration
8 stamp duty
9 Third Party Property insurance

Meaning

A the costs of maintaining a car

B covers the damage to other people's property after an accident in which you are at fault

C an annual charge paid to the Roads and Maritime Services to own a car

D a measure of the level of safety and accident protection a new car offers

E additional costs of buying a new car that must be paid before you drive it

F covers the cost of repairs or replacement to your car and other property damaged after an accident whether or not you are at fault

G a tax paid to the state government when ownership of a vehicle changes hands

H the amount of fuel a vehicle uses as a rate per kilometre travelled

I the type of insurance that covers the cost of injury to passengers and other people involved in an accident in which you are at fault

SOLUTION TO THE CHAPTER PROBLEM

Problem

Sally and Harry are going to drive to Perth to visit Sally's parents. Their SUV (sport utility vehicle) has a fuel consumption of 9.8 litres per 100 kilometres. On some parts of the trip, they will have long distances to travel with few petrol stations and Sally would like to work out how far the car will go on a full tank of petrol. The petrol tank holds 64 litres but, to be safe, she will base her calculations on 90% of this figure. How far can they travel between fuel stops?

Solution

First, calculate 90% of the tank's capacity in order to determine how far the car can travel on this amount.

90% of 64 L = 57.6 L

Convert the fuel consumption rate to L/km.

$$\frac{9.8 \text{ L}}{100 \text{ km}} = 0.098 \text{ L/km}$$

We wish to find the distance (km) travelled on 57.6 L of petrol, so divide by the rate.

Distance = 57.6 ÷ 0.098

$\quad\quad\quad\quad$ = 587.755 …

$\quad\quad\quad\quad$ ≈ 588 km (to the nearest kilometre)

Sally and Harry can travel up to 588 km between fuel stops.

- Give examples of information you learned in this chapter that was new to you.

- Do you think working through this chapter will help when you buy a car? Explain why or why not.

- Is there any part of the calculations you didn't understand? If so, ask your teacher for help.

Copy and complete this mind map of the topic, adding detail to its branches and using pictures, symbols and colour where needed. Ask your teacher to check your work.

1 Jordan purchases a used Toyota Camry for $19 990. Calculate the 3% stamp duty payable on this purchase.

Exercise
13.02

2 Wendi buys a new Holden Commodore priced at $66 990. Calculate how much stamp duty she will pay if it is $1350 plus 5% of the value greater than $45 000.

Exercise
13.02

3 Francis borrows $10 000 at 5.7% p.a. to be repaid monthly over 3 years. The monthly repayment is $313.

   a How much will Francis repay over the 3 years?

   b How much interest will Francis pay?

Exercise
13.03

4 Rajesh is buying a large car and needs to borrow $40 000. He is offered a loan at 6.4% p.a. over 7 years with monthly repayments of $602.

   a How much will Rajesh repay over the 7 years?

   b How much interest will he pay?

Exercise
13.03

5 Rajesh is going to take out comprehensive car insurance with an annual premium of $1562. He is entitled to a no-claim discount of 30%. How much will he have to pay?

Exercise
13.04

6 Klaas drives 524 km, a mixture of highway and city driving, and uses 31.4 L of fuel. Calculate the fuel consumption of Klaas' car in L/100 km, correct to one decimal place.

Exercise
13.05

7 Jayden's Holden Commodore has a fuel consumption of 11.3 L/100 km.

   a How much petrol will his car use on a 423 km trip? Answer correct to one decimal place.

   b What is the cost of fuel for the journey if petrol costs $1.45 per litre.

Exercise
13.05

8 Olivia owns a Subaru Forester, which has a fuel tank of 60 L and fuel consumption of 8.5 L/100 km.

   a How far (correct to one decimal place) can Olivia travel on one tank of petrol?

   b How far do you think it would be safe to travel before filling up the petrol tank?

Exercise
13.05

9 Sunil owns a small car which he mostly uses for driving in town. He records all costs associated with his car for one year. The total is $4400.

   a How much is this per week?

   b Sunil is going to allow for a 10% increase in costs in the coming year. How much per week should he budget for the car with this increase?

Exercise
13.06

10 Monica owns a van which cost her $7030 in running costs in the last year. She knows it is coming up for some major work and that she will be using it in the coming year. She expects her costs to increase by 30%. How much per week should she budget for the van in the coming year?

Exercise
13.06

14.

COMPARING DATA

Chapter problem

Darryl and Andrew are arguing about which local rugby league team is the best. The points scored by their favourite teams in each match over the previous season are listed:

Eagles:	20	10	40	12	17	20	22	20
	34	19	36	18	24	12	38	34
	24	36	32	22	6	7	38	18
Cougars:	14	18	24	39	14	4	4	14
	10	13	28	22	16	18	18	12
	18	28	21	6	10	18	36	12

a To show this data, which is the best graph to use? Construct this graph.

b Which is the better team based on this data? Justify your answer.

c What additional data might you need to have in order to decide which is the better team?

CHAPTER OUTLINE

WHAT WILL WE DO IN THIS CHAPTER?

- Calculate five-number summaries and use them to draw box plots
- Compare data using back-to-back stem-and-leaf plots
- Compare data using parallel box plots
- Consider the shape of data distributions

HOW ARE WE EVER GOING TO USE THIS?

- When comparing products for sale
- When examining performances in a variety of sports
- When interpreting data presented in the media

Box-and-
whisker plots

Box plots:
Graphics
calculator

14.01 Box plots

The three quartiles Q_1, Q_2 and Q_3 and the lowest and highest scores together make a **five-number summary** for data, which can then be graphed on a **box plot**, also called a **box-and-whisker plot**.

The **five-number summary** for a set of data consists of:

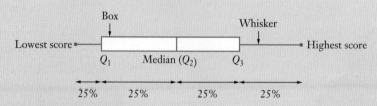

- the lowest score
- the first or lower quartile, Q_1
- the second quartile, the median, Q_2
- the third or upper quartile, Q_3
- the highest score.

EXAMPLE 1

The ages of the 23 people at a cafe are shown here.

33	23	28	36	27	15	32	18
13	13	38	38	27	7	34	27
12	26	33	21	24	39	20	

This data was also used in Example 3 of Chapter 8 on p. 215.

a   Find the five-number summary for this data.

b   Draw a box plot for this data.

Solution

a   Find the median: it is also the 2nd quartile.

7, 12, 13, 13, 15, 18, 20, 21, 23, 24, 26,(27,)27, 27, 28, 32, 33, 33, 34, 36, 38, 38, 39

$Q_2 = 27$

Find the middle of each half of the scores. These are the 1st and 3rd quartiles.

7, 12, 13, 13, 15,⟨18⟩20, 21, 23, 24, 26,⟨27⟩

27, 27, 28, 32, 33,⟨33⟩34, 36, 38, 38, 39

$Q_1 = 18$, $Q_3 = 33$

Include the lowest score (7) and the highest score (39).

The five-number summary is:

7, 18, 27, 33, 39

b   Draw the box plot with the box between 18 and 33, and with a middle bar at 27, and the whiskers extending to 7 and 39.

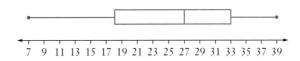

7 9 11 13 15 17 19 21 23 25 27 29 31 33 35 37 39

Exercise 14.01 Box plots

In this exercise, you will use your answers to Exercise 8.02 on page 215. If you can't find your own answers, then look up the answer section at the back of the book.

1 Use your answers to Question **1** about the sale of fish in a fish and chip shop to:
   a   write a five-number summary for the data
   b   draw a box plot for the data.

2 Use your answers to Question **2** about daily maximum temperatures in Cairns to:
   a   write a five-number summary for the data
   b   draw a box plot for the data.

3 Use your answers to Question **3** about student heights to draw a box plot for the data.

4 a   Use your answers to Question **5** about retail theft to write a five-number summary and draw a box plot for each set of data.
   b   In which area do you think it would be safer to open a shop? Justify your answer.

5 The monthly numbers of home burglaries in Emu Springs were recorded over 27 months.

21	25	17	23	16	21	41	22	25
20	22	11	20	12	13	12	6	12
10	19	30	22	21	14	34	33	24

   a   What is the five-number summary for this data?
   b   Draw the box plot for this data.

6 This box plot represents the amount of pocket money in dollars earned by a sample of 60 children.

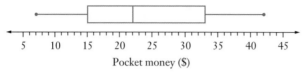

Pocket money ($)

a Find the median.

b Find the range.

c How many children earned between:

   i $33 and $42?   **ii** $15 and $42?

d Find the interquartile range.

7 This stem-and-leaf plot shows the number of students served at the school canteen over a 3-week period.

Stem	Leaf
7	6
8	1 6 8
9	5 7 8
10	1 5 5
11	2 2 4 7
12	4

a Find the five-number summary for this data.

b Draw the box plot for this data.

8 This box plot shows the number of cigarettes smoked per day by a sample of 60 smokers who are trying to quit.

a What is the median number of cigarettes smoked per day?

b What is the interquartile range?

c What is the lowest score?

d How many people smoked between 20 and 25 cigarettes per day?

e How many people smoked fewer than 20 cigarettes per day?

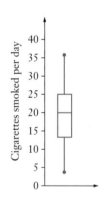

ISBN 9780170413503

14.02 Back-to-back stem-and-leaf plots

We can use **back-to-back stem-and-leaf plots** to compare two data sets.

Madeline is trying to decide which gym to join. She records the ages of people attending classes in the afternoon at two local gyms.

Allfit:	17	22	51	12	27	43	39	20	21	15
	15	43	15	20	32	21	23	16	34	22
Superfit:	19	20	32	46	27	16	11	34	38	21
	13	22	34	35	23	31	20	16	22	27

a   Draw a back-to-back stem-and-leaf plot for this data.

b   Find the median age for each gym.

c   Find the range for each gym.

d   Are there any outliers in either gym? If so, state the outliers.

e   State one similarity between the two sets of data.

f   State one difference between the two sets of data.

Solution

a   The stem is in the middle.

Allfit		Superfit
7 6 5 5 5 2	1	1 3 6 6 9
7 3 2 2 1 1 0 0	2	0 0 1 2 2 3 7 7
9 4 2	3	1 2 4 4 5 8
3 3	4	6
1	5	

b	Each gym has 20 scores, so the middle scores are the 10th and 11th scores.	Median for Allfit $= \dfrac{21+22}{2} = 21.5$
		Median for Superfit $= \dfrac{22+23}{2} = 22.5$
c	Range = highest score – lowest score.	Range for Allfit $= 51 - 12$
		$= 39$
		Range for Superfit $= 46 - 11$
		$= 35$
d	An outlier is an extreme score.	51 is an outlier for Allfit.
e	Look for similar things about the data.	The medians for both sets of data are quite close. Both sets of data are also clustered in the 20s.
f	Look for different things about the data.	Allfit's ages are more spread out than Superfit's. Superfit's ages are spread evenly across the 10s, 20s and 30s, whereas Allfit's ages are clustered in the 10s and 20s.

Exercise 14.02 Back-to-back stem-and-leaf plots

1 These are the results of two Year 10 classes in their final mathematics exam:

10 Green: 84 71 79 82 78 89 71 95 93 81
85 65 70 95 91 89 89 75 62 71
69 88 94 81 85 76 80 67 60

10 Blue: 88 60 66 74 42 59 61 68 54 65
56 46 65 71 57 72 40 65 41 38
35 39 51 47 39 42 39 42

a Draw a back-to-back stem-and-leaf plot for this data.
b How many students are there in 10 Green?
c Find the median score for each class.
d Find the range for each class.
e Are there any outliers in either class? If so, state the outliers.
f If one class is the top class and one class is the middle class, which class is which?

ISBN 9780170413503

2 The school principal Mr Farley is concerned about absences in Years 11 and 12. The following shows the number of absentees per day in Years 11 and 12 over a 4-week period.

Year 11:	30	15	30	23	39	20	31	42	22	41
	30	25	23	30	22	30	29	15	15	44
Year 12:	20	22	12	8	19	13	14	23	7	22
	18	20	13	19	21	15	24	10	26	24

a Draw a back-to-back stem-and-leaf plot for this data.

b Find the median score for each year group.

c Find the range for each year group.

d Are there any outliers in either year group? If so, state the outliers.

e State one similarity between the two sets of data.

f State one difference between the two sets of data.

g Mr Farley believes there is more absenteeism in Year 11 than in Year 12. What other information would we need to evaluate this statement?

3 For a PE assignment, Kayne compares the heights (in centimetres) of males and females in Year 11.

Males:	178	183	167	184	181	170	190	181	181	200
	183	160	165	172	178	178	168	191	181	190
	180	184	180	175	170					
Females:	178	166	166	150	168	166	163	162	167	159
	157	185	176	164	165	164	160	185	176	177
	171	152	173	173						

a Draw a back-to-back stem-and-leaf plot for this data. Use stems of 15, 16, …

b How many males have been measured?

c How many females have been measured?

d Find each of the five-number summaries for the males and the females.

e Are there any outliers in each group? If so, state them.

f State one difference between the two sets of data.

g Write two or three sentences that Kayne could use as a conclusion for his assignment.

4 The daily maximum temperatures for Broome, WA and Kiama, NSW in the month of February are shown below.

Broome:	32.1	32.3	33.1	33.4	31.8	31.0	33.7
	34.3	32.7	32.7	31.0	33.4	33.6	34.2
	32.3	32.5	31.9	32.1	32.8	31.7	29.3
	31.9	31.2	31.7	31.2	32.2	29.5	31.0
Kiama:	33.0	27.9	32.6	27.6	36.3	29.6	21.0
	23.0	22.5	25.0	26.4	22.1	20.7	22.3
	24.9	25.6	25.5	24.5	27.0	32.5	21.8
	20.3	22.5	24.7	25.6	25.9	26.7	25.4

a Draw a back-to-back stem-and-leaf plot for this data. Use stems 20, 21, 22, ... to 36.

b How many days are there in February?

c Find the median for each town.

d Find the range for each town.

e Are there any outliers in the data for either town? If so, state them.

f State one difference between the two sets of data.

g Write two or three sentences comparing the temperatures in the two towns.

h Based on this data, in which town would you prefer to live? Justify your answer.

14.03 Parallel box plots

In statistics, there are many opportunities to compare two sets of data. We can compare sets of data by drawing **parallel box plots** using a common scale.

Double box plots

Comparing word lengths

Investigating young drivers

Comparing sports scores

Comparing city temperatures

EXAMPLE 3

The two five-number summaries below describe the number of rainy days per month in Sydney and Melbourne.

Sydney:	9	11	13	14	15
Melbourne:	7	10	14	16	19

a Draw parallel box plots for these summaries.

b Find the median for each city.

c What is the interquartile range for each city?

d Which city has more rainy days per month?

e If Corrina prefers a more consistent pattern of rainy days, which city would you recommend for her? Justify your answer.

Alamy Stock Photo/Jackie ellis

Solution

a Use one scale for both box plots.

Sydney
Melbourne
7 8 9 10 11 12 13 14 15 16 17 18 19 20

b The median is the middle number in the five-number summary, or use the middle bar in the box plots.

Median for Sydney = 13.

Median for Melbourne = 14.

c IQR = $Q_3 - Q_1$.

IQR for Sydney = 14 − 11 = 3.

IQR for Melbourne = 16 − 10 = 12.

d Compare the medians and boxes of the box plots.

Melbourne has more rainy days per month. Its median is higher and half of its scores are 14 or above compared to $\frac{2}{5}$ of scores for Sydney.

e Consistent means the data is less spread out.

Sydney has the more consistent pattern of rainy days because its range and interquartile range are smaller than Melbourne's. This is shown by its shorter box plot.

Exercise 14.03 Parallel box plots

1 Rigby and Alex are in different mathematics classes. The following five-number summaries are for yearly exams in each class.

Rigby's class:	48	64	75	87	96
Alex's class:	47	57	69	80	97

a Draw parallel box plots for these summaries.

b What is the median for:
  i Rigby's class?   **ii** Alex's class?

c What is the range for:
  i Rigby's class?   **ii** Alex's class?

d Both Alex and Rigby scored 85% in their yearly exams. Who has performed better in relation to their own class? Justify your answer.

e Which class generally performed better in the yearly exams? Justify your answer.

f Can we calculate the mean from the given information? Explain.

2 These are the waiting times in minutes when calling customer service of two phone companies.

Chatphone:	10	7	6	8	7	5	6	9	7
	3	8	8	9	7	9	7	9	8
Oztel:	10	5	9	9	9	10	11	9	8
	7	9	7	7	6	9	8	11	11

a Find the five-number summary for each set of data.

b Construct parallel box plots for this data.

c Find the median for each company.

d Find the mode for each company.

e Find the interquartile range for each company.

f Chatphone claims that its waiting times are generally lower than those of Oztel. Is this correct? Justify your answer.

g Is this sufficient information to decide which company you would choose as your mobile phone service provider? What other information would you need, if any?

3 The mayor of Middleton claims that his town is safer for drivers than the nearby town of Blakewell. To test this, James measured the speed (in km/h) of a sample of 20 cars in each town.

Middleton:	60	65	70	68	62	75	80	83	82	69
	73	75	85	72	67	88	90	85	72	63
Blakewell:	76	64	58	82	72	70	68	75	63	67
	74	70	79	80	73	75	71	68	72	73

a Find the five-number summary for each set of data.

b Construct parallel box plots for this data.

c Find the median for each town.

d Find the mode(s) for each town.

e Find the mean for each town.

f Find the interquartile range for each town.

g Is the mayor of Middleton correct? Justify your answer.

4 This back-to-back stem-and-leaf plot shows the number of points scored in each match by two basketball teams during last season.

Langley Lynx		**Blakely Bears**
6 6 5 4 3	4	4 9
8 8 3 0	5	2 3 3 6 8
8 8 6 6 3 1 1	6	5 6 8 9
7 4 3 0	7	0 0 1 3 6
6 6 5	8	2 5 7 7 9 9
2 2	9	0 3 4

a How many games were played last season?

b Find the range for each team.

c Find the median for each team.

d Find the five-number summary for each team.

e Draw parallel box plots for the two teams.

f Comment on the similarities and differences of the points scored by the two teams.

g Which is the better-scoring team?

5 Nadine surveyed Year 7 and Year 12 students at her school about their favourite drinks and showed the results on this **clustered bar chart**.

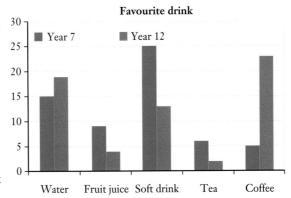

a How many Year 7 students did Nadine survey?

b How many Year 12 students did Nadine survey?

c What is the most popular drink in Year 7?

d What is the most popular drink in Year 12?

e Why do you think the most popular drink is different in Year 7 and Year 12?

CensusAtSchool

INVESTIGATION

COMPARING SETS OF DATA USING *CENSUS AT SCHOOL*

Use data from the **CensusAtSchool** website at the Australian Bureau of Statistics (ABS) to compare two groups.

1 Decide which two groups you will compare; for example, male and female, different states, city versus country, different ages or different years.

2 Decide what numerical data you are going to compare; for example, heights, arm spans, reaction times, foot length.

3 Record 25 values of data for each group you are comparing. Only record sensible measurements and exclude data that is obviously wrong; for example, a height of 60 cm if you are comparing heights of 17-year-olds.

4 Draw a back-to-back stem-and-leaf plot for both sets of data.

5 Find the median for each group.

6 Find the range for each group.

7 State one similarity between the two groups.

8 State one difference between the groups.

9 Find the five-number summary for each data set.

10 Construct a parallel box plot for both sets of data.

11 Write a short paragraph about the similarities and/or differences in the data based on the box plots.

14.04 The shape of a distribution

Shapes of distributions

The shape of a frequency distribution

When there is an overall pattern to the data in a histogram, we can draw a smooth curve through the histogram to represent the data. Statisticians like the smooth curve because it makes calculations easier than working with the histogram. We can also draw a curve around dot plots and stem-and-leaf plots to see the shape of the data.

Here are three examples of smooth curves that represent the general **shape of a distribution**.

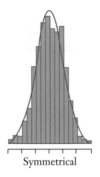

Symmetrical

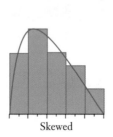

Skewed

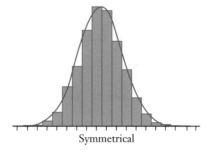

Symmetrical

Symmetrical distributions

The first and third curves above are examples of a **symmetrical distribution**. The curve is symmetrical about the peak or mode. You could fold it down the middle and the two sides would match. A symmetrical curve could represent the heights or weights of all Year 11 students in Australia.

Skewed distributions

When data is not symmetrical but pushed to one side, it is said to be **skewed** (which means 'twisted'). We can identify the type of skew by looking at the 'tail' of the curve. When the tail is on the left, the data are **negatively skewed**. When the tail is on the right, the data are **positively skewed**. One way of remembering the direction of skewness is to note that on a number line, negative numbers are on the left and positive numbers are on the right.

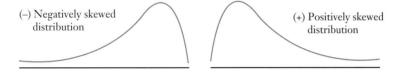

(–) Negatively skewed distribution

(+) Positively skewed distribution

A **negatively skewed distribution** could represent the marks scored by students on an easy test. Most students achieved high marks while comparatively few scored low marks.

A **positively skewed distribution** could represent house prices in a small country town. There are lots of moderate prices for the houses and comparatively few high prices for bigger houses and farms.

Bimodal and multimodal distributions

This graph represents the amount of traffic passing a school per hour over a day. There are two peaks on the curve, corresponding to the morning and afternoon peak hours. Because it has two peaks, this curve represents a **bimodal distribution**. When there is only one peak, it is called a **unimodal distribution**. When there is more than one peak, it is called a **multimodal distribution**.

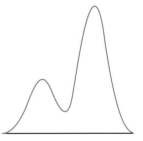

EXAMPLE 4

The distribution below shows the number of car owners by age who claim on their car insurance. Describe the shape of the distribution.

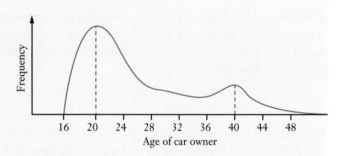

Age of car owner

Solution

'The distribution is skewed. 'The tail points to the right, so it is positively skewed. There are peaks at ages 20 and 40, so the distribution is also bimodal.

'The distribution is positively skewed and bimodal.

Exercise 14.04 The shape of a distribution

1 State whether each distribution is roughly symmetrical. It may help to copy the diagram and draw a smooth curve over it.

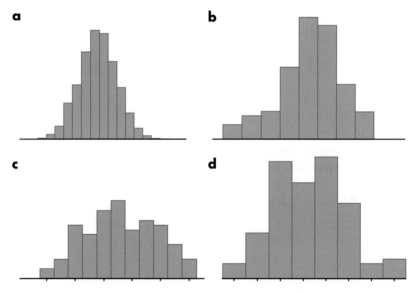

a

b

c

d

2 What type of distribution is shown in each of the following graphs? Choose from symmetrical, positively skewed, negatively skewed, unimodal and bimodal.

a   Age at retirement

b   Australian family income

c   Number of people waiting to catch a train

d   Height of trees in a forest

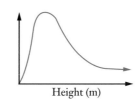

3 Which graphs in Question **2** are unimodal?

4 Sketch a graph to describe each distribution. Remember to label the axes.

a   The age a person dies is negatively skewed, with a mode of 80 years.

b   The percentage marks students scored on a difficult assessment task had a bimodal, positively skewed distribution. A small group scored 90%, but the majority scored around 40%.

c   A bus carrying passengers to a concert includes a large group of dancers and a small group of rugby players. The distribution of the body mass (weights) of the passengers is bimodal.

d   The heights of 2000 randomly selected girls aged 17 are distributed symmetrically.

e   The number of minutes visitors to a museum spend looking at a display that includes a 5-minute video is bimodal. Most visitors spend less than 1 minute looking at the display, but some also watch the entire video and the display. Very few people look at the display for between 1 to 5 minutes.

5 How can you tell that the data in this box plot is positively skewed?

6 Construct a box plot that shows data that are negatively skewed.

7 Construct a dot plot with scores from 1 to 8 that have a bimodal and symmetrical distribution.

8 This multimodal distribution represents the number of customers at a city business at different times of the day.

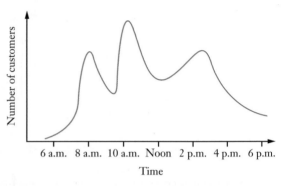

  a This distribution is 'trimodal'. What does this mean?

  b What type of business do you think it could be? Give a reason for your answer.

9 This histogram shows the number of minutes customers at an appliance store had to wait for items to be brought in from the warehouse.

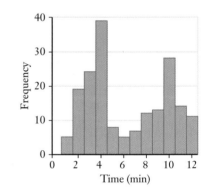

  a Describe the shape of the distribution.

  b Suggest a possible reason for the shape of the distribution.

10 Sometimes, exam and course results are presented to students and parents in the form of a graph showing the position of the score relative to the distribution of the scores of all students. The graph below shows an example for one student, Brock, for his test mark in mathematics.

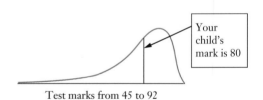

Test marks from 45 to 92

  a Describe the shape of the distribution.

  b Was the mathematics test easy or hard? Justify your answer.

  c Estimate the mode mark.

  d How did Brock perform in the test compared to the other students?

NCM 11. Mathematics Standard (Pathway 1)

ISBN 9780170413503

Statistics crossword

DESCRIPTIONS AND DEFINITIONS

1 In your own words, describe a back-to-back stem-and-leaf plot and a parallel box plot. Draw a rough sketch of what they look like.

2 What is a five-number summary?

3 Match the terms in the left column with their correct meanings in the right column.

a	bimodal	**A**	A rectangular distribution where every score has a similar frequency.
b	skewed	**B**	A distribution with two peaks.
c	unimodal	**C**	The scores that divide the data into four equal parts.
d	quartiles	**D**	A distribution where most scores are to the left or right of centre and there is a tail on the side that doesn't have many scores.
e	uniform	**E**	A distribution with one peak.

SOLUTION TO THE CHAPTER PROBLEM

Problem

Darryl and Andrew are arguing about which local rugby league team is the best. The points scored by their favourite teams in each match over the previous season are listed:

Eagles:	20	10	40	12	17	20	22	20
	34	19	36	18	24	12	38	34
	24	36	32	22	6	7	38	18

Cougars:	14	18	24	39	14	4	4	14
	10	13	28	22	16	18	18	12
	18	28	21	6	10	18	36	12

a To show this data, which is the best graph to use? Construct this graph.

b Which is the better team based on this data? Justify your answer.

c What additional data might you need to have in order to decide which is the better team?

Solution

a This data is best shown on a stem-and-leaf plot.

Eagles		Cougars
7 6	0	4 4 6
9 8 8 7 2 2 0	1	0 0 2 2 3 4 4 4 6 8 8 8 8 8
4 4 2 2 0 0 0	2	1 2 4 8 8
8 8 6 6 4 4 2	3	6 9
0	4	

b The median for the Eagles is 21. The median for the Cougars is 17. The scores for the Cougars are clustered in the 10s. The scores for the Eagles are spread evenly across the 10s, 20s and 30s. Based on this data, the Eagles are the better team.

c You would need to know how many games each team won. It would also help to know how many points were scored against each team.

ISBN 9780170413503

- What parts of this chapter were new to you?
- Give examples of jobs where you might need to compare different data sets.
- Note any difficulties you had with the work in this chapter. Discuss them with your teacher.

Copy and complete this mind map of the topic, adding detail to its branches and using pictures, symbols and colour where needed. Ask your teacher to check your work.

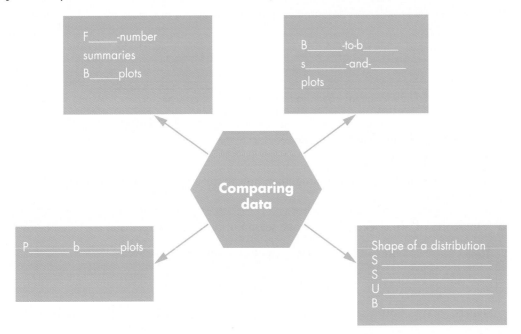

F_____-number summaries
B_____plots

B_____-to-b_____
s_____-and-_____
plots

Comparing data

P_____b_____plots

Shape of a distribution
S _____
S _____
U _____
B _____

14. TEST YOURSELF

1 The following are the test results for Manuel's class.

6	4	2	10	16	12	14	12	6

11	9	10	12	13	18	4	10	14

a What is the five-number summary for this data?

b Draw the box plot for this data.

2 This box plot summarises the number of hours worked in one week by each employee of Café Coffee.

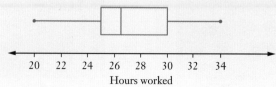

a What is the median for this data?

b Find the range.

c If there are 36 employees altogether, how many employees worked between 30 and 34 hours?

3 The scores of two cricket teams in one season of the local competition are listed below:

Bulls:	83	125	89	113	109	90
	127	159	98	140	114	137

Tigers:	130	144	104	72	139	133
	109	97	138	147	126	139

a Draw a back-to-back ordered stem-and-leaf plot for this data.

b How many matches were played in one season?

c Find the median for each team.

d Are there any outliers in either set of data? If so, state the outliers.

e Which team is the better team? Justify your answer.

4 These are the five-number summaries for the archery scores of Team Magenta and Team Blue.

Exercise
14.03

Team Magenta: 1   5   7   8   10

Team Blue: 0   4   7   9   10

a Draw a parallel box plot for these five-number summaries.

b What is the median for each team?

c What is the interquartile range for each team?

d Shane in Team Blue scored 9 and Adele in Team Magenta scored 9. Who has performed better in relation to their own team? Justify your answer.

5 What type of distribution is shown in each of the following graphs? Choose from symmetrical, positively skewed, negatively skewed, uniform and bimodal.

Exercise
14.04

a

Goals scored

b

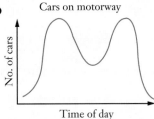

Cars on motorway

c

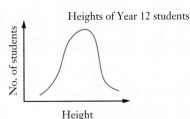

Heights of Year 12 students

6 Sketch a graph for each distribution described below. Remember to label the axes.

Exercise
14.04

a The percentage marks students scored on an easy assessment task – the results are negatively skewed with most scoring between 80 and 90 per cent.

b The masses of all Year 11 students in Queensland are distributed symmetrically.

c When a card is chosen randomly from a deck of cards 500 times, the suit of the card (diamonds, hearts, clubs, spades) has a uniform distribution.

15.

IT'S ABOUT TIME

Chapter problem

The FA Cup soccer final is played at Wembley Stadium in London in May. At this time of year, England is on daylight saving time. The match starts at 1700 local time. At what time should TV viewers in Australia tune in to watch the final?

CHAPTER OUTLINE

WHAT WILL WE DO IN THIS CHAPTER?

- Convert between 12- and 24-hour time
- Calculate time differences
- Use timetables for public transport
- Learn about time zones and daylight saving in Australia and around the world
- Use latitude, longitude and position coordinates to locate cities around the world
- Solve practical problems involving time zones

HOW ARE WE EVER GOING TO USE THIS?

- Planning an Australian or overseas trip
- Reading timetables to catch public transport
- Using 24-hour time to program clocks and timers
- Adjusting the time for daylight saving

12-hour and
24-hour time

24-hour time
on an analogue
clock

15.01 0600 hours

There are two ways to show the time of day:

- **12-hour time:** the conventional way using a.m. and p.m. and the hours 1 to 12

- **24-hour time:** a formal way that uses 4 digits, the hours 00 to 23 and no a.m./p.m.

There are two main types of clock that we use to tell the time:

Analog clock

Digital clock

This table shows the relationship between 12-hour and 24-hour time.

24-hour time	12-hour time		24-hour time	12-hour time
0000	12 a.m. (midnight)		1200	12 p.m. (midday)
0100	1 a.m.		1300	1 p.m.
0200	2 a.m.		1400	2 p.m.
0300	3 a.m.		1500	3 p.m.
0400	4 a.m.		1600	4 p.m.
0500	5 a.m.		1700	5 p.m.
0600	6 a.m.		1800	6 p.m.
0700	7 a.m.		1900	7 p.m.
0800	8 a.m.		2000	8 p.m.
0900	9 a.m.		2100	9 p.m.
1000	10 a.m.		2200	10 p.m.
1100	11 a.m.		2300	11 p.m.

EXAMPLE 1

Express 8:25 a.m. in 24-hour time.

Solution

8:25 a.m. is in the morning. Leave out the dots and put a zero at the front.   8:25 a.m. = 0825

EXAMPLE 2

Write 1615 in 12-hour time.

Solution

1615 is bigger than 1200, so it's in the afternoon (p.m.). For the hour, take 12 away from 16.

$1615 = 4{:}15$ p.m.

$16 - 12 = 4$.

Exercise 15.01 0600 hours

1 Write in 24-hour time:

   a 11:44 a.m.   **b** 6:35 p.m.   **c** 2:51 a.m.   **d** 9:54 p.m.

2 Write in 12-hour time:

   a 0845   **b** 1320   **c** 2331   **d** 1045

3 Write the time shown on each clock in 12-hour time:

  a    **b**    **c**

4 Write the time shown on each clock in 12-hour time. All times are in the morning.

  a    **b**    **c**

5 Write the time shown on each clock in 12-hour time. All times are in the afternoon or evening.

  a    **b**    **c**

6 Show each time on an analog clock face.

   a 8:15 a.m.   **b** 10:40 p.m.   **c** 3:35 p.m.

7 Write the times missing from this table.

12-hour time	5 a.m.	**a**	2:10 p.m.	5:18 p.m.	**b**	**c**	**d**
24-hour time	**e**	0715	**f**	**g**	1730	2150	1120

8 Sue wants to program her DVR to record a television program when she's out. The program starts at 8:30 p.m. and lasts for an hour and a half. Her DVR works in 24-hour time.

   a At what time will Sue need to set her DVR to start recording the program?

   b What time should she set for her finish time? Allow an extra 5 minutes for the recording.

9 The army, navy and air force all use 24-hour time, and sometimes 24-hour time is called 'military time'. Sean rings his wife from the army camp at 1730. What time is this in 12-hour time?

10 Andrew was on police duty one Saturday night from 2200 to 0600 on Sunday morning. He receives a report of a fight at 2245 and again at 0220.

   a How long was Andrew on duty?

   b How long was it between the two incident reports?

   c A patient from the second fight is taken to hospital and arrives 45 minutes after the incident was reported. What time does the patient arrive at hospital?

15.02 How long will it take?

Time
differences

How long do I have before my favourite TV program starts? How long is my stopover in Hong Kong? How long have I worked here? We can answer all these questions by doing a time calculation.

EXAMPLE 3

What is the difference in time between 8:35 a.m. and 3:10 p.m.?

Solution

Use a timeline and count to the next full hour.

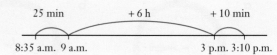

From 8:35 a.m. to 9:00 a.m.	25 minutes
From 9:00 a.m. to 3:00 p.m.	6 hours
From 3:00 p.m. to 3:10 p.m.	10 minutes
Add together.	Total time difference = 6 h + 25 min + 10 min
	= 6 h 35 min

EXAMPLE 4

What is the time 7 hours and 40 minutes after 11:52 p.m.?

Solution

Use a timeline and add the hours first.

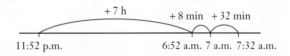

11:52 p.m. + 7 hours	6:52 a.m.
Count to the next full hour and add the number of minutes.	8 minutes to 7 a.m.
To make up 40 minutes, add another 32 minutes: 7 a.m. + 32 min.	7:32 a.m.

EXAMPLE 5

In March 2017, Kane's car was 5 years and 9 months old. When was Kane's car made?

Solution

Subtract 5 years from March 2017.	2017 − 5 = 2012
	March 2012
Count back to the start of 2012 and subtract the number of months.	3 months to start of 2012
To make up 9 months, subtract another 6 months: start of 2012 − 6 months.	June 2011

Exercise 15.02 How long will it take?

1 Calculate the time difference between each pair of times.

    a   7:27 a.m. and 1:12 p.m.   **b**   4:09 a.m. and 9:53 a.m.

    c   3:42 p.m. and 6:02 p.m.   **d**   11:15 p.m. and 3:08 a.m.

2 What is the time difference between each pair of 24-hour times?

    a   0800 and 1100   **b**   0500 and 1500

    c   0940 and 1455   **d**   1340 and 2150

3 Vamsee's flight lands in Hong Kong at 0950. His next flight leaves Hong Kong at 1920. How long does Vamsee have at Hong Kong airport?

4 At S. Thomson College, the school day begins at 8:45 a.m. and ends at 3:10 p.m. How long is the school day?

5 What time will it be:

    a   5 hours after 3:00 p.m.?   **b**   28 minutes after 7:15 p.m.?

    c   3 hours 32 minutes after 9:45 a.m.?   **d**   9 hours 10 minutes after 5:14 p.m.?

6 What time was it:

    a   4 hours before 6:15 p.m.?   **b**   45 minutes before 3:20 a.m.?

    c   2 hours 10 minutes before 1:35 p.m.?   **d**   3 hours 35 minutes before 11:25 a.m.?

7 Ewan has to take his tablets $1\frac{1}{2}$ hours after he finishes lunch. He finished his lunch at 1:45 p.m. What time should Ewan take his medicine?

8 Ramona wants to go for a jog for 50 minutes, but be back in time to watch *Zombie Vampire Athletes* on TV starting at 8:35 p.m. What is the latest time she can leave for her jog?

9 Jamie bought a computer in December 2016. How old is it:

    a   in August 2020?   **b**   in January 2021?   **c**   today?

10 A box of chocolates bought in November 2018 has a 'Use by' date of March 2020. How long, in months, will the chocolates last?

11 In February 2015, Stephanie's TV was $3\frac{1}{2}$ years old. When did Stephanie buy it?

12 Ask your teacher when they started teaching at your school. Work out how long they have been at your school in years, months and days.

13 How old are you today in years, months and days?

15.03 Timetables

Thousands of Australians rely on rail, bus and ferry **timetables** every day.

TV times

Public transport trips

EXAMPLE 6

A section of a train timetable is shown here.

Armadale/Thornlie to Perth

	AM	AM	AM	AM	AM	AM	AM	AM
Armadale	7:16		7:46		8:16		8:31	
Sherwood	7:18		7:48		8:18		8:33	
Challis S	7:20		7:50		8:20		8:35	
Kelmscott	7:22		7:52		8:22		8:37	
Seaforth	7:25		7:55		8:25		8:40	
Gosnells	7:27		7:57		8:27		8:42	
Maddington	7:30		8:00		8:30		8:45	
Kenwick	7:32		8:02		8:32		8:47	
Beckenham	7:34		8:04		8:34		8:49	
Thornlie		7:35		8:05		8:35		8:50
Cannington	7:36	7:39	8:06	8:09	8:36	8:39	8:51	8:54
Queens Park		7:40		8:10		8:40		8:55
Welshpool		7:42		8:12		8:42		8:57
Oats Street	7:39	7:44	8:09	8:14	8:39	8:44	8:54	8:59
Carlisle		7:45		8:15		8:45		9:00
Victoria Park		7:48		8:18		8:48		9:03
Burswood		7:50		8:20		8:50		9:05
Belmont Park				8:22		8:52		9:07
Claisebrook	7:47	7:54	8:17	8:24	8:47	8:54	9:02	9:09
McIver	7:49	7:56	8:19	8:26	8:49	8:56	9:04	9:11
Perth	7:50	7:58	8:20	8:28	8:50	8:58	9:05	9:13

Source: Transperth

Perth to Armadale/Thornlie

	PM	PM	PM	PM	PM	PM	PM	PM
Perth	4:37	4:45	4:52	5:00	5:07	5:15	5:22	5:30
McIver	4:38	4:46	4:53	5:01	5:08	5:16	5:23	5:31
Claisebrook	4:40	4:48	4:55	5:03	5:10	5:18	5:25	5:33
Belmont Park	4:42		4:57		5:12		5:27	
Burswood	4:44		4:59		5:14		5:29	
Victoria Park	4:46		5:01		5:16		5:31	
Carlisle	4:48		5:03		5:18		5:33	
Oats Street	4:49	4:54	5:04	5.09	5:19	5:24	5:34	5:39
Welshpool	4:51		5:06		5:21		5:36	
Queens Park	4:53		5:08		5:23		5:38	
Cannington	4:55	4:58	5:10	5:13	5:25	5:28	5:40	5:43
Beckenham		4:59		5:14		5:29		5:44
Kenwick		5:01		5:16		5:31		5:46
Maddington		5:03		5:18		5:33		5:48
Gosnells		5:07		5:22		5:37		5:52
Seaforth		5:09		5:24		5:39		5:54
Kelmscott		5:12		5:27		5:42		5:57
Challis		5:14		5:29		5:44		5:59
Sherwood		5:16		5:31		5:46		6:01
Armadale		5:19		5:34		5:49		6:04
Thornlie	5:00		5:15		5:30		5:45	

Source: Transperth

Louise catches the 7:52 a.m. train from Kelmscott to get to work at Burswood.
What time does she arrive at Burswood station?

Solution

The 7:52 a.m. train from Kelmscott does not stop at Burswood. She will need to change trains at Oats Street. She can then catch a train to Burswood.

The train to Burswood arrives at the station at 8:20 a.m.

Exercise 15.03 Timetables

Use the timetable above to answer Questions 1 to 7.

1 At what time does the 7:16 a.m. train from Armadale arrive at Perth?

2 When does the 5:03 p.m. train from Carlisle arrive at Thornlie?

3 Ravi lives in Queens Park and works in Perth. He starts work at 8:45 a.m. What is the latest train he can catch at Queens Park to get to work on time?

4 Ravi finishes work at 4:30 p.m. It takes him 10 minutes to walk to the station.

   a What time can he catch the train home?

   b What time does he get to Queens Park?

   c Three days a week, Ravi catches up with friends after work for half an hour. On these days, what time does he catch the train home?

5 Frances catches the 5 p.m. train from Perth. What time will she arrive at Gosnells?

6 Katarina is going to catch the train at Maddington to meet her friend Vimala at Belmont Park. They have agreed to meet for shopping and coffee at 9:15 a.m. What time does Katarina need to catch the train?

7 It takes 1 hour and 10 minutes to travel by bus from Thornlie to Perth.

   a How much longer does it take by bus than by train?

   b Suggest reasons why the bus might take longer.

This is the bus timetable for Adelaide to Alice Springs and return.

ALICE SPRINGS - ADELAIDE

TOWN	CODE		GX850 DAILY EST	PICK-UP AND SET-DOWN POINT
ALICE SPRINGS	ASP	Dep	10:30A	Greyhound Terminal, Shop 2/76 Todd St
Erldunda	ERL		12:35P	Desert Oaks Resort, Cnr Stuart & Lasseter Hwy
Kulgera	KUL	Arr	1:25P	Kulgera Roadhouse Hotel, Stuart Hwy
		Dep	2:00P	
SOUTH AUSTRALIA				
Indulkna Turnoff	IDK		3:20P	Roadside mailbox at turn off
Marla	MBR		4:00P	Traveller's Rest, Stuart Hwy
Cadney Park	CDY		4:55P	Mobil Roadhouse
Coober Pedy	CPD	Arr	6:30P	Terminal, 52-56 Hutchison St
		Dep	7:25P	
Bilgunnia Turnoff	BGN		8:55P	Turn Off
Glendambo	GBO	Arr	10:10P	Mobil Roadhouse
		Dep	10:30P	
Pimba	PIM	Arr	11:45P	Pimba Roadhouse
		Dep	12:01A	
Port Augusta	PUG		1:50A	Post Office, 50 Commerical Road

Port Augusta Meal Break	PUB	Arr	2:00A	Gull Service Station, Lot 8 Highway
		Dep	2:45A	
Port Pirie	PIR	Arr	3:40A	BP Service Station
Port Wakefield	PWF	Dep	5:00A	BP Service Station, 26 Snowtown Rd
Bolivar	BLV		5:45A	BP Service Station
Cavan	CAV		6:00A	Bus Stop 26, Port Wakefield Road
ADELAIDE	ADL	Arr	6:25A	Greyhound Terminal, 85 Franklin St
Subject to change without notice			Effective: 07 April, 2013	

ADELAIDE - ALICE SPRINGS

TOWN	CODE		GX580 DAILY EST	PICK-UP AND SET-DOWN POINT
ADELAIDE	ADL	Dep	6:00P	Greyhound Terminal, 85 Franklin St
Cavan	CAV		6:20P	Bus Stop 26, Port Wakefield Road
Bolivar	BLV		6:30P	Caltex Service Station
Port Wakefield	PWF		7:25P	BP Service Station, 26 Snowtown Rd.
Port Pirie	PIR		8:40P	BP Service Station
Port Augusta Meal Break	PUB	Arr	9:45P	Gull Service Station, Lot 8 Highway
		Dep	10:30P	
Port Augusta	PUG		10:50P	Post Office, 50 Commerical Rd
Pimba	PIM	Arr	12:45A	Shell Roadhouse
		Dep	1:00A	
Glendambo	GBO	Arr	2:15A	Shell Roadhouse
		Dep	2:30A	
Bulgunnia Turnoff	BGN		3:35A	Turn Off
Coober Pedy	CPD	Arr	5:15A	Terminal, 52-56 Hutchison St
		Dep	5:50A	
Cadney Park	CDY	Arr	7:30A	Mobil Roadhouse
Marla	MBR		8:20A	Traveller's Rest, Stuart Hwy
			9:05A	
Indulkna Turn Off	IDK		9:35A	Roadside mailbox at turn off
NORTHERN TERRITORY				
Kulgera	KUL		11:00A	Kulgera Roadhouse Hotel, Stuart Hwy
Erldunda	ERL	Arr	11:45A	Desert Oaks Resort, Cnr Stuart & Lasseter Hwy
		Dep	12:20P	
ALICE SPRINGS	ASP	Arr	2:30P	Greyhound Terminal, Shop 2/76 Todd St

Source: Greyhound Australia

ISBN 9780170413503

8 How long does the trip from Adelaide to Alice Springs take?

9 How many times during the trip does the bus stop for more than 20 minutes? Where are these stops?

10 Harry joins the bus at Port Augusta to travel to Coober Pedy.
   a How long is his bus trip?
   b When Harry returns from Coober Pedy he will travel all the way to Adelaide. How long will it take?

11 Ariana and Axel are two friends. Ariana lives in Adelaide and Axel lives in Alice Springs. They are planning to meet 'in the middle' to spend the weekend together.
   a Where is the closest stop to 'the middle'?
   b Taking into account the time of day and the arrival time of each person, where would be the best place to meet along this route? Justify your answer.

12 The train from Adelaide to Alice Springs leaves Adelaide at 12:20 p.m. on Sunday and arrives in Alice Springs at 1:45 p.m. on Monday.
   a How long does it take to travel by train from Adelaide to Alice Springs?
   b How does this time compare to the bus trip?
   c Would you prefer to travel by bus or train from Adelaide to Alice Springs? Justify your answer.

This timetable is for CityCat ferries operating on the Brisbane River. Use it for Questions 13 to 15 next page.

Departs Terminal:	a.m.	a.m.	a.m.	a.m.	a.m.	a.m.	a.m.	a.m.	p.m.
Northshore Hamilton	.....	.....	.....	.....	.....	.....	.....	.....	.....
Apollo Wharf	10:25	10:37	10:50	11:02	11:15	11:27	11:40	11:52	12:05
Bretts Wharf	10:28	10:40	10:53	11:05	11:18	11:30	11:43	11:55	12:08
Bulimba	10:34	10:46	10:59	11:11	11:24	11:36	11:49	12:01	12:14
Teneriffe	10:37	10:49	11:02	11:14	11:27	11:39	11:52	12:04	12:17
Hawthorne	10:41	10:53	11:06	11:18	11:31	11:43	11:56	12:08	12:21
New Farm Park	10:46	10:58	11:11	11:23	11:36	11:48	12:01	12:13	12:26
Mowbray Park	10:50	11:02	11:15	11:27	11:40	11:52	12:05	12:17	12:30
Sydney Street	10:53	11:05	11:18	11:30	11:43	11:55	12:08	12:20	12:33
Riverside	11:01	11:13	11:26	11:38	11:51	12:03	12:16	12:28	12:41
QUT Gardens Point	11:09	11:21	11:34	11:46	11:59	12:11	12:24	12:36	12:49
South Bank 2	11:12	11:24	11:37	11:49	12:02	12:14	12:27	12:39	12:52
North Quay	11:16	11:28	11:41	11:53	12:06	12:18	12:31	12:43	12:56
Regatta	11:24	11:36	11:49	12:01	12:14	12:26	12:39	12:51	1:04
Guyatt Park	11:28	11:40	11:53	12:05	12:18	12:30	12:43	12:55	1:08
West End	11:31	11:43	11:56	12:08	12:21	12:33	12:46	12:58	1:11
University of Queensland	11:35	11:47	12:00	12:12	12:25	12:37	12:50	1:02	1:15

Source: CityCat ferrie

13 How long does the ferry take to travel from Bretts Wharf to South Bank 2?

14 a Phillip needs to be at work at a restaurant at Riverside by 11:30 a.m. What is the latest time he can catch the ferry at Teneriffe?

b After work, he travels to the University of Queensland to meet up with friends. How long does the ferry trip take?

c At the end of the evening he returns home. How long will this trip take if he chooses to go by ferry? Assume the reverse journey will take the same length of time as the forward journey.

d To travel from the University of Queensland to Teneriffe by bus takes approximately 55 minutes. Why might people choose to travel by bus rather than by ferry?

15 Why is one part of the timetable grey and one part green?

16 The City Explorer Bus stops at places of interest in the city.

a How many buses are needed to meet the following City Explorer Bus timetable? Explain how you arrived at your answer.

Depart									
Explorer depot	10:00	10:25	10:50	11:15	11:45	12:00	12:25	12:50	1:15
City cathedral	10:08	10:33	10:58	11:23	11:53	12:08	12:33	12:58	1:23
Railway station	10:15	10:40	11:05	11:30	12:00	12:15	12:40	1:05	1:30
Parliament	10:24	10:49	11:14	11:39	12:09	12:24	12:49	1:14	1:39
Museum	10:35	11:00	11:25	11:50	12:20	12:35	1:00	1:25	1:50
City square	10:45	11:10	11:35	12:00	12:30	12:45	1:10	1:35	2:00
Zoo	11:00	11:25	11:50	12:15	12:45	1:00	1:25	1:50	2:15
Dockland shops	11:12	11:37	12:02	12:27	12:57	1:12	1:37	2:02	2:27
Arts centre	11:19	11:44	12:09	12:34	1:04	1:19	1:44	2:09	2:34
Water gardens	11:30	11:55	12:20	12:45	1:15	1:30	1:55	2:20	2:45
Hall of fame	11:38	12:03	12:28	12:53	1:23	1:38	2:03	2:28	2:53
Arrive									
Explorer depot	11:50	12:15	12:40	1:05	1:35	1:50	2:15	2:40	3:05

Source: City Explorer Bus

b Vo, Binh and Vicki came to the city by train, arriving at the station at 11:42 a.m. They caught the City Explorer Bus to the zoo. What is the earliest time they could expect to arrive at the zoo? Explain your answer.

c Manuel and Sofia are dropped off by car at the City cathedral at 10:25am. They arrange to meet their hosts at the Hall of fame at 2:45pm. They want to spend at least half an hour at the museum, photograph the City square and do some souvenir shopping at the Dockland shops. Plan a list of times for them to catch the City Explorer Bus to do these things and meet their hosts on time.

d In summer, extra City Explorer Bus tours leave the depot at 11:30 a.m., 1:30 p.m. and 2:30 p.m. Make a list of departure times that appear in the timetable for each of these tours.

15.04 Australian time zones

Australia has three time zones:

- **Australian Eastern Standard Time (AEST)**, covering Queensland, New South Wales, Australian Capital Territory, Victoria and Tasmania.

- **Australian Central Standard Time (ACST)**, covering South Australia and Northern Territory, $\frac{1}{2}$ an hour behind AEST.

- **Australian Western Standard Time (AWST)**, covering Western Australia, 2 hours behind AEST.

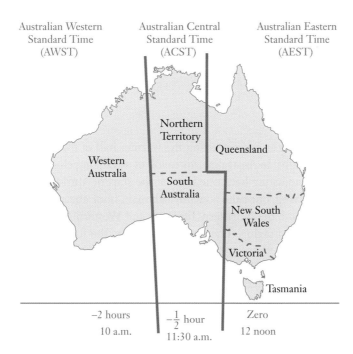

The best way to represent time zones is on a timeline with your location being 0 hours.

EXAMPLE 7

Show the Australian time zones on a timeline with Western Australia as zero hours.

Solution

If Western Australia is 0, then Eastern Australia is +2 hours and Central Australia is $+1\frac{1}{2}$ hours.

0		+2
AWST	ACST	AEST
0	$+1\frac{1}{2}$	+2

When moving east, add on the time difference.

When moving west, subtract the time difference.

EXAMPLE 8

a It is 9 a.m. in Perth. What time is it in Adelaide?

b It is 2 p.m. in Brisbane. What time is it in Adelaide?

Solution

a Moving east from Perth to Adelaide: add the time difference.

$9 \text{ a.m.} + 1\frac{1}{2} \text{ hours} = 10{:}30 \text{ a.m.}$

b Moving west from Brisbane to Adelaide: subtract the time difference.

$2 \text{ p.m.} - \frac{1}{2} \text{ hour} = 1{:}30 \text{ p.m.}$

Daylight saving

Many countries adopt **daylight saving** during the warmer half of the year to take advantage of the extra hours of sunlight. In Australia, this involves advancing clocks by one hour at the start of October and adjusting back one hour at the start of April ('spring' forward, 'fall' back). During daylight saving, our time zones are an extra hour ahead of the standard time zones, except in Queensland, the Northern Territory and Western Australia, the states that do not participate in daylight saving. The map below shows the five time zones in Australia during daylight saving.

EXAMPLE 9

When it is 10:20 a.m. in December in Adelaide, what time is it in Perth?

Solution

In December, Adelaide is on daylight saving time. This means it is $2\frac{1}{2}$ hours ahead of Perth instead of the usual $1\frac{1}{2}$ hours.

Time in Perth $= 10:20$ a.m. $- 2\frac{1}{2}$ hours
$= 7:50$ a.m.

Exercise 15.04   Australian time zones

1 Show the time zones on a timeline with the Northern Territory as zero hours.

2 State whether each location is ahead of, behind, or has the same time as Adelaide.
   a Sydney   **b** Melbourne
   c Darwin   **d** Canberra
   e Geraldton (WA) **f** Mt Isa (Qld)
   g Ballarat (Vic) **h** Ceduna (SA)
   i Hobart   **j** Broome (WA)

3 It is 10:30 a.m. in Adelaide. Find the time in:
   a Cairns (Qld)   **b** Freemantle (WA)   **c** Alice Springs (NT)
   d Launceston (Tas)   **e** Bendigo (Vic)   **f** Broome (WA)

4 What is the time in each city when it is 11 p.m. in NSW?
   a Brisbane   **b** Adelaide   **c** Perth
   d Darwin   **e** Hobart   **f** Canberra

Example 7

Example 8

5 The AFL Grand Final started in Melbourne at 2:15 p.m. and was shown live in Adelaide. What time did the game start on Adelaide TV?

6 a Mick's flight from Sydney to Perth will take 4 hours. He leaves Sydney at 2 p.m. What time does he land in Perth? Give your answer as Perth local time.

b When Mick flies home, he leaves Perth at 11:30 a.m. What time does he land in Sydney? Give your answer as Sydney local time.

7 When Joanna was holidaying at Margaret River in WA, she phoned her parents in Canberra. She rang at 6:30 p.m. Western Standard Time. What time was it in Canberra?

8 Simone, in Townsville, Qld, uses the Internet to talk with her cousins in Broome, WA. At what time should Simone log on to catch her cousins at 8:00 p.m. Broome time?

9 a Use the Internet to find out when daylight saving begins and ends in NSW and Victoria?

b Why do some states have daylight saving?

c How does daylight saving affect the different time zones?

d When it is 12:30 p.m. in Western Australia (not on daylight saving), what time is it in NSW on Eastern Standard Daylight Saving Time?

10 Broken Hill in NSW operates on Central Standard Time rather than Eastern Standard Time like the rest of NSW. Suggest a reason why the people of Broken Hill might have chosen to do this. (Hint: look at a map.)

11 For this question, assume daylight saving is operating.

a When it is 8:30 p.m. in Darwin, what time is it in Hobart?

b What time is it in Adelaide when it is 11:45 a.m. in Melbourne?

c When it is 1:05 p.m. in Sydney, what time is it in Perth?

d What time is it in Brisbane when it is 8:20 a.m. in Darwin?

12 Ben, working in the mines in Western Australia, rings his wife in Victoria during summer. He knows she will have the children in bed by 8 p.m. After what time in Western Australia can he ring so that he can talk to her after the children are in bed?

13 Janine, who lives in Queensland, wants to ring her daughter Alexa in Adelaide before she goes to work to start a new job. She knows Alexa intends to leave for work at 7:45 a.m., so wants to talk to her at 7:15 a.m. At what time in Queensland should Janine ring her?

PLANNING A TRIP

You are visiting one of the state capitals in Australia. You are planning a trip to the central business district (CBD) and to an island off the coast.

1 Choose which capital city you are going to visit.

2 You need to travel from the airport to the CBD. Choose a hotel in the capital city at which to stay. Find out the best way to travel from the airport to the hotel, how long it will take and what it will cost.

3 You are going to visit one of the islands off the coast near the city you have chosen (Rottnest, Kangaroo, Phillip or Magnetic, for example). Find out how to get there from your hotel (train, bus and/or ferry), how long it will take and what it will cost.

4 Choose another tourist attraction in your chosen capital city and find out how to get there from your hotel.

5 Find out if there are any special public transport deals available in your chosen capital city that could save you money.

15.05 Latitude and longitude

Street maps use grid references, such as D7, to locate streets, parks and other features. To locate positions on the Earth's surface, we use imaginary latitude and longitude lines on the globe as a grid.

Parallels of **latitude** are imaginary lines that are parallel to the **Equator**. The North Pole has a latitude of 90° north (90°N).

The Equator is at 0° latitude.

The South Pole's latitude is 90°S.

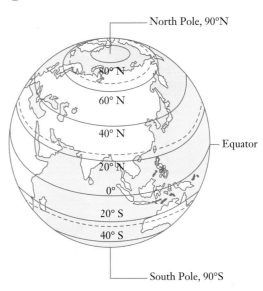

Meridians of **longitude** are imaginary lines running around the Earth from pole to pole. The 0° meridian of longitude runs through the Greenwich Observatory, near London, and is called the **prime meridian** or **Greenwich meridian**.

Greenwich is pronounced 'gren-itch'.

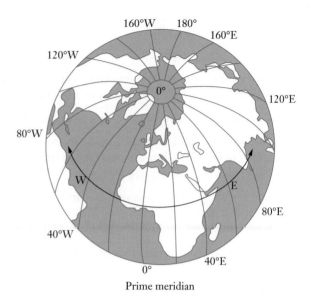

Prime meridian

Longitude is measured in degrees east and west of the prime meridian. The 180°E and 180°W meridians are the same line, on the opposite side of the Earth.

EXAMPLE 10

This map shows the lines of latitude and longitude running across North and South America, Europe and Africa.

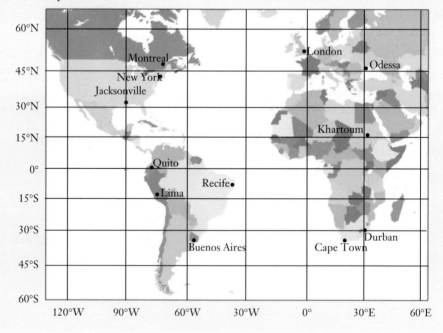

a Is the city of Buenos Aires in Argentina north or south of the Equator?

b Is Buenos Aires east or west of Greenwich?

c What are the position coordinates of Odessa, in the Ukraine?

Solution

a Buenos Aires is below the Equator (0° horizontal line), so it is south.

Buenos Aires is south of the Equator.

b Buenos Aires is to the left of the Greenwich meridian (0° vertical line), so it is west.

Buenos Aires is west of Greenwich.

c Odessa is above the Equator at 45°N and right of Greenwich at 30°E.

The coordinates of Odessa are approximately 45°N, 30°E.

When we write position coordinates we always give the **latitude** (**north** or **south**) first and **longitude** (**east** or **west**) second.

Exercise 15.05 Latitude and longitude

Use the map from the previous page for Questions 1 to 5.

1 Which of the cities Cape Town, Montreal, New York and Recife are:

Example 10

   a north of the Equator?   **b** south of the Equator?

   c east of Greenwich?   **d** west of Greenwich?

2 Write two cities with approximate latitude 45°N.

3 Write three cities with approximate longitude 30°E.

4 Give the approximate latitude and longitude of each city.

   a Jacksonville   **b** Buenos Aires   **c** Recife

   d Khartoum   **e** Greenwich (London)

5 Name the city with each pair of coordinates given.

   a 45°N, 67°W   **b** 40°N, 70°W   **c** 32°S, 26°E

   d 30°S, 30°E   **e** 15°S, 70°W   **f** 0°, 75°W

6 Use this globe diagram to write the position coordinates of:

   a New Caledonia
   b Mt Isa
   c Manila
   d Mauritius
   e Bangkok

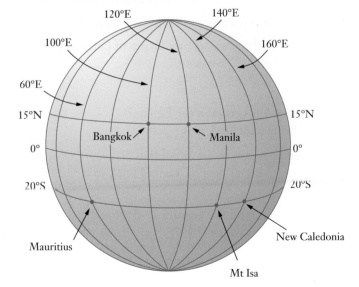

7 When Neil was scuba diving, he discovered the wreck of an old Spanish ship at 15°S, 70°E. Is the wreck closest to New Caledonia, Manila, Mauritius or Bangkok?

World Atlas, GPS coordinates

INVESTIGATION

WHERE ON EARTH ARE YOU?

1 Go to the World Atlas website and search for the Latitude and longitude finder. Use it to find the coordinates of the capital city of each Australian state and territory.

2 Go to the GPS coordinates website to find:

   • the position coordinates of your home and school

   • the places with position coordinates 28°S, 153.42°E and 40.7°N, 74°W.

International time zones

15.06 International time zones

Around the world, the local time of a place is determined by its longitude. Different parts of the world operate on different **time zones**, which are measured relative to the Greenwich Observatory. This map shows international time zones, related to the time along the Greenwich meridian called **Coordinated Universal Time (UTC)** or **Greenwich Mean Time (GMT)**.

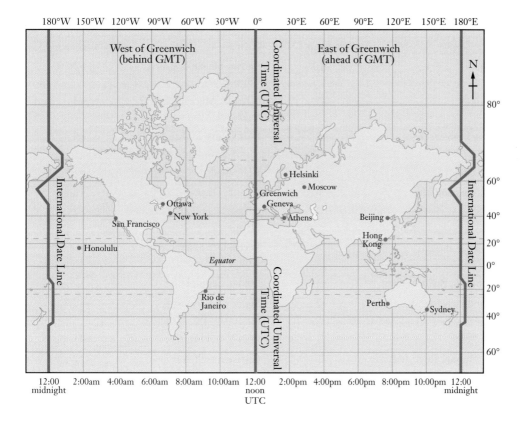

Places west of England (such as the USA) are behind UTC and places east of England (such as Asia) are ahead of UTC.

West (–)   UTC   East (+)

Australian Eastern Standard Time is UTC + 10 hours.

EXAMPLE 11

State whether each country is ahead of or behind UTC.

a Australia   b Canada   c Chile   d China

Solution

a On the map, Australia is east (to the right) of England.   Australia is ahead of UTC.

b USA is west (to the left) of England.   USA is behind UTC.

c Chile is in South America, west of England.   Chile is behind UTC.

d China is in Asia, east of England.   China is ahead of UTC.

Table of
time zones

This table shows the time in various cities in relation to UTC.

City	Hours from UTC	Cities	Hours from UTC
Honolulu	−10	Beirut, Helsinki, Cairo, Cape Town	+2
Anchorage	−9	Nairobi, Mecca, Baghdad, Moscow	+3
Las Vegas, Los Angeles, Vancouver	−8	Port Louis, Dubai	+4
Banff, La Paz	−7	Islamabad	+5
Chicago, Mexico City	−6	Astana, Dhaka	+6
Atlanta, New York, Lima	−5	Jakarta, Bangkok, Hanoi	+7
Bridgetown, Santiago	−4	Perth, Denpasar, Hong Kong, Singapore	+8
Rio de Janeiro, Buenos Aires	−3	Tokyo, Seoul, Dili	+9
South Sandwich Islands	−2	Sydney, Melbourne, Port Moresby, Hagatna	+10
Azores	−1	Port Vila, Honiara	+11
London, Dublin, Reykjavik	0	Christchurch, Suva	+12
Florence, Algiers, Paris, Berlin	+1	Nuku'alofa	+13

EXAMPLE 12

a   What is the time difference between Rio de Janeiro (Brazil) and Christchurch (New Zealand)?

b   When it is 2 a.m. in Rio de Janeiro, what time is it in Christchurch?

c   What time is it in Rio de Janeiro when it is 7 a.m. in Christchurch?

Solution

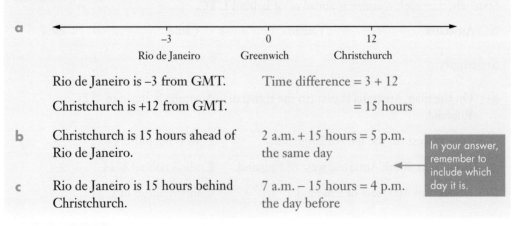

a

−3	0	12
Rio de Janeiro	Greenwich	Christchurch

Rio de Janeiro is −3 from GMT.   Time difference = 3 + 12

Christchurch is +12 from GMT.   = 15 hours

b   Christchurch is 15 hours ahead of   2 a.m. + 15 hours = 5 p.m.
Rio de Janeiro.   the same day

In your answer, remember to include which day it is.

c   Rio de Janeiro is 15 hours behind   7 a.m. − 15 hours = 4 p.m.
Christchurch.   the day before

Exercise 15.06 International time zones

Use the map of time zones and the table of cities to complete this exercise.

1 For each country, state whether it is ahead of or behind Greenwich Mean Time.

   a Alaska   **b** Japan   **c** India

   d Argentina   **e** Hawaii   **f** Saudi Arabia

2 What is the time difference between the following cities?

   a Mecca (Saudi Arabia) and Las Vegas (USA)

   b Lima (Peru) and Melbourne

   c Honolulu (Hawaii) and Buenos Aires (Argentina)

   d Helsinki (Finland) and Suva (Fiji)

   e Banff (Canada) and Florence (Italy)

> Remember a line diagram can help you work it out!

3 Use your answers to Question **2** to answer the following.

   a When it is 9 a.m. in Las Vegas, what time is it in Mecca?

   b When it is 4 p.m. in Mecca, what time is it in Las Vegas?

   c What time is it in Sydney when it is 3 a.m. in Lima?

   d When it is 10 a.m. in Sydney, what time is it in Lima?

   e What time is it in Honolulu when it is 2 p.m. in Buenos Aires?

   f What time is it in Buenos Aires when it is 10 p.m. in Honolulu?

   g When it is 10 a.m. in Suva, what time is it in Helsinki?

   h When it is 10 p.m. in Helsinki, what time is it in Suva?

   i When it is 8 p.m. in Florence, what time is it in Banff?

   j What time is it in Florence when it is 11 a.m. in Banff?

4 a When it is 5 a.m. in Mexico City, what time is it in Helsinki?

   b When it is 4 p.m. in Hong Kong, what time is it in Vancouver (Canada)?

   c What time is it in Algiers (Algeria) when it is 10 a.m. in Honolulu?

   d When it is 1 p.m. in La Paz (Bolivia), what time is it in Reykjavik (Iceland)?

   e If it is 9 p.m. in Atlanta (USA), what time is it in Dhaka (Bangladesh)?

   f When it is 11 a.m. in Chicago (USA), what time is it in Port Vila (Vanuatu)?

5 At 9:00 a.m. in Bridgetown, Wesley phones his cousin in Cairo. What time is it in Cairo?

6 James is playing an online game at 10 p.m. in Perth. He is multi-playing with Juan in Santiago (Chile). What time is it in Santiago?

7 Michelle caught a flight from Melbourne to Christchurch. This flight left Melbourne at 1030.

    a What time was it in Christchurch when Michelle's flight left Melbourne?

    b The flight arrived in Christchurch at 1800 (New Zealand time). How long did Michelle's flight take?

8 Carol flew from New York to Honolulu. Her flight left New York at 1520 on Saturday and took 11 hours to get to Honolulu. At what local time did she arrive in Honolulu?

9 Simon flies from Brisbane to Los Angeles. He leaves Brisbane at 1030 on Saturday. He arrives in Los Angeles at 0900 on the same day (Saturday).

    a How is this possible?

    b How long is the flight from Brisbane to Los Angeles?

10 Michael is conducting business in Melbourne with a company in Jakarta, Indonesia. Melbourne is on daylight saving, while Indonesia is not. Michael needs to organise a video conference for 2 p.m. Indonesian time. What time will it be in Melbourne?

11 Joe is visiting Ouagadougou in Burkina Faso in West Africa. This city does not observe daylight saving. He wants to connect to his friend Steve in Sydney via the Internet. He logs on at 6 p.m. local time.

    a What time will it be in Sydney?

    b Is this a good time for Joe to try to talk to his friend? Why or why not?

12 Alycia is studying in Adelaide. Her brother is working in Baku in the former Russian region of Azerbaijan. It is February and daylight saving is operating in Adelaide. Between what times in Adelaide can Alycia ring him so that she rings between 6 p.m. and 10 p.m. in Baku?

INVESTIGATION

THE INTERNATIONAL DATE LINE

Use the Internet to research the following questions.

1 What is the International Date Line?

2 How is it related to the prime meridian through Greenwich?

3 What happens when you cross the International Date Line travelling west; for example, flying from the USA to Australia?

4 What happens when you cross the International Date Line travelling east; for example, flying from Australia to the USA?

5 Name an island on each side of the International Date Line.

6 Two countries have recently changed where they are in relation to the International Date Line. What are they and why did they make the change?

7 Find and copy a map showing the International Date Line.

ISBN 9780170413503

15.07 Watching international sport

Major sporting events such as the tennis at Wimbledon, the World Cup soccer, Super Rugby and Formula 1 motor races are watched by millions of people in many different countries. We can use time zones to work out when different events will be shown live on TV in Australia.

Exercise 15.07 Watching international sport

Use international time zones to answer the questions.

1 Formula 1 Grand Prix races are held all over the world, 19 races in all. What time do you need to turn on the TV to see the start of each of the following races?

 a Sakhir in Bahrain in April, starting at 6 p.m. local time.

 b Shanghai in China in April, starting at 3 p.m. Shanghai time.

 c Monza in Italy in September, starting at 2 p.m. local time

 d Austin in the USA in November, starting at 2 p.m. local time.

 > Remember to take into account whether daylight saving is operating in Australia or in the country in question.

2 Wimbledon, the oldest tennis tournament in the world, takes place in London at the end of June and the start of July.

 a Each day, matches start at 11 a.m. local time. What time is this in NSW?

 b The semi-finals start at 2 p.m. local time. What time will you need to switch on the TV to watch them?

 c The Men's Final starts at 2:30 p.m. local time. One final lasted 3 hours and 35 minutes. At what time, in NSW, did this final finish?

3 The FIFA World Cup was held in Brazil in June and July 2014. Australia played three other countries in the Group rounds.

 a The Australia versus Chile match was scheduled for 6 p.m. local time. What time was this in NSW?

 b Australia's other two matches against the Netherlands and Spain were scheduled for 1 p.m. local time. What time was this in NSW?

 c The final was played at 4 p.m. local time. The broadcast lasted 2 hours and 45 minutes. What time did it finish in NSW?

4 There are 18 teams in the Super Rugby competition; five from Australia, five from New Zealand, six from South Africa, one from Japan and one from Argentina. For the Australian teams, some games are played in Australia and some are played in the other countries. New Zealand is GMT +12 and South Africa is GMT +2.

 a The Queensland Reds will play the South African Sharks in Brisbane with kick-off at 1605 local time. What time is this for fans in South Africa?

 b The Western Force play the New Zealand Highlanders in Dunedin, New Zealand. Kick-off is at 1935 local time. What time is this in Western Australia for fans who want to watch the game?

 c The South African Bulls play the Melbourne Rebels in Pretoria, South Africa with kick-off at 1910 local time. At what time in Melbourne does the game start?

5 Every four years in July, the Australian Cricket team plays the Ashes Series against England in England. At this time of year, England is on daylight saving time and Australia isn't. The daily schedule is:

- Start of play: 10:30 a.m.
- Lunch: 12:30 p.m. to 1:10 p.m.
- Tea: 3:10 p.m. to 3:30 p.m.
- Close of play: 5:30 p.m.

List the daily schedule in local NSW time.

6 The US Masters Golf Tournament is played each year at Augusta, Georgia in the USA in April. Augusta is GMT −5. In April, Augusta is on daylight saving time.

   a Practice rounds are broadcast from 12 noon to 7 p.m. local time. Peter lives in Hobart and wants to watch practice. Between what times should he be watching?

   b Each of the four days of competition are broadcast from 3 p.m. to 7 p.m. Between what times can you watch the daily rounds?

   c Play finishes at 7 p.m. each day, but, if there is a tie at the end of the competition, extra play-off holes have to be played on the last day. The play-off takes an extra 45 minutes. In your time zone, what time will the play-off finish on the last day?

World Clock

INVESTIGATION

WHERE?

In this investigation, use the World Clock website to answer the following questions.

1 Find three locations that have the same time as you.

2 Find three locations that are 5 hours behind you.

3 Find three locations that are 2 hours ahead of you.

4 What town or city is as far behind you in time as possible?

5 What town or city is as far ahead of you in time as possible?

6 Choose three places you would like to visit and find the time in those places when it is 12 noon where you are.

ISBN 9780170413503

BACKWARDS CROSSWORD

There are no clues to this crossword. Your task is to copy the crossword and write the words below in their correct places in the crossword. The number of letters in each word or phrase is shown in brackets. A different crossword called *World crossword* can be downloaded by your teacher from NelsonNet.

UTC [3]   TIME ZONE [8]   GREENWICH [9]   POSITION
                                                      COORDINATES [19]

EAST [4]   TIMELINE [8]   LONGITUDE [9]   INTERNATIONAL
                                                      DATE LINE [21]

WEST [4]   LATITUDE [8]   SUMMER TIME [10]   STANDARD TIME [12]
NORTH [5]   PARALLEL [8]   24-HOUR TIME [10]
SOUTH [5]   MERIDIAN [8]   DAYLIGHT SAVING [14]
EQUATOR [7]   TIMETABLE [9]

SOLUTION <small>TO THE</small>
CHAPTER PROBLEM

Problem

The FA Cup soccer final is played at Wembley Stadium in London in May. At this time of year, England is on daylight saving time. The match starts at 1700 local time. At what time should TV viewers in Australia tune in to watch the final?

Solution

In London, daylight saving is operating, so the local time is UTC + 1.

NSW and the other Eastern states in Australia are on UTC + 10.

The time difference is 9 hours.

When it is 1700 in London, the time in NSW is 1700 + 9 hours = 0200 or 2 a.m. the next day.

South Australia and Northern Territory are on UTC + 9.5, which is half an hour behind NSW, so their local time will be 1:30 am.

Western Australia is on UTC + 8, which is 2 hours behind NSW, so their local time will be 12 midnight.

TV viewers of the FA Cup soccer final need to tune in at:

- 12 midnight   for Western Australia
- 1:30 am   for South Australia, Northern Territory
- 2 a.m.   for NSW, Queensland, Victoria, ACT and Tasmania

- When have you had to calculate or find out the time at a different location? Why did you need to do it?

- Write a short paragraph describing a funny incident when someone made a mistake at the beginning or end of daylight saving.

- Is there any part of this chapter that you don't understand? If yes, ask your teacher to explain it to you.

Copy and complete this mind map of the topic, adding detail to its branches and using pictures, symbols and colour where needed. Ask your teacher to check your work.

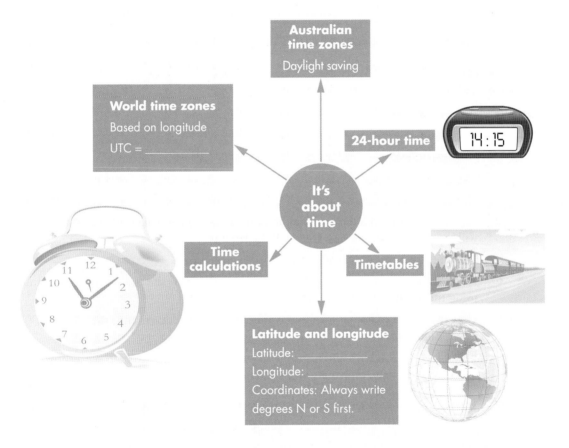

Australian time zones
Daylight saving

World time zones
Based on longitude
UTC = _____

24-hour time

14:15

It's about time

Time calculations

Timetables

Latitude and longitude
Latitude: _____
Longitude: _____
Coordinates: Always write degrees N or S first.

1 Write each time in 24-hour time.

   a   4:20 p.m.   **b**   8:35 a.m.

2 Write each time in 12-hour time.

   a   0725   **b**   1355

3 Calculate the time difference between each pair of times.

   a   8:25 a.m. and 12 noon   **b**   10:36 a.m. and 3:40 p.m.

   c   10:45 p.m. and 2:30 a.m.   **d**   0730 and 1530

4 This timetable shows the train service from Goulburn to Central station.

Goulburn	05:31			07:40	13:20 (Bus)		
Moss Vale	06:29			08:42	14:44	14:53	
Macarthur	07:35	07:38		09:48		16:09	16:13
Campbelltown (arrive)	07:38	07:41		09:51		16:13	16:16
Campbelltown (depart)		07:42		09:52			16:17
International airport		08:20					17:02
Domestic airport		08:23					17:05
Central		08:34	10:34				17:16

   a   Judy is going to catch the 7:40 a.m. train from Goulburn to Central. At what time will the train arrive and how long will the trip take?

   b   How long will this train wait at Campbelltown station?

   c   Judy needs to be in the city before 10 a.m. Describe how she could get there in time using the train.

   d   How much longer does it take to travel by train from Macarthur to the international airport in the afternoon than in the morning at 07:38?

   e   Keith lives in Goulburn and he wants to travel to the city by train to see a show starting at 7:30 p.m. How long will the trip take?

5 a   What time is it in Western Australia when it is 6 a.m. in NSW?

   b   It is 10:15 a.m. in South Australia. What time is it in Tasmania?

   c   Determine the time in the Northern Territory when it is 12 noon in WA.

   d   Sally is talking to her family at home on the phone from her holiday location. At Sally's holiday location it is 2 p.m. and it is 1:30 p.m. at home. Where could Sally's home be?

6 In Australia, Queensland and Western Australia don't follow daylight saving. Calculate the time in each city when the New Year starts in Sydney (that is, at 12 midnight).

   a Melbourne   **b** Brisbane   **c** Perth   **d** Hobart

7 Use the map from Example 10 on page 442 to find:

   a the position coordinates of Jacksonville

   b the city with coordinates 8°S, 35°W.

8 It is 11 a.m. in Canberra. What is the local time in each of the following cities?

   a Christchurch   **b** Bangkok   **c** Atlanta

   d Cape Town   **e** Beirut   **f** New Delhi (UTC + 5.5)

9 a Calculate the time difference between Atlanta and Beirut.

   b What is the time in New Delhi when it is 1340 in Bangkok?

10 Ella likes to record English soccer matches. The game starts at 6 p.m. Friday local time in the UK. Ella lives in Wollongong and it is January. At what time should she set her TV to record the start of the game in the UK?

Practice set 3

Section A: Multiple-choice questions

For each question, select the correct answer A, B, C or D.

Exercise
11.01

1 Given the rule $y = -x - 4$, what is the value of y when $x = -3$?

A   −1   **B**   −7   **C**   7   **D**   12

Exercise
15.02

2 What is the time difference between 10:42 a.m. and 2:13 p.m.?

A   3 h 31 min   **B**   4 h 55 min   **C**   8 h 29 min   **D**   12 h 55 min

Exercise
14.04

3 Describe the shape of the data in this graph.

A   symmetrical

B   bimodal

C   negatively skewed

D   positively skewed

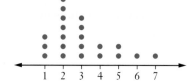

Exercise
12.03

4 Calculate the simple interest earned on $600 at 5% p.a. for 4 years.

A $20   **B** $30   **C** $120   **D** $3000

Exercise
13.04

5 Bill is charged $83.78 per month for comprehensive car insurance. What is the cost per week?

A $19.33   **B** $20.95   **C** $26.66   **D** $29.33

Exercise
15.04

6 Ursula lives in Brisbane and rings her friend in Perth at 1:15 p.m. What time is it in Perth?

A   11:15 a.m.   **B**   12:45 p.m.   **C**   1:45 p.m.   **D**   3:15 p.m.

Exercise
12.01

7 A department store has a mark-up of 200% on clothing. The store buys a vest for $12. What will be its selling price after the mark-up?

A $24   **B** $36   **C** $212   **D** $224

Exercise
14.01

8 Calculate the interquartile range for this set of data.

10   3   5   8   4   2   10

A   3   **B**   5   **C**   7   **D**   10

NCM 11. Mathematics Standard (Pathway 1)   ISBN 9780170413503

9 What is the equation of this line?

A $y = -x + 2$

B $y = x - 2$

C $y = 2 - 2x$

D $y = 2x - 2$

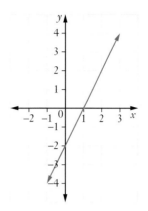

Exercise
11.02

10 Vehicle stamp duty in NSW is 3% for the first $45 000 and 5% on the remainder. What is the stamp duty on a car costing $56 500?

A $1695   B $1925   C $2025   D $2825

Exercise
13.02

11 Use the table in Section 15.06 on page 446 to find the time in Hong Kong when it is 9 p.m. Saturday in Rio de Janeiro.

A 10 a.m. Saturday   B 4 p.m. Saturday

C 2 a.m. Sunday   D 8 a.m. Sunday

Exercise
15.06

12 Danielle invests $4300 at 3% p.a. for 6 months simple interest. How much money will she have at the end of 6 months?

A $774   B $4305.38   C $4364.50   D $5074

Exercise
12.03

Section B: Short-answer questions

1 Graph each linear function on a number plane.

a $y = x - 3$ b $y = 3 - x$

Exercise
11.01

2 Name three countries or cities that are:

a east of Greenwich   b west of Greenwich.

Exercise
15.05

3 a Name two factors that make car insurance more expensive?

b Explain what is meant by a 'no-claim discount'?

Exercise
13.04

4 Terry and Andrea purchased a block of land for $133 000. Six years later, they sell it for $164 000. Calculate their percentage profit correct to two decimal places.

Exercise
12.02

Exercise
14.02

5 Year 11 students completed two tests for their assessment. Both tests were marked out of 50. These are their marks:

Test 1: 48 19 17 45 39 27 40 41 30 23

 38 32 30 27 31 34 36 20 25 22

 40 41 30 46 27 34 31 23 8 38

Test 2: 39 30 20 47 35 35 27 36 34 44

 11 11 47 31 28 32 3 38 7 28

 29 21 32 46 19 30 31 49 17 23

a Draw a back-to-back stem-and-leaf plot for this data.

b Find the range for each test.

c Are there any outliers in either test? If so, state the outliers.

Exercise
14.03

6 Use the data in Question **5** to answer these questions.

a Find the five-number summary for each test.

b Draw a parallel box plot for these two sets of data.

c Write a paragraph describing the similarities and differences between the two sets of results.

Exercise
11.05

7 Use the graph in Test yourself 11 on page 351 to answer these questions.

a Convert $50 to euros.

b Convert €20 to Australian dollars.

c Kate buys a bus ticket in Berlin for €35. How much is this in Australian dollars?

d Calculate how many dollars you would get for €300.

Exercise
13.02

8 Haroon buys a new car for $49 990. How much stamp duty does he pay on his purchase?

Exercise
15.03

9 Use the train timetable in Chapter 15, pages 431–432 to answer these questions.

a Anna catches the 7:57 a.m. train from Gosnells to go to work near Oats Street. What time does she arrive at Oats Street?

b Scott lives in Sherwood. He goes to Cannington to work. He is allowed to arrive at any time between 7 a.m. and 9 a.m. What times could he catch the train?

c Deidre catches the 5:07 p.m. train from Perth. What time can she arrive at Seaforth?

10 Melina measured the height of the 23 students in her Year 11 class (in centimetres):

170   165   159   167   183   174   185   174   168   152   161   163

163   176   186   169   170   149   169   172   186   151   173

a   Find the median of these heights.

b   Find the upper and lower quartiles of these heights.

c   State the five-number summary for this data.

d   Draw a box plot for this data.

11 The fuel consumption for Vicky's car is 8.4 L/100 km.

a   How much fuel is used in travelling 620 km?

b   How far can Vicky travel on 40 L? Answer correct to the nearest whole number.

12 Draw a graph that shows a:

a   symmetrical distribution   **b**   negatively skewed distribution

c   bimodal distribution   **d**   positively skewed distribution.

13 a   A rugby league match starts in Sydney at 4 p.m. What time is this in:

i   Adelaide?   **ii**   Perth?

b   An AFL match starts in Darwin at 11.30 a.m. What time is this in:

i   Canberra?   **ii**   Fremantle, WA?

14 Anita is setting up a plumbing business and purchases new tools costing $1700.

a   Using a straight-line depreciation rate of 22%, how much is the depreciation each year?

b   What is the salvage value of the tools after 3 years?

c   After how many years will the salvage value of the tools be zero?

15 Simon and Maddy are organising the catering for an outdoor wedding reception. The caterer charges $400 plus $55 per person.

a   Copy and complete the table.

Number of wedding guests, n	0	50	100	150	200
Cost of catering, C dollars					

b   Write a formula relating total cost C of catering for n wedding guests.

c   Use your formula to calculate:

i   the cost of catering for 125 wedding guests

ii   the number of wedding guests when the cost was $9200.

d Draw a graph from the table in part **a**.

e What is the gradient of the line in part **d** and what does it represent?

Exercise
11.04

16 Marina works at Newsentry Mall. She is paid according to the formula $W = 23.8h$ where W = wages and h = hours worked.

a How much is Marina paid for 5 hours work?

b What is the constant of variation in this variation equation?

c What does the constant of variation represent in this context?

d How much is Marina paid in a week when she works 37 hours?

e How many hours does Marina need to work to be paid $523.60?

Exercise
15.06

17 The French Open tennis is played in Paris at the end of May and early June. At this time of year, Paris is on daylight saving time.

a Qualifying matches start at 10 a.m. local time. What time is this in NSW?

b You are watching the tennis on TV and the semi-finals start at 10 p.m. What time is it in Paris?

c The Women's Final starts at 2:30 p.m. on Saturday in Paris. The final lasts 2 hours and 35 minutes. At what time in NSW did this final finish?

Exercise
12.06

18 Jude invested $12 000 at 4% p.a. interest compounded annually for 3 years.

a Copy and complete this table to calculate the final value of his investment.

	Interest	Balance
End of the 1st year	$I = Prn$ $= \$12\ 000 \times 0.04 \times 1$ $= \$_____$	$\$12\ 000 + \$\_\_\_\_ = \$_____$
End of the 2nd year	$I = Prn$ $= \$_____ \times 0.\_\_\_ \times 1$ $= \$\_\_\_\_$	$\$\_\_\_\_ + \$\_\_\_\_ = \$\_\_\_\_\_$
End of the 3rd year	$I = Prn$ $= \$\_\_\_\_ \times 0.\_\_\_\_ \times 1$ $= \$\_\_\_\_$	$\$\_\_\_\_ + \$\_\_\_\_ = \$_____$

b How much interest did Jude earn on his investment?

ANSWERS

Chapter 1

Exercise 1.01

1
  a 500 000 people or half a million
  b Sydney, 4 900 000
  c Darwin, 140 000
  d The scale means we can only estimate to the nearest 100 000, but the actual population is more exact than that.
  e Adelaide
  f Teacher to check

2
  a 1.3%
  b ACT
  c Victoria and Northern Territory
  d No, it only tells you the percentage, not the actual numbers.
  e 5174
  f Teacher to check

3
  a Peanut butter
  b Biscuits
  c Cereal
  d 28%
  e Peanut butter
  f Biscuits, cereal
  g 14 g
  h Cereal and biscuits

4
  a Motor vehicle thefts in the country town are decreasing over these 6 years.
  b 2015   c 2013   d 53
  e Teacher to check

5
  a 20°C   b 212°F   c 32°F
  d 38°C

6
  a $120   b $200   c $280
  d $40
  e To cover his business costs, e.g. tools

Exercise 1.02

1
  a Green 31.5c, Octas 33c, System Two 32.5c
  b 1c   c the first graph
  d The scale on the second graph starts at zero; the scale on the first graph only goes from 30 to 33.
  e Green company in order to make their calls seem cheap

2
  a Brisbane   b Darwin   c 116c
  d about 10c
  e Ashwin used a vertical scale from 110c to 130c only and this makes the gaps look bigger.

3
  a The first graph shows the prices on the vertical scale increasing by $2, a large amount for each unit, so it makes the big increases in price look smaller. The second graph shows the 'Increase in price of coffee since 2012' rather than the price, with the vertical scale increasing by 50c, so it makes the increases in price look bigger.
  b The first graph makes the price increase look small, so it may be used by the coffee company. The second graph makes the price increase look big, so it may be used by a competing coffee company.
  c Small: graph the actual population; big: graph the increase in the population.

4
  a 107c   b $69
  c The cost of petrol is in *cent*s, the cost per barrel is in *dollar*s: vertical scale is different for each graph.
  d The two graphs move up and down in a similar way: this suggests the two things are related.

5
  a Australia's average weekly wages are *much* greater than in Malvolia.
  b No vertical scale, using a 3D symbol instead of a column

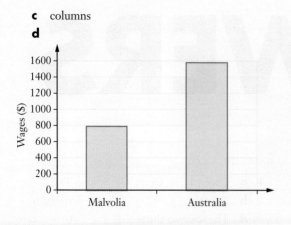

c columns

d

3 Teacher to check

4 Teacher to check

5 a 81   **b** 57   **c** 24

6 a 32   **b** 19

7 a 99   **b** 47   **c** 52

8 a 48   **b** 27

9 a No. He only surveys people who work in the centre of Sydney and are willing to stop and answer questions. He misses country people, people who work elsewhere or people who are too busy to stop.

    b Teacher to check

Exercise 1.03

1 a sample   **b** census   **c** sample

   d sample   **e** sample   **f** census

   g census   **h** sample   **i** sample

   j census   **k** census   **l** sample

2 a students in NSW schools

   b people of voting age in NSW

   c Listeners of a radio station, or readers of a magazine, or musicians

   d students at my school

   e people of NSW

   f the general population

   g Year 12 students

   h wealthy people

   i people who shop at supermarkets

   j the Australian cricket team

Exercise 1.04

1 a systematic   **b** random   **c** stratified

   d self-selected   **e** random   **f** systematic

   g systematic   **h** self-selected   **i** stratified

   j systematic   **k** random   **l** self-selected

   m stratified   **n** random   **o** stratified

2 a Put 400 names in a box and draw 40 out

   b Find how many students are in each Year group and choose an appropriate fraction of the 40 from each Year group

   c Choose every 10th name from a list of the families

   d Send letter or email all parents asking for volunteers to be interviewed

Exercise 1.05

Teacher to check

Exercise 1.06

1 a C   **b** N   **c** N   **d** C   **e** N

   f N   **g** N   **h** N   **i** C   **j** C

2 a N   **b** O   **c** N   **d** N   **e** O

   f O   **g** N   **h** O   **i** N   **j** O

3 a C   **b** C   **c** D   **d** D   **e** D

   f C   **g** C   **h** D   **i** D   **j** C

Exercise 1.07

1 a

Car	Frequency
Holden	8
Ford	7
Toyota	7
Mitsubishi	3
Subaru	3
Other	2
Total	30

   b categorical and nominal

2 a

No. of hamburgers	Frequency
17	5
18	4
19	7
20	4
21	1
22	0
23	1
24	1
25	1
26	1
27	3
28	3
Total	31

b numerical and discrete

3 a 10   **b** 11   **c** 6 letters
   d 55   **e** numerical and discrete

4 a 7   **b** 14   **c** 15
   d 29 months   **e** numerical and discrete

5 a

Heights	Frequency
150–154	2
155–159	4
160–164	3
165–169	5
170–174	5
175–179	4
180–184	2
Total	25

b numerical and continuous

6 a

No. of incidents	Frequency
36–40	4
41–45	6
46–50	5
51–55	12
56–60	2
61–65	2
66–70	4
71–75	0
76–80	0
81–85	0
86–90	1
Total	36

b 90   **c** numerical and discrete

7 a 40   **b** 21   **c** 6
   d numerical and continuous

8 a 50   **b** 3   **c** 13
   d 12   **e** numerical and discrete

Exercise 1.08

1

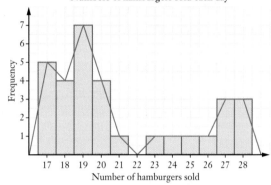

Numbers of hamburgers sold each day

2

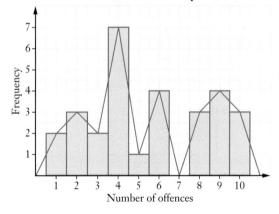

Break-and-enter offences per month

3

Heights of Year 11 students

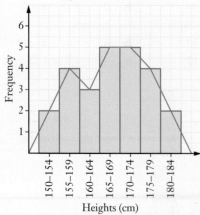

4

Speeds of cars passing school

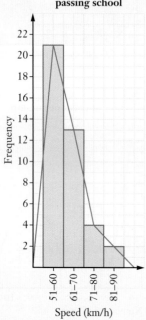

Exercise 1.09

1 a

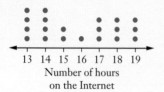

Number of hours on the Internet

b 20   **c** 4   **d** 9

2 a

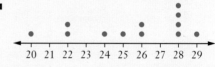

b 3   **c** 25%   **d** 5

3 a

b 12   **c** $28   **d** 5

4 a

Stem	Leaf
7	6
8	1 6 8
9	5 7 8
10	1 5 5
11	2 2 4 7
12	4

Key: 7|6 = 76

b 53%

5 a

Stem	Leaf
1	1 2 2 3 3 4 6 7 7 9
2	0 0 0 2 3 3 4 5 5 5 6 7
3	0 1 3 3 3 4 5 9
4	1 2 8 8
5	5 5

Key: 1|1 = 11

b 8   **c** 22%

6 a

Stem	Leaf
0	0 0 5 8 9
1	1 4 5 6 7
2	1 2 4 5 6 7 9
3	1 3 5
4	2

Key: 1|1 = 11

b 42 hours   **c** 5   **d** 11

e 21   **f** 52%

7 a traffic

b traffic, childcare problems, public transport

c Teacher to check

8 a administration, quality of product

b 75%

c Teacher to check   **d** True

9 a

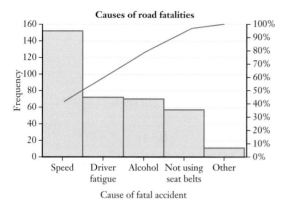

b speed, driver fatigue

Test yourself 1

1 a 2 300 000   **b** 2 200 000

2 Teacher to check

3 Sample, teacher to check reasons

4 a 75   **b** 59   **c** 16

5 a Teacher to check   **b** Teacher to check

6 a C   **b** N   **c** N

   d C   **e** N   **f** C

7 a

Temperature	Tally	Frequency
15	II	2
16		0
17	I	1
18	II	2
19	III	3
20	III	3
21	IIII I	6
22	IIII	4
Total		**21**

b 21 days   **c** 8 days

d, e

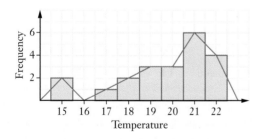

8 a

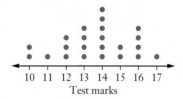

b 23   **c** 7

9 a

Stem	Leaf
2	5 6 8 8 9 9
3	2 3 3 5 5 6 9
4	1 7 8
5	0 0 2 5 6

b 21   **c** 56   **d** 42.9%

10 a hard drive, monitor

    b approximately 80%

    c operating system, keyboard

    d 25

Chapter 2

Exercise 2.01

1 a $6y$ **b** $6h$

   c $12ab$ **d** $7b^2 + b^3$

   e $11p + 2$ **f** $5 + x$

   g $9w - 2$ **h** $25 - 3t$

   i $9k$ **j** $5f + 5f^2$

   k $4a - b$ **l** $4z + 5$

   m $16ab + 2a + b$ **n** $5x^2 - x^3$

   o $10k - 2$

2 a $15xy$ **b** $12h$ **c** $15st$ **d** $\dfrac{5g}{b}$

   e $\dfrac{3t}{2w}$ **f** $\dfrac{2p}{t}$ **g** $\dfrac{6ab}{c}$ **h** $\dfrac{8ab}{5y}$

   i $\dfrac{2pq}{a}$ **j** $15m^3$ **k** $3m^2$ **l** $4p^2$

   m $3l$ **n** $60abc$ **o** $2a$ **p** $2h^2$

   q $2b$ **r** $\dfrac{2}{9}$

Exercise 2.02

1 a $2x + 10$ **b** $6a - 15$

   c $8x + 12y + 8$ **d** $2a^2 + 3ab - 4a$

   e $8a - 12ab$ **f** $3b^3 + 3b^2 + 21b$

2 a $6 + 5b$ **b** $8 + 3x$ **c** $7d + 20$
  d $6x + 24$ **e** $20p + 11q$ **f** $4w + 32$
3 a $-2x - 5$ **b** $-3a + 1$
  c $-4 - 5b$ **d** $-1 + x$
  e $-4p + 3q - 10$ **f** $-x + y - z$
4 a $-6x - 15$ **b** $-8x + 12$ **c** $2ax + 10x$
  d $7a + 14$ **e** $-9 + x$ **f** $-x^2 + 3x$
  g $6d + 16$ **h** $y^2 + 2y$ **i** $2k$

Exercise 2.03

1 a 3000 **b** 1372 **c** 5760
2 1266 **3** 43 **4** 42
5 40 **6** 588.98 **7** 30
8 8000 **9** 5735.16 **10** 10.9
11 25 **12** 10.1

Exercise 2.04

1 3 mL
2 177°C
3 a 285 km **b** 71.25 km
4 a Young's rule: 10.3 mL
   Clark's rule: 7.5 mL
  b Clark's rule because it's based on weight and
     Isabella is underweight for her age.
5 a 0.06 **b** $8119.60
6 51%
7 4140 m³
8 82 km/h

Exercise 2.05

1 $a = 53$ **2** $x = 5$ **3** $p = 33$
4 $d = 4$ **5** $m = 9$ **6** $y = 2$
7 $x = 4$ **8** $a = 3$ **9** $y = 80$
10 $g = 36$ **11** $t = 22.5$ **12** $h = 4$
13 $g = 2.5$ **14** $p = 7.5$ **15** $t = 12.5$
16 $y = 17.5$ **17** $a = 2.4$ **18** $p = 3\frac{2}{3}$
19 $x = 0.12$ **20** $p = \frac{15}{32}$ **21** $d = -4\frac{2}{3}$

Exercise 2.06

1 48.5 cm correct to 1 decimal place
2 $a = 3$

3 25 km
4 18.5 cm
5 150 volts
6 30 mL
7 a 30°C **b** 120
8 a 241 228 km/day
  b 3600 hours, 15 278 km/h
9 23 days
10 a 3 **b** 41
11 a $1368.03 **b** $27 045
12 1.8 s

Exercise 2.07

1 a $5t = 55, t = 11$ **b** $10x = 4.8, x = 48c$
  c $2n = 110, n = 55$ **d** $y - 4 = 6, y = 10$
2 a B, $N = 299$ beetles **b** A, 1140 mL
  c A, 39 boxes **d** D, $N = 53$
3 a $5n - 130 = 300, n = 86$
  b $(a + 15) \times 7 = 294, a = 27$
  c $2l + 34 = 100, l = 33$ cm
  d $(y - 13) \times 6 + 5 = 95, y = 28$
  e $11d \div 2 = 44, d = 8$ cm
  f $200 + x \div 5 = 750, x = 2750

Exercise 2.08

1 $S = \dfrac{D}{T}$ **2** $h = \dfrac{V}{lw}$ **3** $A = \dfrac{150D}{M}$

4 $R = S + 5n$ **5** $r = \dfrac{C}{2\pi}$ **6** $x = \dfrac{y - c}{m}$

7 $d = \dfrac{C - 2}{2.1}$ **8** $W = \dfrac{P - 2L}{2}$ **9** $I = \dfrac{V}{P}$

10 $h = \dfrac{2A}{b}$ **11** $n = \dfrac{A + 360}{180}$

Keyword activity

1 quantities **2** like **3** divide
4 expressions **5** expand **6** term
7 negative **8** formula **9** subject
10 pronumeral **11** value **12** same

Test yourself 2

1 a $13p + 3$ **b** $15 + 4x$ **c** $2w - 12$
 d $4t + t^2$ **e** $80ab + 2$ **f** $4m + 8n$
 g $19m$ **h** $4pq$ **i** $\dfrac{3ab}{y}$
 j 72 **k** $\dfrac{2pq}{t}$ **l** 9

2 a $8a + 6$ **b** $-8 + y$
 c $-x^2 + 5x$ **d** $15p + 27$

3 a 179.6 **b** 25 **c** 11.8

4 a 460 km **b** 28.75 km

5 $1\ 355\ 000$

6 a $x = 44$ **b** $p = 22$
 c $x = 4.36$ to 2 decimal places

7 135.4 m

8 32

9 $T = \dfrac{100I}{PR}$

10 $n = \dfrac{C - 340}{50}$

Chapter 3

Exercise 3.01

1 a $14\ 800$ kJ **b** 22%
2 a 8000 kJ **b** 2400 kJ
3 a 5000 kJ **b** 1000 kJ
4 a 2100 **b** 85.7 **c** 285.6
 d 5.95 **e** 585.7 **f** 2571
5 a 2310 **b** 45
6 a boys **b** 2 to 3 years **c** 3350 kJ
 d 4 to 8 years **e** $56\ 000$ kJ
 f No, 1100 kJ too many **g** increase
7 a 2000 kJ **b** 2 hours
 c **i** $10\ 500$ kJ
   ii No
   iii Eat more or exercise less

Exercise 3.02

1 a 2659 kJ **b** 1463.5 kJ
 c 5691 kJ **d** 725 kJ
2 a Breakfast 4491 kJ, lunch 3180 kJ, dinner 6608 kJ, daily total = 14 279 kJ
 b Unless he is very, very active, he will put on weight.

3 Teacher to check
4 a 3450 kJ **b** 115 minutes
5 a 690 kJ **b** 765 kJ **c** 1020 kJ
 d 3975 kJ **e** 1920 kJ **f** 8050 kJ
6 a 306 kJ **b** 888 kJ **c** 1152 kJ
 d $14\ 754.6$ kJ **e** 76 kg

Exercise 3.03

1 a $T = 5$ **b** $D = 240$
2 a 180 km **b** 4 h
3 275 km
4 8 km
5 a 0.15 h **b** 9 min
6 a 2 km
 b 30 is minutes not hours **c** 4 km/h
7 a 279 m **b** 4.5 s
8 a 5400 m **b** $11\dfrac{1}{9}$ s
9 a 750 km **b** 10 h
 c 76 km/h **d** 5 a.m. Tuesday
10 A

Exercise 3.04

1 32.3 m
2 a 20.2 m **b** 1.7 m **c** slow down
3 a Yes, he could stop in 16.8 m
 b Yes, he would take 18.3 m to stop
4 a 51.0 m **b** 24.0 m
5 He should leave about 34 m, as he needs that distance to stop.
6 Agree. Teacher to check reasoning.
7 a 79.2 km/h **b** 16 km/h
8, 9 Teacher to check

Exercise 3.05

1 0.058 **2** 0.066 **3** 0.204
4 0.0139 **5** 4 drinks
6 a 3 **b** 6 **c** 4 **d** 5
7 2 h
8 12:20 a.m.
9 0.11; No, she will still be under the influence of alcohol.

10 a 6   **b** 0.078   **c** 5 h 12 min

11 a 0.118   **b** 0.194   **c** 0.046

12 a decrease   **b** increase   **c** increase

      d increase   **e** increase

Keyword activity

1 Reaction time   **2** BAC

3 Speed   **4** Velocity

5 Calorie   **6** Kilojoule

7 Stopping distance   **8** Distance

9 Standard drinks   **10** Energy

11 Road surface index

Test yourself 3

1 a 2016 kJ   **b** 667 calories

2 3264 kJ

3 35 km/h

4 a 447 m   **b** 298 m

5 5 standard drinks

Chapter 4

Exercise 4.01

1 a $1330   **b** $2660   **c** $69 160

2 a $6875   **b** $3173.08   **c** $1586.54

3 a $67 600

    b A month is longer than 4 weeks

    c $5633.33   **d** $37.14

4 a $249.76   **b** 4.5 hours

5 a $879.23   **b** $1170.43

6 $15.68

7 $4160

8 a $178   **b** 11

    c Teacher to discuss

9 a Job 3, the salary

    b Teacher to discuss

Exercise 4.02

1 a $25.80   **b** $34.40

    c $21.54   **d** $28.72

    e $36.90   **f** $49.20

    g $46.88   **h** $62.50

2 $123

3 $122.40

4 $272.50

5 a 2   **b** $1060.20

6 a **i** $19   **ii** $28.50   **iii** $38

      iv $399   **v** $114   **vi** $228

    b $741

7 a $18   **b** $81

8 $105

9 a $135.20   **b** 7:30 a.m.   **c** 1 hour

    d 3:30 p.m.   **e** $101.40   **f** 11 a.m.

Exercise 4.03

1 a $962.15   **b** $2290.88

2 $611.25   **3** $1002

4 $1113.30   **5** $791.21

6 $576.25

7 a **i** 8   **ii** 7

      iii 4   **iv** 5

      v $12.32   **vi** $18.48

      vii $18.48   **viii** $24.39

      ix $98.56   **x** $129.36

      xi $73.92   **xii** $121.95

    b $423.79

Exercise 4.04

1 a $61.25   **b** $221.20

    c $581.88   **d** $1206.63

2 a $486.50   **b** $2655.50

3 $3492.10

4 a $60 644.40   **b** $4664.95   **c** $816.37

5 $6536.62

6 $5963

7 $376

8 a $363.69   **b** $5420.89

9 $448

10 a $78.75   **b** Teacher to check

Exercise 4.05

1 a $2250   **b** $40   **c** $6000

    d $12 500   **e** $6570   **f** $570

2 $117.25

3 $4603.50

4 $554.40

5 $10 600

6 a $18   **b** $61   **c** $159.50

7 a $18   **b** $48.91   **c** $214

8 $2235   **9** $88

10 $412 700

11 a $20 440   **b** $43 400

12 a $16.80   **b** $62.40   **c** 417

13 $2875

14 a $204   **b** $945.20   **c** $2363

Exercise 4.06

1 $82

2 $398

3 $112.80

4 $2376.27

5 a $20 745.40   **b** $204.20

6 a $745.30   **b** $658.80

7 Age pension $769 155.40, annual phone allowance $4850.40

8 The pension received ($5235.10) is the correct amount (not $3614.25) and Rose has not been overpaid.

Exercise 4.07

1 a $5800   **b** $111.54

2 a

Income		Expenses	
Office	$620	Rent	$280
Club	$215	Food	$60
		Mobile phone	$20
		Travel	$60
		Total fixed expenses	$420
		Entertainment	$60
		Savings	$315
		Clothes	$40
Total income	$835	Total expenses	$835

Other answers possible for entertainment, savings, clothes.

b

Income		Expenses	
Office	$620	Rent	$280
Club	$215	Food	$60
		Mobile phone	$20
		Car	$175
		Entertainment	$50
		Savings	$210
		Clothes	$40
Total income	$835	Total expenses	$835

3 a $312   **b** $41 080   **c** $16 224

4 a Income = $420, Expenses = $300

   b $1440   **c** Teacher to check

5 a $158   **b** $223

   c Approx. 113 weeks

   d Teacher to check

Keyword activity

1 J   **2** F   **3** G   **4** I

5 K   **6** L   **7** A   **8** C

9 E   **10** H   **11** D   **12** B

Test yourself 4

1 a $656.25   **b** $1312.50   **c** $34 125

2 a $8000   **b** $3692.30

3 a $30.36   **b** $283.36

4 $21   **5** $115.85

6 $875   **7** $6227.50

8 $40 500   **9** $782

10 $21 015.80

11 She should budget. If she saves $80 per fortnight and puts it into a special account she will have a little more than she needed this year to pay these expenses.

Chapter 5

Exercise 5.01

1 a $595.50   **b** $4394

2 a $688   **b** $38   **c** $1175

   d $1121   **e** $311

3 a $617.75   **b** $552.75

4 $186.62

5 a $674.88   **b** $2402.04

6 a $112.40   **b** $574.13

7 =D2-D3

8 $26 936

Exercise 5.02

1 a $70 880   **b** $69 920

2 $415

3 a $105   **b** Refund of $669.50

4 a $2035   **b** $52 910

   c He will get a bill because he paid $260 × 26 = $6760 in PAYG tax, which is less than the $7897.75 tax payable.

5 a $21 112   **b** $20 887

   c $2444   **d** $194 refund

6 She will have to pay $54 more.

7 $77 295.20

8 a $85 020   **b** $25 801.36

   c $84 821

   d Yes, because she paid more tax ($25 801.36) than was required ($19 113.84).

Exercise 5.03

1

	Income tax	Medicare levy	Total tax
a	$6237	$904	$7141
b	$3564.40	$739.20	$4303.60
c	$7675.13	$992.50	$8667.63
d	$32 957	$2450	$35 407
e	$56 932	$3720	$60 652

2 a $38 710   **b** $4127.75   **c** $774.20

   d No   **e** $381.95

3 a $30 274.50   **b** $3745.63   **c** $1440.63

4 a $520   **b** $32

5 Wrong! Australian income tax is progressive. The more your taxable income, the greater proportion you pay in tax. Jacob's tax is $7797 and Garth's tax is $43 132 which is more than 5.5 times Jacob's tax.

6 a Tax = $16 637, Medicare levy = $1544, total $18 181

   b $7881

7 a $2700   **b** $900

   c It's $65 cheaper to have the basic health insurance.

8 $64 250

Exercise 5.04

1 a $4.60   **b** $50.60

2 $380

3 a $7.50   **b** $82.50   **c** $85

   d $23   **e** $1.50   **f** $16.50

4 a $26.40   **b** $264

5 a $5   **b** $50   **c** $15.40

   d $154   **e** $9.60   **f** $96

   g $0.15   **h** $1.50   **i** $2

   j $20

6 $5200

7 C4=C3/11

   C5=C4*10   OR =C3-C4   OR =C3/11*10

8 $7486.36

9 a Multiplying by 1.15 is the same as adding 15%.

   b $NZ120

10 a 2800 euros   **b** 560 euros

11 1200 Danish krone

Keyword activity

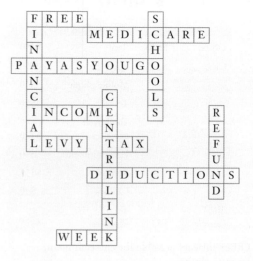

Test yourself 5

1 a $702 **b** $604

2 $89 585

3 Taxable income = $64 600, Medicare levy = $1292

4 a $18 359 **b** $1577

5 Income tax = $20 562, income tax after paying $25 000 into her superannuation account = $12 347, tax saving = $8215.

6 a $52.80 **b** $39.60

  c $25 **d** $180

7 a 49.2 euros **b** 23.2 euros

Practice set 1

Section 1

1 A **2** D **3** A **4** B

5 C **6** B **7** A **8** B

9 C **10** C **11** D **12** D

Section 2

1 a $4ab + 3bc$ **b** $6x^2 - 13x + 2$

  c $6e$ **d** $5p$

2 Teacher to check all parts

3 $2x + 4 = 38, x = 17$

4 $1826

5 a

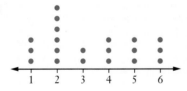

  b 9 **c** 45%

6 a $4192.31 **b** $1172.50

7 $23

8 a

Waiting time	Tally	Frequency
1–5	II	2
6–10	IIII IIII IIII IIII	20
11–15	IIII IIII IIII III	18
16–20		0
21–25	II	2

  b 18 **c** 4.8%

9

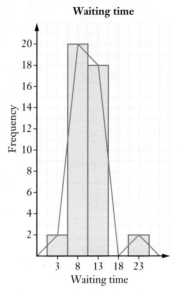

10 $77 210

11 a $17 384.64 **b** $16 640.45

  c Refund, $744.39

12 a $p = 5$ **b** $a = 60$

13 a 1405 kJ **b** 42.15 minutes

14 a $10 150 **b** 140

  c $n = \dfrac{C - 350}{49}$

15 $4362.72

16 29.216 m

17 a $3880 **b** $679 **c** $4559

18 $8991

19 a 225

  b installation, payroll section

  c about 75%

Chapter 6

Exercise 6.01

1 $\dfrac{19}{30}$

2 a $\dfrac{2}{5}$ **b** $\dfrac{3}{5}$

3 $\dfrac{3}{4}$ **4** $\dfrac{1}{2}$ **5** $\dfrac{5}{9}$

6 a i $\dfrac{5}{20} = \dfrac{1}{4}$ **ii** $\dfrac{12}{20} = \dfrac{3}{5}$ **b** $\dfrac{7}{20}$

7 $\dfrac{2}{7}$

8 a 18   **b** $\dfrac{24}{60}=\dfrac{2}{5}$ **c** $\dfrac{18}{60}=\dfrac{3}{10}$

9 a $\dfrac{2}{5}$ **b** $\dfrac{1}{100}$ **c** $\dfrac{81}{100}$

10 a $\dfrac{36}{100}$ **b** $\dfrac{4}{100}$

11 $\dfrac{19}{20}$

12 110% = 1.1 and the maximum value for probability is 1.

13 a $\dfrac{6}{15}=\dfrac{2}{5}$ **b** $\dfrac{3}{5}$

14 a $\dfrac{3}{20}$ **b** $\dfrac{17}{20}$

Exercise 6.02

1 a $\dfrac{1}{5}$ **b** $\dfrac{1}{10}$ **c** 27 cm

2 a 0.08   **b** $\dfrac{17}{25}$

3 a 24   **b** $\dfrac{16}{24}=\dfrac{2}{3}$ **c** $\dfrac{1}{3}$

4 a $\dfrac{5}{40}=\dfrac{1}{8}$ **b** $\dfrac{35}{40}=\dfrac{7}{8}$

5 a $\dfrac{19}{20}$ **b** 0.35

c $\dfrac{1}{10}$ **d** yes, no, no

e Special interest groups who have non-representative views.

6 a $\dfrac{10}{75}=\dfrac{2}{15}$

b 0.28, 0.16, 0.13, 0.15, 0.16, 0.12

c It could be biased. Number 1 appears to occur nearly twice as often as it should. More trials are required before a conclusion can be reached.

7 Answers will differ according to the students' simulations.

Exercise 6.03

1 $\dfrac{2}{5}$ **2** $\dfrac{1}{12}$ **3** 0.65   **4** $\dfrac{2}{3}$

5 a $\dfrac{19}{20}$ **b** $\dfrac{14}{25}$

6 75%   **7** $\dfrac{1}{20}$ **8** 20%

Exercise 6.04

1 a

	1	2	3	4	5	6	7	8
Head	Head, 1	Head, 2	Head, 3	Head, 4	Head, 5	Head, 6	Head, 7	Head, 8
Tail	Tail, 1	Tail, 2	Tail, 3	Tail, 4	Tail, 5	Tail, 6	Tail, 7	Tail, 8

b **i** $\dfrac{1}{16}$ **ii** $\dfrac{9}{16}$

iii $\dfrac{4}{16}=\dfrac{1}{4}$ **iv** $\dfrac{4}{16}=\dfrac{1}{4}$

2 a

	1	2	3	4	5	6
1	2	3	4	5	6	7
2	3	4	5	6	7	8
3	4	5	6	7	8	9
4	5	6	7	8	9	10
5	6	7	8	9	10	11
6	7	8	9	10	11	12

b 36

c **i** $\dfrac{5}{36}$ **ii** $\dfrac{5}{36}$

iii $\dfrac{3}{36}=\dfrac{1}{12}$ **iv** $\dfrac{1}{36}$

d total of 9   **e** 1.5

3 a

	0	1	2	3	4	5
1	1	2	3	4	5	6
1	1	2	3	4	5	6
3	3	4	5	6	7	8
3	3	4	5	6	7	8
4	4	5	6	7	8	9
6	6	7	8	9	10	11

b **i** $\dfrac{5}{36}$ **ii** $\dfrac{1}{36}$ **iii** $\dfrac{9}{36}=\dfrac{1}{4}$

c 6

4 a

	1	2	3	4	5	6
1	0	1	2	3	4	5
2	1	0	1	2	3	4
3	2	1	0	1	2	3
4	3	2	1	0	1	2
5	4	3	2	1	0	1
6	5	4	3	2	1	0

b $\frac{6}{36} = \frac{1}{6}$ **c** 0   **d** 1

5 a

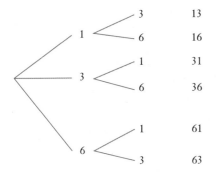

	1	2	3	4	5	6
1	1	2	3	4	5	6
2	2	2	3	4	5	6
3	3	3	3	4	5	6
4	4	4	4	4	5	6
5	5	5	5	5	5	6
6	6	6	6	6	6	6

b $\frac{5}{36}$ **c** 6   **d** 5

6 $\frac{16}{36} = \frac{4}{9}$

Exercise 6.05

1 a

First digit Second digit Outcomes

```
         ┌─ 3    13
    1 ───┤
         └─ 6    16
         ┌─ 1    31
    3 ───┤
         └─ 6    36
         ┌─ 1    61
    6 ───┤
         └─ 3    63
```

b $\frac{3}{6} = \frac{1}{2}$

2 a

First set   Second set   Outcomes
of lights   of lights

```
         ┌─ stop         stop, stop
  stop ──┤
         └─ not           stop, not stop
            stop
         ┌─ stop         not stop, stop
  not ───┤
  stop   └─ not           not stop, not stop
            stop
```

b $\frac{1}{4}$ **c** $\frac{3}{4}$

3 a

First pen   Second pen   Outcomes

```
              ┌─ red      yellow, red
   yellow ────┤
              └─ blue     yellow, blue
              ┌─ yellow   red, yellow
   red ───────┤
              └─ blue     red, blue
              ┌─ yellow   blue, yellow
   blue ──────┤
              └─ red      blue, red
```

b $\frac{1}{6}$ **c** $\frac{2}{6} = \frac{1}{3}$

d

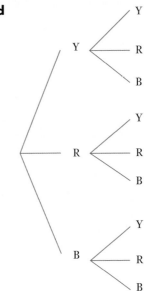

```
              ┌─ Y
        Y ────┤─ R
              └─ B
              ┌─ Y
        R ────┤─ R
              └─ B
              ┌─ Y
        B ────┤─ R
              └─ B
```

e $\frac{3}{9} = \frac{1}{3}$

4 a Teacher to check   **b** $\frac{1}{9}$

5 a Teacher to check

b 6   **c** $\frac{2}{6} = \frac{1}{3}$

Exercise 6.06

Teacher to discuss answers with the class

Exercise 6.07

1 a 0.1

b

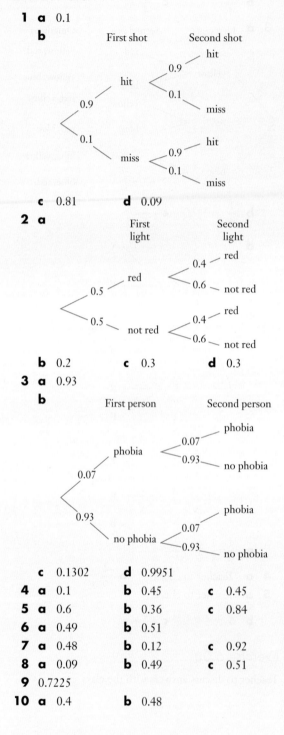

c 0.81 **d** 0.09

2 a

b 0.2 **c** 0.3 **d** 0.3

3 a 0.93

b

c 0.1302 **d** 0.9951

4 a 0.1 **b** 0.45 **c** 0.45

5 a 0.6 **b** 0.36 **c** 0.84

6 a 0.49 **b** 0.51

7 a 0.48 **b** 0.12 **c** 0.92

8 a 0.09 **b** 0.49 **c** 0.51

9 0.7225

10 a 0.4 **b** 0.48

Keyword activity

Theoretical probability is about the long-term chance that something will happen. The theoretical probability of getting a head when we toss a coin is $\frac{1}{2}$. This doesn't mean that every second time we toss a coin we will get a head. Neither does it mean that when we toss a coin 100 times we will get a head 50 times. It also doesn't mean that if we get 8 tails in a row, then the next toss of the coin is more likely to be a head. Coins can't remember what's happened in the past and the chance of getting a head in the future doesn't change because we've had lots of tails. What the probability of a head being $\frac{1}{2}$ does mean is that if we toss the coin thousands of times, about half of the time we'll get a head, but on no individual future occasion can we know what is going to happen.

Test yourself 6

1 a $\frac{1}{8}$ **b** $\frac{3}{8}$ **c** $\frac{3}{8}$ **d** 0

2 $\frac{1}{3}$

3 $\frac{19}{20}$

4 a Teacher to check, 24 **b** $\frac{1}{6}$

5 Teacher to check

6 a 3 or 4

b There's nothing guaranteed in probability. Expectation is a long-term average and is only an indication in the short term.

c No, it's still a $\frac{1}{4}$ probability.

7 a $\frac{9}{25}$ **b** $\frac{4}{25}$ **c** $\frac{21}{25}$

Chapter 7

Exercise 7.01

1 a centimetres or metres

b minutes **c** kilograms

d kilometres **e** days

f millilitres **g** metres

h litres **i** millimetres

j hours **k** grams

l kilolitres **m** tonnes

n seconds **o** milligrams

2 a 1.8 m   **b** 9500 m

  c 13.65 m   **d** 19 800 m

3 a 12 200 L   **b** 6.34 L

  c 0.55 L   **d** 980 L

4 a 1920   **b** 0.55   **c** 9.05

  d 45 000   **e** 15 200   **f** 48

  g 4   **h** 180   **i** 192 000

  j 65.7   **k** 960   **l** 12.3

5 a 3 200 000 g   **b** 4320 min

  c 7.65 km   **d** 0.0568 kg

6 a 500 000   **b** 10 800   **c** 0.45

  d 3.1945   **e** 980 000   **f** 5.46

  g 2   **h** 1 200 000   **i** 3.5

7 a B   **b** C   **c** C   **d** B

8 a 6.6, 13, 8.8, 3.5

  b beans 0.792 m, corn 1.56 m, tomato 1.056 m, silver beet 0.42 m

9 6.75

10 a 720 g   **b** 17.28 kg

  c 345.6 kg, 0.3456 t

11 a 1.5 L   **b** 36

12 2.8 h or 2 h 48 min

Exercise 7.02

1 a 3000   **b** 12 000   **c** 1.5   **d** 2.4

  e 850 000   **f** 0.9   **g** 2500   **h** 0.5

2 a 10 000   **b** 10

3 a 4 mg   **b** 54 g   **c** 2160 mg

4 a 135 g   **b** 135 000 mg   **c** 270 mg

5 20

6 Teacher to discuss answers with the class.

7 360 mg

8 a 11 pounds

  b Yes, its mass is less than 12 pounds.

9 Sum of the dimensions = 40.4 inches, mass = 52.8 pounds. The bag is small enough but it is too heavy.

10 a 1350 kg   **b** 10.8 t

11 1600 g

12 Approximately $41 398

Exercise 7.03

1 a i 1 m   **ii** 4 m

  iii ±0.5 m   **iv** ±12.5%

  b i 0.1 cm   **ii** 1.8 cm

  iii ±0.05 cm   **iv** ±2.8%

  c i 5 km/h   **ii** 50 km/h

  iii ±2.5 km/h   **iv** ±5.0%

  d i 0.5°C   **ii** 38°C

  iii ±0.25°C   **iv** ±0.7%

  e i 5 m/s   **ii** 45 m/s

  iii ±2.5 m/s   **iv** ±5.6%

  f i 500 rpm   **ii** 7000 rpm

  iii ±250 rpm   **iv** ±3.6%

2 a ±0.5 min   **b** ±30 s   **c** ±7.14%

3 a i ±0.5 L   **ii** 17.5 L   **iii** 18.5 L

  b i ±0.5 cm   **ii** 77.5 cm   **iii** 78.5 cm

  c i ±2.5 kg   **ii** 32.5 kg   **iii** 37.5 kg

  d i ±0.05 s   **ii** 9.35 s   **iii** 9.45 s

  e i ±5 km   **ii** 725 km   **iii** 735 km

4 a ±5 m   **b** 235 m and 245 m

  c 245 m

5

	Item	Absolute error	Limits of accuracy
a	Fuel	2.5 L	277.5 L to 282.5 L
b	Fence posts	25 posts	325 to 375 posts
c	Fencing wire	10 m	530 m to 550 m
d	Fertiliser	5 kg	865 kg to 875 kg

6 a 12.16 m   **b** 15.8 kg   **c** 77.86 L

Exercise 7.04

1 370 000

2 a $110 000   **b** $100 000   **c** $105 000

3 a $500 000   **b** $480 000   **c** $482 000

4 a 0.0046   **b** 0.026

  c 0.020   **d** 0.000 47

5 a 4 000 000

  b Any value from 4 245 000 to 4 254 999

  c 3 740 000

6 a 3000   **b** 3   **c** 10 000

  d 0.005   **e** 20   **f** 0.7

7 a 0.67   **b** 0.0055   **c** 1.3
  d 0.087   **e** 790 000   **f** 7.3
  g 24   **h** 2.4   **i** 34
8 a 352 000 kg   **b** 4190 m   **c** 67.1 km/h
  d 14.8 mL   **e** 150 000 000 km

Exercise 7.05

1 a 8.6×10^5 **b** 9.14×10^9
  c 2.01×10^3 **d** 3.6×10^{-4}
  e 1.8×10^{-9} **f** 1.01×10^{-4}
2 a 3 520 000   **b** 35 000
  c 8700   **d** 645 000 000
  e 0.0061   **f** 0.000 019 3
  g 0.000 000 5   **h** 0.01
3 1×10^9
4 1×10^{-7}
5 6×10^7, 53 000 000
6 a 1×10^{12} **b** 640
  c 2.166×10^{10} **d** 351 000
  e 1100   **f** 6×10^7
  g 3×10^9 **h** 3.76×10^{10}
7 1.419×10^{16}
8 4.6×10^{18}
9 Bacteria, 3.4998×10^{-3}
10 111 111 111
11 a 5.3×10^7 **b** 1.5×10^5 **c** 2.5×10^3
  d 4.6×10^{-4} **e** 2.7×10^{-3} **f** 1.0×10^{-1}
12 5.17×10^{11}

Exercise 7.06

1 8.4 m   **2** 23.8 m   **3** 5.831 m
4 152 km/h   **5** 12.37 m   **6** 733 m
7 44.4 m   **8** 148 km
9 Teacher to check

Exercise 7.07

1 a 24 m   **b** 20 m   **c** 34 m
2 a 14 m
  b one 5 m roll and one 10 m roll
  c $48.45
3 30 m
4 a 14.9 cm   **b** 17 cm

5 Any two lengths that add to 18 m
6 a 25 m   **b** 24 cm   **c** 24 mm
7 a 21 m   **b** $98
8 gutter $525.15, pipes $89.80, total = $614.95

Exercise 7.08

1 a 75.4 mm   **b** 47.1 cm
  c 29.8 m   **d** 18.1 mm
2 a 4524 m   **b** 3.5 laps
3 a 46.3 m   **b** 36.8 cm   **c** 17.6 cm
4 46.3 cm
5 336 m

Keyword activity

1 metric   **2** length   **3** kilometres
4 one-thousandth   **5** perimeter   **6** mass
7 tonne

Test yourself 7

1 a 120   **b** 0.75   **c** 8.5
  d 25 000   **e** 5600   **f** 18
  g 2   **h** 110   **i** 266 000
  j 95.7   **k** 900 000   **l** 28 800
  m 0.85   **n** 4.5   **o** 980 000
  p 7.55   **q** 3   **r** 4 200 000
2 a 5600   **b** 18 000
  c 4.5   **d** 6.2
  e 930 000   **f** 0.4
  g 4100   **h** 0.6
3 0.5 g
4 62.5 g and 67.5 g
5 a 740 000   **b** 8 300 000 000
  c 4030   **d** 0.000 72
  e 0.000 003 5   **f** 0.000 21
6 a 7.4×10^5 **b** 8.34×10^9
  c 4.03×10^3 **d** 7.2×10^{-4}
  e 3.4×10^{-6} **f** $2.06 \, 10^{-4}$
7 a 5 600 000   **b** 91 000
  c 0.0041   **d** 0.0001
8 a 18.7   **b** 12.8
9 a 27.5 m   **b** 54 m   **c** 48 cm
10 a 24.7 cm   **b** 42.6 m

Chapter 8

Exercise 8.01

1 a 14   **b** 14   **c** 13.8

2 a 4   **b** 1, 3, 4, 7   **c** 4.3

3 All three measures are lower in the country region. There are possibly fewer people on bail in the country region.

4 a 22   **b** no mode

   c 22.5. The mean best represents the data.

5 a i 31   **ii** 36   **iii** 29.9

   b Each measure is found in a different way; for example, the mode is the most frequent score. The best measure for this set of data is the mean because there are no outliers and every score is involved in calculating the mean.

6 a Mean = 51.1, median = 47.5, modes = 45, 50

   b Mean = 93.1, median = 100, mode = 100

   c Mean = 38.3, median = 39, modes = 39, 41

7 a Median, because it's not affected by outliers

   b Median or mode, because they're not affected by outliers

   c Modes represent the middle scores

8 a B

   b No, as the data is not numerical

9 a $128 167   **b** $50 500

   c Median, as the mean is greater than all but two salaries

   d Mean, as it's higher

   e Every score is different. There isn't a most frequent score.

10 a Mean = $49 400, median = $61 000

   b Median, as in four out of five years the profit was higher than the mean

   c Wanting to minimise tax paid

11 a Mean = $431 125, median = $61 000

   b Median, to attract more customers at a cheaper price

   c Mean

12 a Simon 43.9, Daniel 44.3

   b Daniel

   c Simon 45, Daniel 37

   d Simon, he is more consistent

Exercise 8.02

1 a range = 11

   b $Q_1 = 18, Q_2 = 19, Q_3 = 25$

   c 7

2 a range = 7   **b** $Q_1 = 29, Q_3 = 31$

   c 2

3 a range = 31

   b $Q_1 = 159.5, Q_2 = 169, Q_3 = 174.5$

   c 15

4 a range = 6   **b** $Q_1 = 4, Q_2 = 5.5, Q_3 = 8$

   c 4

5 a Inner Sydney

     i 43

     ii $Q_1 = 20, Q_2 = 25.5, Q_3 = 33$

     iii 13

   South Coast

     i 38

     ii $Q_1 = 52, Q_2 = 60, Q_3 = 64.5$

     iii 12.5

   b South Coast quartiles are much higher, but interquartile range much the same.

Exercise 8.03

1 a $30 000   **b** $44 000

   c 2009, 2011   **d** $36 000 to $64 000

   e 2010

2 a $30 000   **b** 78th percentile

   c $10 000   **d** Median   **e** 9%

3 a 60%   **b** 90   **c** 66 or 67

   d 3rd and 4th deciles

   e Easy, as the median score is 64% and more students scored at the higher end.

4 a 6   **b** October

   c August   **d** five times

   e Poor, as the below average months are very poor. Other answers are possible.

5 a 25%   **b** 95%

   c 188 cm   **d** 5 years

   e i 178 cm or 179 cm   **ii** 182 cm

Exercise 8.04

1 a 28   **b** 1   **c** 12   **d** 94

2 a i $Q_1 = 14$, $Q_3 = 19$, IQR = 5
   ii 6.5, 26.5   **iii** Yes (28 > 26.5)

   b i $Q_1 = 5.5$, $Q_3 = 8$, IQR = 2.5
   ii 1.75, 11.75   **iii** Yes (1 < 1.75)

   c i $Q_1 = 29.5$, $Q_3 = 37$, IQR = 7.5
   ii 18.25, 48.25   **iii** Yes (12 < 18.25)

   d i $Q_1 = 35$, $Q_3 = 49$, IQR = 14
   ii 14, 70   **iii** Yes (94 > 70)

3 a 0, 2, 16   **b** 0, 2, 16

4 a 30, 47, 48, 48, 49, 54, 59, 59, 63, 64, 68, 68, 68, 80

   b 30, 80   **c** 18, 98   **d** No

5 a

Stem	Leaf
2	5
16	1 5 7 7 9 9
17	0 0 1 3 5 6 6 9
18	0 0 2 2 4 5 6
19	7 8
23	0

   b 25, 230

   c 25 cm – definitely a wrongly recorded measurement, as no Year 11 student would be this short. 230 cm – probably a wrongly recorded measurement, as a Year 11 student is unlikely to be this tall.

6 a $1 800 000

   b Reasonable – this could be paid for a particularly large/luxurious house.

   c $696 500   **d** $767 273

   e Median – it is more typical of most prices.

   f $664 000, much closer to the median

7 a $275 000, $890 000

   b Reasonable – this could be the salary of the CEO or overall business manager.

   c $154 000   **d** $70 500

   e Median, 10 out of 12 scores are close to the median.

   f Mean, it's higher   **g** $68 300, Yes

8 a 204 deliveries

   b This doctor could be a specialist in difficult pregnancies and deliveries.

c Mean = 30.65, median = 21.5

d Mean = 21.5, median = 21

e The outlier has a big effect on the mean but only a very small effect on the median.

Exercise 8.05

1 a

Numbers of computers in household	Frequency	Cumulative frequency
0	6	6
1	8	14
2	2	16
3	1	17
4	2	19
5	1	20
Total	20	

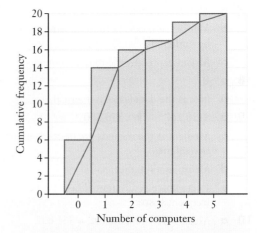

b

Hours of sleep	Frequency	Cumulative frequency
6	3	3
7	9	12
8	11	23
9	6	29
10	5	34
11	2	36
Total	36	

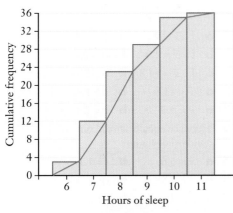

c

Hours spent training	Frequency	Cumulative frequency
1–3	1	1
4–6	3	4
7–9	6	10
10–12	10	20
13–15	14	34
16–18	6	40
Total	40	

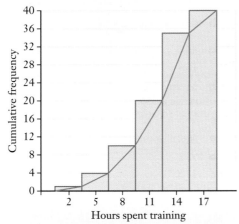

2 a

Class centre	Frequency	Cumulative frequency
10	8	8
30	4	12
50	40	52
70	20	72
90	8	80

b 54 **c** 22

3 a

Class centre	Frequency	Cumulative frequency
175	30	30
225	30	60
275	20	80
325	15	95
375	15	110
425	5	125
475	5	130

b 260 **c** 145

4 a 11.5 **b** 6

Exercise 8.06

1 a

Score, x	Frequency, f	Cumulative frequency	fx
5	4	4	20
6	3	7	18
7	8	15	56
8	4	19	32
Totals	19		126

b i 7 **ii** 7 **iii** 6.6 **iv** 3

2 a

Number of texts, x	Frequency, f	Cumulative frequency	fx
0	8	8	0
1	4	12	4
2	10	22	20
3	10	32	30
4	15	47	60
5	3	50	15
Totals	50		129

b i 3 **ii** 4 **iii** 2.58 **iv** 5

3 a 50

b Mode = 6, median = 5, mean = 4.4, range = 6

c Mean, as it is the lowest.

ISBN 9780170413503

4 a 34

b

Price ($)	Class centre, x	Frequency	Cumulative frequency	fx
5000–<6000	5500	3	3	16 500
6000–<7000	6500	5	8	32 500
7000–<8000	7500	6	14	45 000
8000–<9000	8500	12	26	102 000
9000–<10 000	9500	8	34	76 000
Totals		34		272 000

c i 8000–<9000 **ii** 8000–<9000
iii 8000

5 a 36

b

Fuel cost ($)	Class centre, x	Frequency	Cumulative frequency	fx
10–<20	15	9	9	135
20–<30	25	11	20	275
30–<40	35	5	25	175
40–<50	45	3	28	135
50–<60	55	2	30	110
60–<70	65	1	31	65
70–<80	75	2	33	150
80–<90	85	3	36	255
Totals		36		1300

c i 20–29 **ii** 20–29 **iii** 36.1
6 a 30

b

Marks (%)	Class centre	Frequency	Cumulative frequency	Frequency × class centre
40–49	44.5	4	4	178.0
50–59	54.5	7	11	381.5
60–69	64.5	10	21	645.0
70–79	74.5	3	24	223.5
80–89	84.5	5	29	422.5
90–99	94.5	1	30	94.5
Totals		30		1945.0

c i 60–69 **ii** 60–69 **iii** 64.8

Exercise 8.07

1 a i $\bar{x} = 9$ **ii** $\sigma = 1.41$
b i $\bar{x} = 4.4$ **ii** $\sigma = 2.33$
c i $\bar{x} = 14.8$ **ii** $\sigma = 1.72$
2 a $\bar{x} = 13.8$ **b** $\sigma = 2.4$
3 a $\bar{x} = 4.3$ **b** $\sigma = 2.4$
4 They are equally spread out: same standard deviation.
5 a $\bar{x} = 6.6$ **b** $s = 1.1$
6 a $\bar{x} = 2.58$ **b** $\sigma = 1.52$
7 a $\bar{x} = 4.4$ **b** $s = 1.63$

Exercise 8.08

1 a 15.9%, 14.8%, 107, 29.3% **b** $\dfrac{58}{365}$
c Phone use, distraction, medical emergency
2 a 0.00   0.00   0.02   0.04   0.05   0.07
0.09   0.09   0.12   0.15   0.18   0.20
b Mean = 0.08, modes = 0.00 and 0.09, median = 0.08, range = 0.20
c 3 **d** $Q_1 = 0.03$, $Q_3 = 0.135$
e 0.105 **f** $\dfrac{5}{6}$ **g** 25%
h 0.09, 0.09, 0.12

3 a 46   98   102   102   110   146
168   183   194   208   210   305
b Mean = 156, median = 157, mode = 102, range = 259
c 99 **d** 66.9
e Mean or median, teacher to check reasons
4 a 924 **b** 67%
c 17% **d** Teacher to check
5 a

Class	Class centre, x	Frequency, f	Cumulative frequency	fx
0–<10	5	2	2	10
10–<20	15	11	13	165
20–<30	25	8	21	200
30–<40	35	5	26	175
40–<50	45	3	29	135
50–<60	55	3	32	165
60–<70	65	4	36	260
Totals		36		1110

b 30.8 km **c** 20–<30
d 62 **e** 10–<20

Keyword activity

1 mode   **2** number
3 range   **4** quartiles
5 average   **6** often
7 outlier   **8** central tendency
9 interquartile range
10 order
11 spread
12 ten
13 cumulative frequency
14 percentiles
15 histogram, polygon
16 mean

Test yourself 8

1 a i no mode   **ii** 913   **iii** 830
  b i $40   **ii** $37   **iii** $35.82
  c i 12°C, 20°C, 23°C, 25°C   **ii** 20°C
    iii 19°C
  d i no mode   **ii** 125.5 mm   **iii** 128 mm
2 a 61%
  b 56%, 58%, 59%, 61%, 62%
  c 62.3%
  d Modes not useful. Mean or median are a good reflection of the data.
3 a i 826   **ii** 554, 1064.5   **iii** 510.5
  b i $12   **ii** $32, $40   **iii** $8
  c i 14°C   **ii** 14°C, 23.5°C   **iii** 9.5°C
  d i 168 mm   **ii** 89 mm, 150.5 mm
    iii 61.5 mm
4 a Computer fraud, rainfall, student income (no outliers)
  b Temperatures (outliers)
5 a 30%   **b** 39
  c 8th and 9th deciles
6 a 6
  b September, November, March and June
  c About 210 mm

7 a 300   **b** 962.5
  c It lowers the mean.
8 a 59, 202, 227
  b Mean = 116.9 mm, median = 124 mm
  c Raise the mean
  d Little effect – the medians are close together.
  e Either one of the medians is a good measure of central tendency, as most scores are close to them.

9 a

Number of cars	Frequency	Cumulative frequency
0	2	2
1	7	9
2	8	17
3	2	19
4	1	20
Total	20	

  b

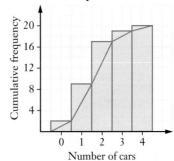

Cars per household

10 a

Class centre	Frequency	Cumulative frequency
3	9	9
8	10	19
13	11	30
18	12	42
23	10	52
28	8	60

  b i 15.5   **ii** 14

11 a

Number of drinks	Frequency	Cumulative frequency	Frequency × score
0	7	7	0
1	5	12	5
2	13	25	26
3	4	29	12
4	1	30	4
5	2	32	10
6	1	33	6
Totals	33		63

    b i 2   **ii** 2   **iii** 1.9   **iv** 6

12 a

Number of seedlings	Class centre	Frequency	Cumulative frequency	Frequency × class centre
1–5	3	3	3	9
6–10	8	6	9	48
11–15	13	6	15	78
16–20	18	9	24	162
21–25	23	7	31	161
26–30	28	5	36	140
31–35	33	4	40	132
Totals		40		730

    b i 16–20   **ii** 16–20   **iii** 18.25

13 5.05

14 a 1.91   **b** 1.50

Chapter 9

Exercise 9.01

1 a 36 cm^2 **b** 55 m^2 **c** 104 m^2
  d 100 m^2 **e** 70 m^2 **f** 96 m^2

2 a 184.8 mm^2 **b** 1386 cm^2 **c** 55 m^2
  d 45 m^2 **e** 20.72 m^2
  f 4.0194 cm^2

3 7.14 cm^2 and 5.4 cm^2

4 Many answers are possible, for example:

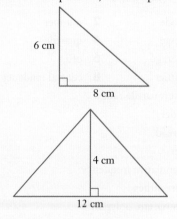

The product of the base and the height must be 48.

5 a 25.44 cm^2 **b** 203.52 cm^2
  c 216.73 cm^2

6 a 176.8 cm^2 **b** 19.84 m^2
  c 875 mm^2 **d** 1000 cm^2

7 a 156 m^2 **b** 500 mm^2
  c 71 m^2 **d** 201.19 km^2

8 a 80 m^2 **b** $576

Exercise 9.02

1 a 452.39 cm^2 **b** 49.87 cm^2
  c 47.78 cm^2 **d** 3019.07 cm^2
  e 117.29 cm^2 **f** 151.32 cm^2
  g 760.27 cm^2 **h** 173.11 cm^2

2 a 2.02 m^2 **b** 1.818 m^2

3 a 113.1 m^2 **b** 56.5 m^2 **c** 84.8 m^2
  d 139.3 m^2 **e** 94.6 m^2 **f** 198.5 m^2

4 a 37.7 m^2 **b** 272.4 m^2

5 251 cm^2

6 3.39 cm

Exercise 9.03

1 a 50 m^2 **b** 140 m^2
  c 140 m^2 **d** 33 m^2

2 Approximately 11.25 km^2

3 a 8 m   **b** 4 m

Exercise 9.04

1 a 29 400 cm^2 **b** 600 mm^2

2 6125 cm^2

3 1.215 m^2

4 a 348 cm^2 **b** 11.33 m^2

5 $12\,680 \text{ cm}^2$

6 a 89.25 m^2 **b** $414

   c 29.25 m^2 **d** $460

7 12.72 m^2

8 a 254 cm^2 **b** 279.4 cm^2

9 71 m^2

10 a 45 cm by 25 cm

   b 1125 cm^2 **c** 3925 cm^2

Exercise 9.05

1 a 604.76 cm^2 **b** 865.70 cm^2

   c 672.36 cm^2 **d** 36.32 m^2

2 345.58 cm^2

3 367.6 cm^2

4 a 154 cm^2 **b** 77 cm^2

5 18 m^2

6 a $514\,718\,540 \text{ km}^2$

   b $360\,302\,980 \text{ km}^2$

   c $154\,415\,560 \text{ km}^2$

7 161 m^2

8 a 3421.19 cm^2 **b** 2376 tiles

9 a 1728 cm^2 **b** 39.27 cm^2

Exercise 9.06

1 48 m^2

2 38.4 m^2

3 6.36 m^2

4 a 2617 cm^2 **b** 2879 cm^2

5 a 7.875 m^2 **b** 40.23 m^2

6 2300 cm^2

Exercise 9.07

1 a 111 cm^3 **b** 2.5 m^3

   c $15\,370 \text{ cm}^3$ **d** 91.125 cm^3

2 a $16\,065 \text{ cm}^3$ **b** $1\,285\,200 \text{ cm}^3$

3 a 5000   **b** 5

4 2000

5 400

6 1.5 L

7 1000 cm^3

8 1250 cm^3

9 a 16.2 m^2 **b** 29.16 m^3 **c** $30\,000$ L

10 a $42\,875 \text{ cm}^3$ **b** $471\,625 \text{ cm}^3$

   c approximately 330 books

11 a 84 cm^3 **b** 240 cm^3

12 a 9.2 m^2 **b** 5.98 m^3 **c** 5980 L

13 a 126 pavers   **b** $4\,500\,000 \text{ mm}^3$

   c 5.67×108 mm

   d 13   **e** $737.10

14 200 m^3

Exercise 9.08

1 a $127\,234.50 \text{ mm}^3$ **b** $36\,945.13 \text{ cm}^3$

   c 1.42 m^3

2 707 cm^3

3 a 3053.6 cm^3 **b** 659.6 mm^3

   c 7238.2 cm^3 **d** 1150.3 mm^3

4 a 0.39 m^3 **b** 390 L

5 $1.1 \times 10^{12} \text{ km}^3$

6 a 431 cm^3 **b** $13\,520 \text{ cm}^3$

   c 3176 cm^3 **d** 23.49%

7 a $28\,148.67 \text{ cm}^3$

   b 5387.83 cm^3 **c** $22\,760.84 \text{ cm}^3$

8 4.6 m^3

9 a 6795 cm^3 **b** 6.8 L

10 a 0.13 m^3 **b** 130 L   **c** 3250 L

   d 2312 L

   e No. He will need to make two trips.

11 a 523.6 m^3 **b** 523 600 L

12 a 1047.2 cm^3 **b** radius 5 cm, height 20 cm

   c 1570.8 cm^3

Keyword activity

Across

3 COMPOSITE   **7** DIMENSION

9 TRAPEZOIDAL   **11** VOLUME

12 LITRE

Down

1 SURFACE   **2** CAPACITY

3 CUBIC   **4** PRISM

5 RECTANGULAR   **6** CYLINDER

8 NET   **10** SPHERE

Test yourself 9

1 a 81 cm^2 **b** 96 mm^2
    c 9.12 cm^2 **d** 150 mm^2

2 a 834.7 mm^2 **b** 47.7 cm^2 **c** 420.5 m^2

3 6000 m^2

4 a 125.5 m^2 **b** 720 m^2

5 a 36 835 mm^2 **b** 1521 cm^2

6 1576 m^2

7 a 768.8 mm^3 **b** 27 720 cm^3

8 a 542.9 mm^3 **b** 137.3 m^3

Chapter 10

Exercise 10.01

1 a $530   **b** $13 780

2 a $432   **b** $22 464

3 a $510   **b** $2210

4 a $780   **b** $1112
    c $1240   **d** $1160

5 $1380

6 a $220   **b** $475

7 a i $950   **ii** $1900
    b $3325   **c** $1108.33   **d** $790
    e $262.33   **f** $456.67
    g She should think carefully. She needs to calculate how much she needs for other expenses such as savings, clothes, transport.

Exercise 10.02

1 Greater Middleton Council

2 10 Feb 2018

3 9 Mar 2018

4 $71.81

5 30 789 999

6 Yes (no arrears)

7 26 Oct 2017 to 24 Jan 2018

8 90 days

9 43 kL

10 478 L

11 Water consumption is highest in summer and then decreases during the year. Highest bill in January, lowest bill in October.

12 Watering gardens, topping up swimming pools, playing water games

13 In person, by post, by credit card (over the phone), using BPAY, at the post office

14 11% p.a. interest accrued daily

15 20 mm

16 0.8 kL = 800 L, Less

17 Water Tariff 1

18 $1.67/kL

19

Average daily use	Tariff	Cost of supply
1.67	$2.25	$337.50
10.89	$1.67	$1636.60
5	$1.67	$751.50
0.96	$2.25	$193.50
6	$1.67	$901.80
1.83	$1.67	$275.55

20 a 5200 L   **b** $8.68

21 a 20 L/min   **b** 900 L
    c 46 800 L   **d** $78.16

Exercise 10.03

1 a 15 L/min   **b** 180 L   **c** 54 L
    d 25 L   **e** 259 L   **f** 94.535 kL
    g 189.07 kL

2 a 33 L   **b** 12.045 kL

3 a 540 L   **b** 77.1 L   **c** 28.08 kL

4 a 5400 L   **b** 771.4 L   **c** 280.8 kL

5 a

	Litres/day	Kilolitres/year
Bathroom	518	189.07
Kitchen	33	12.045
Laundry	77.1	28.08
Outside	771.4	280.8
Total	1399.5	509.995

    b $1300.49   **c** $50.02

Exercise 10.04

1 a 30 L   **b** 90 L   **c** 18 L   **d** 12.5 L

2 12.5 L

3 a 180 L   **b** 25.7 L

4 a 1800 L   **b** 257.1 L

5 a

	Litres/day	Kilolitres/year
Bathroom	120.5	43.9825
Kitchen	12.5	4.5625
Laundry	25.7	9.3805
Outside	257.1	93.8415
Total	415.8	151.767

b $387.01

6 reducing length of shower, water-efficient showerhead, dual-flush toilet, aerator on tap, use dishwasher less, front-loading washing machine, use recycled grey water in the garden

Exercise 10.05

1 $350.83   **2** Yes

3 $269.34   **4** $26.93

5 $296.27   **6** 9.09%

7 Decreased, 3.8 kWh/day

8 a 10 976 kWh   **b** 9012 kWh

9 90

10 a 757 kWh   **b** 922 kWh

11 a $0.063100   **b** $0.189300

12 a $47.77   **b** $174.53

13 domestic

14 $3.36

Exercise 10.06

1

	Appliance	Power rating (W)	kWh per day	Daily cost (cents)	Monthly cost ($)
a	Fridge (600 L)	800	19.2	518	155.40
b	Clothes dryer	2400	3.6	97	29.10
c	Washing machine	900	0.45	12	3.60
d	Bathroom/fan/ heater/light	1100	1.1	30	9.00

e	Incandescent light globe (100 W)	100	0.6	16	4.80
f	Food processor	380	0.19	5	1.50
g	Electric kettle	1500	1.5	41	12.30
h	Stove hotplate (max. setting)	1500	0.375	10	3.00
i	Dishwasher	1900	1.9	51	15.30
j	Toaster	650	0.325	9	2.70
k	Vacuum cleaner	950	0.2375	6	1.80
l	Stereo system	40	0.2	5	1.50
m	Hair dryer	1400	0.7	19	5.70
n	TV	550	3.3	89	26.70
o	Iron	950	0.7125	19	5.70

2–4 Teacher to check

Exercise 10.07

1

	Appliance	Quarterly usage (kWh)	Quarterly cost @ 27c/kWh
a	Fridge	225	$60.75
b	Hot water system	275	$74.25
c	Space heating	864	$233.28
d	Lighting	675	$182.25
e	Washing machine	62.5	$16.88
f	Clothes dryer	37.5	$10.13
g	Oven	324	$87.48
h	Dishwasher	50	$13.50
i	Microwave	60.75	$16.40
j	Toaster	27	$7.29
k	Stove hotplate	216	$58.32
l	Other kitchen	90	$24.30
m	Iron	72	$19.44
n	Hair dryer	81	$21.87
o	TV	118.8	$32.08
p	Other	90	$24.30
q	Total	3268.55	$882.51

2 a

Quarterly usage (kWh)	Quarterly cost @ 19.8c/kWh
275	54.45

b $19.80 **c** 2.2%

3 a i $45.56 **ii** 5.2%

b i

Quarterly usage (kWh)	Quarterly cost @ 27c/kWh
135	$36.45

ii $145.80

iii 16.5%

4 a

Quarterly usage (kWh)	Quarterly cost @ 27c/kWh
324	$87.48

b $145.80

c 16.5%

5 a

	Appliance	Quarterly usage (kWh)	Quarterly cost @ 27c/kWh
i	Clock radio	8.64	$2.33
ii	Computer monitor	10.8	$2.92
iii	Cordless phone	6.48	$1.75
iv	Microwave oven	8.64	$2.33
v	Computer	4.32	$1.17
vi	Printer	17.28	$4.67
vii	Stereo	21.6	$5.83
viii	Television	21.6	$5.83
ix	DVD player	17.28	$4.67
	Total	116.64	$31.49

b $31.49 **c** 3.6% **d** $15.75

6 Off-peak hot water, turning lights off, using CFL globes, reverse-cycle air conditioner, not leaving items on standby

Test yourself 10

1 a $12 220 **b** $1018.33 **c** $470

2 $1410

3 a 11 Nov 2018 **b** 12 Dec 2018

c $277.25 **d** 11c/kL

4 153 L

5 a 720 L **b** 102.9 L

6 51 L

7 a 720 L

b No, it is the same water usage.

8 a $203.91 **b** $18.54

c $185.37 **d** $43.74

9 a 2.75 kWh **b** $0.74

10 a 7.14 kWh **b** $1.93

11 a 47c per day **b** $171.55

12 a $1.10 per day **b** $401.50

Practice set 2

Section 1

1 C **2** B **3** B

4 D **5** D **6** A

7 C **8** A **9** D

10 A **11** B **12** C

Section 2

1 a 6400 mm^2 **b** 49.5 cm^2

2 $7858.50

3 a 14.8 cm **b** 10.8 cm

4 $1555

5 a 6.75 **b** 1.25

6 a 7 **b** 4 **c** 7

7 a i 268.08 cm^3 **ii** 201.06 cm^2

b i 1436.76 cm^3 **ii** 615.75 cm^2

8 a 4.23×10^5 **b** 3.4×10^{-4} **c** 2.4×10

9 a 81 000 **b** 0.00299 **c** 500

10 a 74 kL **b** $207.20 **c** 813 L

11 a 7, 9 **b** $Q_1 = 1, Q_3 = 3, IQR = 2$

c i −2 **ii** 6

d Yes, they are both greater than 6.

12 a

b $\frac{1}{4}$

13 a 135 cm^3 **b** 64 m^3 **c** 264 mm^3

14 a 198 cm^2 **b** 96 m^2 **c** 312 mm^2

15 a 0.167 **b** 0.244

c Teacher to check **d** 20

16

Measurement	Absolute error	Lower limit of accuracy	Upper limit of accuracy
55 cm correct to the nearest cm	0.5 cm	54.5 cm	55.5 cm
75 kg correct to the nearest 5 kg	2.5 kg	72.5 kg	77.5 kg
280 km correct to the nearest 10 km	5 km	275 km	285 km

17 a $2, 3, 6, 9, 9, 15, 22, 26, 27$

b

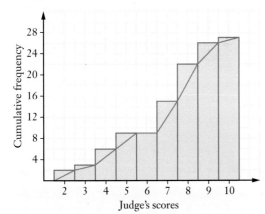

c 7

18 a 235.62 mm^2 **b** 16.08 cm^2

19 61.42 mm

20 a

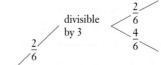
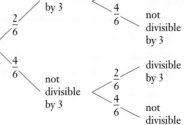

b $\frac{1}{9}$ **c** $\frac{4}{9}$

21 a 691 kWh **b** $\$164.46$ **c** 8 kWh

22 a $\$8984$ **b** $\$9883$ **c** $\$824$

d Teacher to check

Chapter 11

Exercise 11.01

1 a

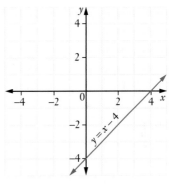

b

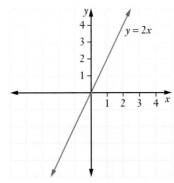

c

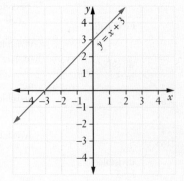

d

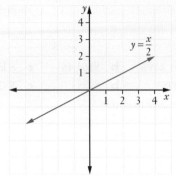

e

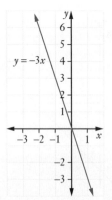

f

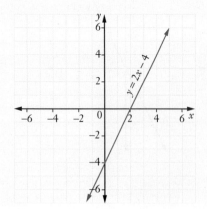

g

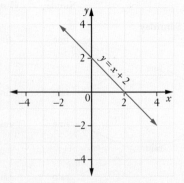

h

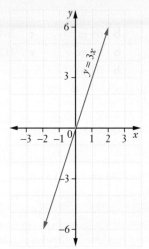

i

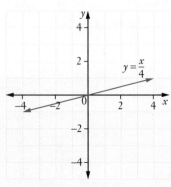

Exercise 11.02

1

	Gradient	y-intercept	Equation
a	$\frac{1}{2}$	2	$y = \frac{1}{2}x + 2$
b	3	−1	$y = 3x - 1$
c	1	6	$y = x + 6$
d	$\frac{1}{2}$	1	$y = \frac{1}{2}x + 1$

2 a $y = -2x + 5$ **b** $y = -\frac{1}{2}x + 4$

c $y = -\frac{1}{2}x + 3$ **d** $y = -x + 5$

3 A

4 a i 3 **ii** 2

b i −2 **ii** 3

c i −1 **ii** 7

d i 4 **ii** 0

e i $\frac{1}{4}$ **ii** −2

5 a positive, $m = 1$ **b** positive, $m = 2$

c positive, $m = 1$ **d** positive, $m = \frac{1}{2}$

e negative, $m = 3$ **f** positive, $m = 2$

g negative, $m = 1$ **h** positive, $m = 3$

i positive, $m = \frac{1}{4}$

6 a C **b** B

c D **d** A

7 a D **b** C

c A **d** B

Exercise 11.03

1 a 40

b The number of chirps when the temperature is zero

c 8

d The additional number of chirps per minute when the temperature increases by one degree Celsius

e $n = 8T + 40$

f 296

g 15°C

h 840, not realistic, teacher to check

2 a 0 **b** 3.5

c Number of degrees cooler at which water boils for every extra km in altitude

d $T = -3.5h + 100$

e about 90°C

f 72, teacher to check

3 a Laura $m = 14$, Bevan $m = 8$, Jacob $m = 3$

b speed

c Laura $D = 14t$, Bevan $D = 8t$, Jacob $D = 3t$

d Jacob; he has the smallest gradient, or he didn't travel as far as the others in 1 hour

e Laura bike, Bevan jog, Jacob walk

4 a $h = 0.6l + 2.95$

b No, the formula says the head circumference should be about 35.08 cm, which it is.

5 a $40 **b** $\frac{1}{4}$

c Additional cost per ticket

d $p = \frac{1}{4}n + 40$ **e** $165

f $D = \frac{1}{2}n + 110$

6 a 100, 130, 160, 190, 210, 240

b Teacher to check

c 70 **d** Delivery charge

e 30 **f** Cost per additional tonne

g Teacher to check: $70 + 20 \times 30 = 670$

h $30 per tonne plus the delivery fee

i $C = 27n + 80$

j The size of the delivery trucks

7 a 48, 57, 66, 75, 84, 93, 102

b Teacher to check: $95 \times 0.9 + 48 = 56.55$

c $C = 0.9n + 48$

d i $161.40 **ii** 147

e Teacher to check

f 48, the fixed cost of the oil

g 0.9, the cost to make a container of chips

h number $\times$ $2.50

i Teacher to check

j Break-even point

k 30

l Teacher to check

8 a $50 **b** $1 **c** $C = n + 50$

d Teacher to check

1 a

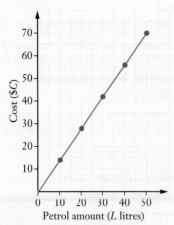

b 1.4   **c** 1.4
d The cost of 1 L of petrol
e $C = 1.4L$ **f** $105
2 a 11 kg   **b,c** Teacher to check
d $\dfrac{1}{6}$ **e** $\dfrac{1}{6}$
f $M = \dfrac{E}{6}$ **g** 264 kg
3 a $148.80   **b** 21.2
c Hourly rate of pay, $21.20
d $784.40   **e** 17 h
4 a $400   **b** 450   **c** 2
d The cost of a ticket
e $A = 2n$ **f** $2750   **g** 6400
5 a

Distance (d km)	100	200	300	400	500
Petrol used (P L)	8	16	24	32	40

b

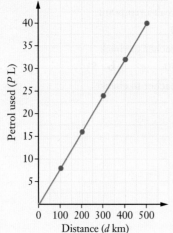

c $\dfrac{2}{25}$ or 0.08   **d** $\dfrac{2}{25}$ or 0.08
e Amount of petrol used to travel 1 kilometre
f $P = 0.08d$ **g** 20 L
h Teacher to check
6 a,b

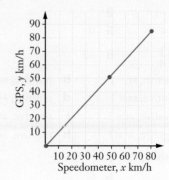

c $\dfrac{17}{16}$ **d** $y = \dfrac{17x}{16}$
e 106.25 km/h
f Yes, he was speeding at 41.4 km/h
g Drive slower than the speed limit

Exercise 11.05

In this set of exercises, the answers from graphs are approximate only.

1 a Teacher to check
b 2.5 m
2 a Teacher to check
b 20 L/100 km
c 2.5 gallons/100 miles
3 a Teacher to check
b 20 euros   **c** $13.50
4 a 0, 7.6, 76, 152
b Teacher to check
c $32.90   **d** $304
5 a 48 km/h   **b** 1.6
c $k = 1.6m$ **d** 160 km/h

Keyword activity

AXIS   GRADIENT   VARIATION
CONVERSION   INTERCEPT   *Y*-INTERCEPT
DIRECT   LINEAR   CONSTANT
EQUATION   MODELLING   TABLE OF VALUES

Teacher to check meanings

Test yourself 11

1 a

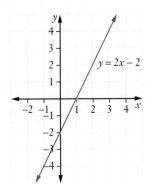

b

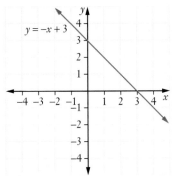

2 a i 1   **ii** −3   **iii** $y = x − 3$

  b i −2   **ii** 1   **iii** $y = −2x + 1$

3 a i 4   **ii** −3

  b i −1   **ii** 2

  c i $\frac{1}{3}$ **ii** 0

4 a 3 min   **b** 2

  c Time to pack a box

  d 33 min   **e** $T = 2n + 33$

5 a 0, 4, 8, 12, 16

  b $C = 4n$ **c** 0

  d Income for 0 mugs of coffee

  e 4   **f** Income per mug

6 a $3600   **b** Teacher to check

  c Teacher to check

  d 30   **e** 30

  f Cost per square metre

  g $R = 30S$ **h** $55 500

7 a €10.5   **b** $70   **c** $50

  d Yes, as €25 ≈ $35   **e** €154

Chapter 12

Exercise 12.01

1 a 152 kg   **b** $2650   **c** 157.5 m

  d 13.3 L   **e** $64.50   **f** 1440 L

  g 105.05 kg   **h** 714 students

2 $35.37   **3** $679.15   **4** $234

5 $20 790   **6** 1157

7 a $22.50   **b** $877.50

8 $158 080

9 $563.33

10 a $39.57   **b** $1939.13

11 $172.01

12 $1032

13 a $516.25   **b** $567.88

  c Because prices are usually rounded to the nearest dollar or $10.

14 a i $670   **ii** $549.40

  b $494.46

Exercise 12.02

1 a 28.6%   **b** 29.6%

  c 83.1%   **d** 55.8%

2 a 18.6%   **b** 40.2%

  c 12.0%   **d** 15.4%

3 87.5%

4 45.5%

5 a Profit, $15   **b** 11.1%

6 a Loss, $500   **b** 11.8%

  c Keira might be happy as she used the car for a year.

7 $272.25

8 a $23 985   **b** $21 100

  c Loss   **d** 12%

9 a $315   **b** $324

  c Profit   **d** 2.9%

10 45.8%

Exercise 12.03

1 **a** $130.20 **b** $109.20
  c $412.25 **d** $34.88
2 **a** $45.83 **b** $442.74
  c $17.55 **d** $1.66
3 0.39%
4 **a** 0.7% **b** 0.162% **c** 4.2%
  d 0.023% **e** 0.323% **f** 2.1%
5 **a** $864 **b** $3264
6 **a** $6.69 **b** $371.69
7 $18.75
8 **a** $3610 **b** $4455 **c** $150.50
9 $294 150
10 **a** $30
  b **i** $4770
     ii It was probably a scam and he would lose his $30 000.

Exercise 12.04

Answers from graphs are approximate.

1 **a** $24 **b** $6
  c $92
  d A little lower than the 8% line.
  e The bigger the gradient, the bigger the interest rate.
2 **a** $X = 6\%, Y = 5\%, Z = 4\%$
  b $2200 **c** Yes **d** 6
3 $A = 120 $B = 240 $C = 360
4 **a** $360 **b** 12% p.a.
5 **a** $P = 500$ $r = 0.08$
  b $500 \times 0.08 = 40$
  c 0, 40, 80, 120, 160, 200, 240
  d Teacher to check
6 **a** $160 **b** $60
  c $25 **d** It's the principal.

Exercise 12.05

1 **a** 2400 **b** 2810
2 **a** 6400 **b** 7 **c** 57
3 **a** $6250 **b** $75 000
  c 12 **d** 18 years
4 **a** $111 525 **b** $297 400 **c** 7

5 **a** $900 **b** $2700 **c** 3
6 20%
7 **a** $20 \times 10 = 200$
  b 5% **c** $50
8 **a**

Age in years, n	Yearly depreciation, D	Salvage value, S
0	$0	$2400
1	$240	$2160
2	$240	$1920
3	$240	$1680
4	$240	$1440
5	$240	$1200

b

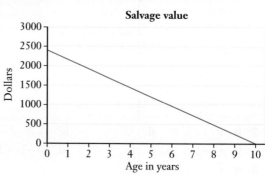

Salvage value

c 8 years **d** The graph is a straight line.

Exercise 12.06

1

	Interest	Balance
End of the 1st year	$I = Prn$ $= $8000 \times 0.06 \times 1$ $= 480	$8000 + $480 $= $8480
End of the 2nd year	$I = Prn$ $= $8480 \times 0.06 \times 1$ $= 508.80	$8480 + $508.80 $= $8988.80
End of the 3rd year	$I = Prn$ $= $8988.80 \times 0.06 \times 1$ $= 539.33	$8988.80 + $539.33 $= $9528.13

2 a $480   **b** $12 480

c $12 979.20   **d** $13 498.37

e $1498.37

3 a $4260.10   **b** $260.10   **c** $4.10

4 a

	Interest	Balance
End of the 1st month	$I = Prn$ $= \$14\,000 \times 0.007 \times 1$ $= \$98$	$\$14\,000 + \$98 =$ $\$14\,098$
End of the 2nd month	$I = Prn$ $= \$14\,098 \times 0.007 \times 1$ $= \$98.69$	$\$14\,098 + \$98.69 =$ $\$14\,196.69$
End of the 3rd month	$I = Prn$ $= \$14\,196.69 \times 0.007 \times 1$ $= \$99.38$	$\$14\,196.69 + \$99.38$ $= \$14\,296.07$

b $296.07

5 a $9600   **b** 0.8%

c 9.6%   **d** $154.21

Exercise 12.07

1 a $9142   **b** $1642

c No, it will be bigger because the amount in Claire's account is bigger. Claire will get 'interest on her interest'.

2 $1552

3 a It multiplies the value in B9 (the principal) by the interest rate.

b Adds the interest in C9 to the principal in B9

c either =B29-B9 or =SUM(C9:C29)

4 a

Number of years	0	1	2	3	4	5	6
Interest earned during the year	0	$240	$257	$267	$278	$289	$300
Value of the investment at the end of the year	$6000	$6420	$6677	$6944	$7222	$7511	$7811

b

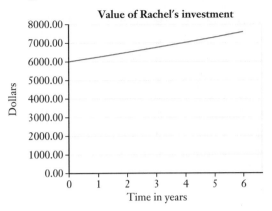

c

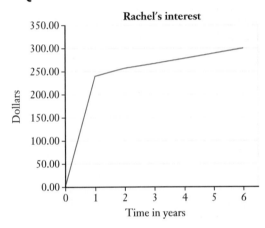

5 a Simple interest graphs are straight lines and compound interest graphs curve up to the right.

b 7.5 years

6 a $400   **b** Nick X, Adam Y

c Nick, by $35

7 The amount of interest increases each year. The increase has to be the same each year for the graph to be a straight line.

ISBN 9780170413503

Keyword activity

1 D **2** A **3** F
4 B **5** C **6** E

Test yourself 12

1 a $504 **b** $296.40
2 $18.43
3 a $3250 **b** 88%
4 72%
5 a $46.20 **b** $270
6 a

Year	0	1	2	3	4	5	6
Amount in the account at the end of the year	$2800	$2884	$2968	$3052	$3136	$3220	$3304

b

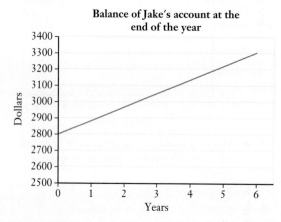

Balance of Jake's account at the end of the year

7 $2860
8 a $75 **b** $1.88

Chapter 13

Exercise 13.01

Teacher to check

Exercise 13.02

1 $414
2 $899.70
3 $1110
4 $1890

5 a $2099.95 **b** $3165.95
6 $20 099.50
7 a

Vehicle	Vehicle price	Stamp duty
1	$5000	$150
2	$20 000	$600
3	$35 000	$1050
4	$45 000	$1350
5	$55 000	$1850
6	$65 000	$2350
7	$80 000	$3100

b

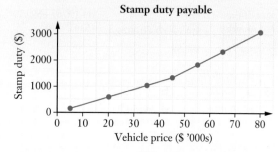

Stamp duty payable

8 a $420 **b** $15 145
9 a $30 **b** $420
 c More money is collected

Exercise 13.03

1 a $16 092 **b** $2092
2 a $45 **b** $1620
3 a $73 320 **b** $23 320
4 a $63 504 **b** $13 504 **c** $9816
5 a $628 **b** $30 144 **c** $5144
6

	Interest rate	Monthly repayment	Total amount repaid	Interest
a	7% p.a.	$396	$23 761	$3761
b	9% p.a.	$415	$24 910	$4910
c	11% p.a.	$435	$26 091	$6091

The higher the interest rate, the higher the total amount repaid.

7 a

Loan term (years)	Monthly repayment	Total repaid	Interest
4	$445	$21 378	$3378
5	$371	$22 262	$4262
6	$322	$23 169	$5169

b Teacher to check

Exercise 13.04

1 $1252.80

2 $168.35

3 a $1604.25   **b** $962.55

   c No, 50% would be $891.25

4–6 Teacher to check

Exercise 13.05

1 8.4 L/100 km

2 a 7 L/100 km   **b** 4.50 L/100 km

   c 9.50 L/100 km

   d 9.01 L/100 km

   e 11.16 L/100 km

3 a 8.75 L/100 km

   b 7.27 L/100 km

   c Yes

4 a 39.405 L   **b** $63.87

5 a 40.95 L   **b** 36.45 L   **c** 33.75 L

   d 22.5 L   **e** 30.6 L

6 a 136.5 L   **b** $197.93

   c 1092 L   **d** $1583.40

7 800 km

8 a 736 km   **b** 820 km   **c** 855 km

   d 470 km   **e** 887 km

9 a i 927.5 L   **ii** $1344.88

   b i 14 L   **ii** 264 km

10 a i 957 km   **ii** 692 km

   b i 705 L   **ii** 975 L   **iii** $405

   c 9 years

11 a 2000 L   **b** $3000   **c** 2592 L

   d $2332.80   **e** $667.20

   f 2.6 years ≈ 3 years

Exercise 13.06

Teacher to check

Keyword activity

1 D   **2** F   **3** I

4 A   **5** H   **6** E

7 C   **8** G   **9** B

Test yourself 13

1 $599.70

2 $2449.50

3 a $11 268   **b** $1268

4 a $50 568   **b** $10 568

5 $1093.40

6 6.0 km/100 L

7 a 47.8 L   **b** $69.31

8 a 705.9 km   **b** about 650 km

9 a $84.62   **b** $93.08

10 $175.75

Chapter 14

Exercise 14.01

1 a 17, 18, 19, 25, 28

   b

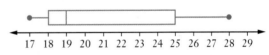

2 a 25, 29, 31, 31, 32

   b

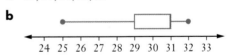

3

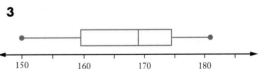

4 a Inner Sydney: 12, 20, 25.5, 33, 55
South Coast: 36, 52, 60, 64.5, 74

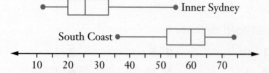

b Inner Sydney, lower risk of theft.

5 a 6, 13, 21, 24, 41

b

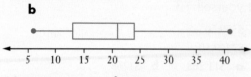

6 a 22   **b** 35
c i 15   **ii** 45
d 18

7 a 76, 88, 101, 112, 124

b

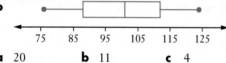

8 a 20   **b** 11   **c** 4
d 15   **e** 30

Exercise 14.02

1 a

10 Green		10 Blue
	3	5 8 9 9 9
	4	0 1 2 2 2 6 7
	5	1 4 6 7 9
9 7 5 2 0	6	0 1 5 5 5 6 8
9 8 6 5 1 1 1 0	7	1 2 4
9 9 9 8 5 5 4 2 1 1 0	8	8
5 5 4 3 1	9	

b 29   **c** 10 Green: 81, 10 Blue: 56.5
d 10 Green: 35, 10 Blue: 53
e an outlier in 10 Blue (88)
f 10 Green: top, 10 Blue: middle

2 a

Year 11		Year 12
	0	7 8
5 5 5	1	0 2 3 3 4 5 8 9 9
9 5 3 3 2 2 0	2	0 0 1 2 2 3 4 4 6
9 1 0 0 0 0 0	3	
4 2 1	4	

b Year 11: 29.5, Year 12: 19
c Year 11: 29, Year 12: 19
d No outliers
e Clustered in the 20s
f Medians very different
g Number of students in each year

3 a

Male		Female
	15	0 2 7 9
8 7 5 0	16	0 2 3 4 4 5 6 6 6 7 8
8 8 8 5 2 0 0	17	1 3 3 6 6 7 8
4 4 3 3 1 1 1 1 0 0	18	5 5
1 0 0	19	
0	20	

b 25   **c** 24
d Males: 160, 171, 180, 183.5, 200
Females: 150, 162, 166, 173, 185
e 200 is an outlier in the male group
f Male median is higher than the female median.
g Teacher to check

4 a

Broome		Kiama
	20	3 7
	21	0 8
	22	1 3 5 5
	23	0
	24	5 7 9
	25	0 4 5 6 6 9
	26	4 7
	27	0 6 9
	28	
5 3	29	6
	30	
9 9 8 7 7 2 2 0 0 0	31	
8 7 7 5 3 3 2 1 1	32	5 6
7 6 4 4 1	33	0
3 2	34	
	35	
	36	3

b 28

c Broome: 32.15°, Kiama: 25.45°

d Broome: 5°, Kiama: 16°

e Broome: 29.3°, 29.5°; Kiama: 36.3°, 33.0°, 32.6°, 32.5°

f Broome temperatures are close together, Kiama temperatures are widely spread out

g Teacher to check

h Teacher to check

Exercise 14.03

1 a

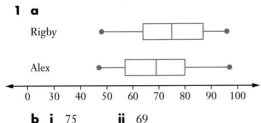

b i 75   **ii** 69

c i 48   **ii** 50

d Alex; his mark is 16% above the median, whereas Rigby's is only 10% above the median.

e Rigby's class; more than 50% of the scores are above the median for Rigby's class.

f No, we don't have any scores or the numbers of students in each class.

2 a Chatphone: 3, 7, 7.5, 9 10; Oztel: 5, 7, 9, 10, 11

b

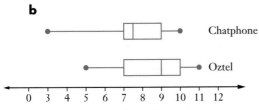

c Chatphone: 7.5, Oztel: 9

d Chatphone: 7, Oztel: 9

e Chatphone: 2, Oztel: 3

f Yes, Chatphone's median is 1.5 minutes less than Oztel's and 75% of Chatphone's callers wait less than 50% of Oztel's callers.

g Teacher to check

3 a Middleton: 60, 67.5, 72.5, 82.5, 90
Blakewell: 58, 68, 72, 75, 82

b

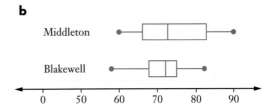

c Middleton: 72.5 km/h, Blakewell: 72 km/h

d Middleton: 72, 75, 85; Blakewell: 68, 70, 72, 73, 75

e Middleton: 74.2 km/h, Blakewell: 71.5 km/h

f Middleton: 15 km/h, Blakewell: 7 km/h

g No, the mean in Blakewell is lower and 25% of the sample in Middleton was faster than the entire Blakewell sample.

4 a 25

b Langley Lynx: 49, Blakely Bears: 50

c Langley Lynx: 66, Blakely Bears: 70

d Langley Lynx: 43, 51.5, 66, 75.5, 92
Blakely Bears: 44, 57, 70, 87, 94

e

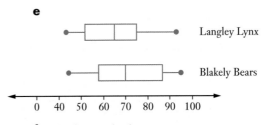

f Teacher to check

g Blakely Bears

5 a 60   **b** 61   **c** Soft drink

d Coffee   **e** Teacher to check

Exercise 14.04

1 a symmetrical **b** not symmetrical

c symmetrical **d** not symmetrical

2 a unimodal, negatively skewed

b unimodal, positively skewed

c bimodal **d** unimodal, positively skewed

3 a, b and **d**

4 Many answers are possible. Teacher to discuss.

a

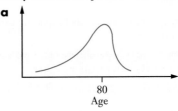

b

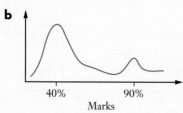

Marks

c

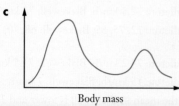

Body mass

d

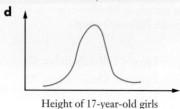

Height of 17-year-old girls

e

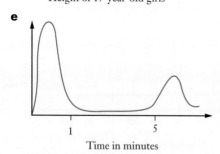

Time in minutes

5 The tail extends on the right to 29, and half of the data lies from 5 to 9.

6 Teacher to discuss. The diagram shows an example answer.

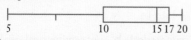

7 Many answers possible. Teacher to check

8 a Has three peaks

b Many possible answers. It could be a coffee shop. Peak times correspond to people going to work, plus morning and afternoon tea times.

9 a Bimodal

b Many possible reasons, including the three given below.

- There are two warehouse employees and one works faster than the other.

- The information is from two different times; for example, weekdays and weekends. The weekend employees are part-time only and can't locate items as quickly as the full-time employees.

- Some items are more difficult to access than others.

10 a Negatively skewed

b Easy, most people got high marks

c Around 85

d Average to just below average, but below the mode.

Keyword activity

1,2 Teacher to check

3 a B   **b** D   **c** E

   d C   **e** A

Test yourself 14

1 a 2, 6, 10.5, 13, 18

b

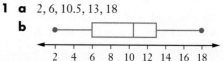

       2   4   6   8   10 12 14 16 18

2 a Median = 26.5

b 14

c 9 employees

3 a

Bulls		Tigers
	7	2
9 3	8	
8 0	9	7
9	10	4 9
4 3	11	
7 5	12	6
7	13	0 3 8 9 9
0	14	4 7
9	15	

b 12

c Bulls: 113.5, Tigers: 131.5

d Bulls: 159, Tigers: 72

e Tigers; higher median and scores clustered in the 130s.

4 a

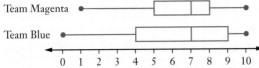

Team Magenta
Team Blue

0 1 2 3 4 5 6 7 8 9 10

b Median is 7 for both teams.

c Team Magenta: 3, Team Blue: 5

d Adele; she is in the top 25% of her team, whereas Shane is at the 75% mark in his team.

5 a positively skewed

b bimodal

c symmetrical

6

a

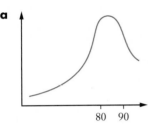

80   90

b

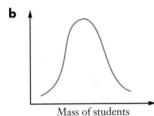

Mass of students

c

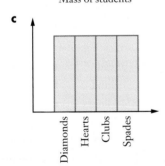

Diamonds Hearts Clubs Spades

Chapter 15

Exercise 15.01

1 a 1144   **b** 1835   **c** 0251   **d** 2154

2 a 8:45 a.m.   **b** 1:20 p.m.

  c 11:31 p.m.   **d** 10:45 a.m.

3 a 11:23 a.m.   **b** 5:54 p.m.   **c** 1:16 p.m.

4 a 5:20 a.m.   **b** 10:05 a.m.   **c** 2:45 a.m.

5 a 3:15 p.m.   **b** 8:31 p.m.   **c** 11:55 p.m.

6 a   **b**

c

7 a 7:15 a.m.   **b** 5:30 p.m.   **c** 9:50 p.m.

  d 11:20 a.m.   **e** 0500   **f** 1410

  g 1718

8 a 2030   **b** 2205

9 5:30 p.m.

10 a 8 hours

  b 3 hours 35 minutes   **c** 0305

Exercise 15.02

1 a 5 h 45 min   **b** 5 h 44 min

  c 2 h 20 min   **d** 3 h 53 min

2 a 3 h   **b** 10 h

  c 5 h 15 min   **d** 8 h 10 min

3 9 h 30 min

4 6 h 25 min

5 a 8:00 p.m.   **b** 7:43 p.m.

  c 1:17 p.m.   **d** 2:24 a.m.

6 a 2:15 p.m.   **b** 2:35 a.m.

  c 11:25 a.m.   **d** 7:50 a.m.

7 3:15 p.m.

8 7:45 p.m.

9 a 3 years 8 months

  b 4 years 1 month

  c Teacher to check

10 1 year 4 months

11 August 2011

12,13 Teacher to check

Exercise 15.03

1 7:50 a.m.

2 5:15 p.m.

3 8:10 a.m.

4 **a** 4:52 p.m.   **b** 5:08 p.m.   **c** 5:22 p.m.

5 5:22 p.m.

6 8:45 a.m.

7 **a** 47 min longer
   b Teacher to check

8 20 h 30 min

9 There are four stops longer than 20 min at 9:45 p.m., 5:15 a.m. (next day), 8:20 a.m., 11:45 a.m.

10 **a** 6 h 25 min   **b** 11 h

11 **a** approximately Bulgunnia Turnoff
    b Teacher to check

12 **a** 25 h 25 min   **b** Longer
    c Teacher to check

13 44 min

14 **a** 11:02 a.m.   **b** 34 min
    c 58 min   **d** Teacher to check

15 green is a.m. times, grey is p.m. times

16 **a** Five buses, bus 1 can do the 10 a.m. and noon trips
    b 12:45 p.m., first bus after 11:42 a.m. is 12 noon
    c,d Teacher to check

Exercise 15.04

1

	NT	
AWST	ACST	AEST
$-1\frac{1}{2}$	0	$+\frac{1}{2}$

2 **a** ahead   **b** ahead
   c same   **d** ahead
   e behind   **f** ahead
   g ahead   **h** same
   i ahead   **j** behind

3 **a** 11 a.m.   **b** 9 a.m.   **c** 10:30 a.m.
   d 11 a.m.   **e** 11 a.m.   **f** 9 a.m.

4 **a** 11 p.m.   **b** 10:30 p.m.   **c** 9 p.m.
   d 10:30 p.m.   **e** 11 p.m.   **f** 11 p.m.

5 1:45 p.m.

6 **a** 4 p.m.   **b** 5:30 p.m.

7 8:30 p.m.

8 10 p.m.

9 **a–c** Teacher to check   **d** 3:30 p.m.

10 Teacher to check

11 **a** 10 p.m.   **b** 11:15 a.m.
    c 10:05 a.m.   **d** 8:50 a.m.

12 after 5 p.m.

13 6:45 a.m.

Exercise 15.05

1 **a** Montreal, New York
   b Recife, Cape Town
   c Cape Town
   d Montreal, New York, Recife

2 Montreal, Odessa

3 Odessa, Khartoum, Durban

4 **a** 32°N, 90°W **b** 34°S, 55°W **c** 8°S, 40°W
   d 15°N, 30°E   **e** 50°N, 0°

5 **a** Montreal   **b** New York
   c Cape Town   **d** Durban
   e Lima   **f** Quito

6 **a** 20°S, 160°E   **b** 20°S, 140°E
   c 15°N, 120°E   **d** 20°S, 60°E
   e 15°N, 100°E

7 Mauritius

Exercise 15.06

1 **a** behind   **b** ahead   **c** ahead
   d behind   **e** behind   **f** ahead

2 **a** 11 hours   **b** 15 hours   **c** 7 hours
   d 10 hours   **e** 8 hours

3 **a** 8 p.m. same day   **b** 5 a.m. same day
   c 6 p.m. same day   **d** 7 p.m. day before
   e 7 a.m. same day   **f** 5 a.m. next day
   g Midnight   **h** 8 a.m. next day
   i Midday same day   **j** 7 p.m. same day

4 **a** 1 p.m. same day   **b** Midnight
   c 9 p.m. same day   **d** 8 p.m. same day
   e 8 a.m. next day   **f** 4 a.m. next day

5 3 p.m. same day

6 10 a.m. same day

7 a 1230   **b** $5\frac{1}{2}$ hours

8 2120

9 a Los Angeles is 18 hours behind Brisbane
   b 16 hours 30 minutes

10 6 p.m.

11 a 5 a.m.   **b** Probably not; too early

12 11:30 a.m. to 3:30 p.m.

Exercise 15.07

1–3 Teacher to check

4 a 0805   **b** 1535
   c 0310 next day

5 Teacher to check

6 a 2 a.m. to 9 a.m. the next day
   b, c Teacher to check

Keyword activity

Across

7 East   **9** Timetable

11 Longitude   **13** Time zone

15 Greenwich   **16** Daylight saving

17 UTC   **18** South

19 West   **20** Parallel

Down

1 International Date Line

2 Latitude

3 Meridian

4 Timeline

5 Position coordinates

6 24-hour time

8 Summer time

10 North

12 Standard time

14 Equator

Test yourself 15

1 a 1620   **b** 0835

2 a 7:25 a.m.   **b** 1:55 p.m.

3 a 3h 35 min   **b** 5h 4 min
   c 3h 45 min   **d** 8 h

4 a 10:34 a.m., 2 hours 54 minutes
   b 1 minute
   c Catch the 5:31 a.m. train and change trains at either Macarthur or Campbelltown
   d 7 minutes   **e** 3 hours 54 minutes

5 a 4 a.m.   **b** 10:45 a.m.
   c 1:30 p.m.   **d** NT or SA

6 a midnight   **b** 11 p.m.
   c 9 p.m.   **d** midnight

7 a 32°N, 90°W   **b** Recife

8 a 1 p.m.   **b** 8 a.m.
   c 8 p.m. previous day
   d 3 a.m.   **e** 3 a.m.   **f** 6:30 a.m.

9 a 7 hours   **b** 1210

10 5 a.m. Saturday

Practice set 3

Section 1

1 A   **2** A   **3** D

4 C   **5** A   **6** A

7 B   **8** C   **9** D

10 B   **11** D   **12** C

Section 2

1 a

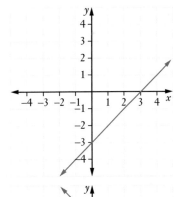

b

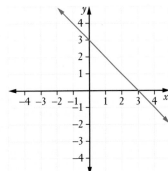

2 Teacher to check

3 Teacher to check

4 23.31%

5 **a**

Test 1	Stem	Test 2
8	0	3 7
9 7	1	1 1 7 9
7 7 7 5 3 3 2 0	2	0 1 3 7 8 8 9
9 8 8 6 4 4 2 1 1 0 0 0	3	0 0 1 1 2 2 4 5 5 6 8 9
8 6 5 1 1 0 0	4	4 6 7 7 9

b Test 1: 40, Test 2: 46

c Test 1: 8, Test 2: none

6 **a** Test 1: 8, 25, 31, 39, 48

Test 2: 3, 21, 30.5, 36, 49

b

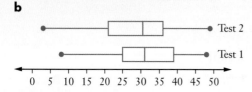

c Teacher to check

7 **a** €35　　**b** $30　　**c** $50　　**d** $420

8 $1599.50

9 **a** 8:09 a.m.

b 7:18 a.m., 7:48 a.m., 8:18 a.m.

c 5:39 p.m.

10 **a** 169 cm　　**b** $Q_L = 163$ cm, $Q_U = 174$ cm

c 149, 163, 169, 174, 186

d

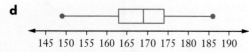

11 **a** 52.08 L　　**b** 476 km

12 Teacher to check all parts

13 **a** **i** 3:30 p.m.　　**ii** 2 p.m.

b **i** 12 noon　　**ii** 10 a.m.

14 **a** $374　　**b** $578　　**c** 5 years

15 **a**

Number of wedding guests, n	0	50	100	150	200
Cost of catering, C dollars	400	3150	5900	8650	11 400

b $C = 400 + 55n$

c **i** $7275　　**ii** 160

d

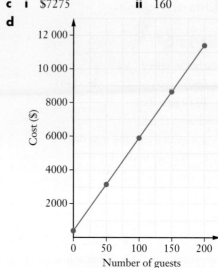

e 55, the number of guests

16 **a** $119　　**b** 23.8

c amount paid per hour

d $880.60　　**e** 22 hours

17 **a** 6 p.m.　　**b** 2 p.m.

c 1:05 a.m. Sunday

18 **a**

	Interest	Balance
End of the 1st year	$I = Prn$ $= \$12\,000 \times 0.04 \times 1$ $= \$480$	$\$12\,000 + \$480$ $= \$12\,480$
End of the 2nd year	$I = Prn$ $= \$12\,480 \times 0.04 \times 1$ $= \$499.20$	$\$12\,480 + \$499.20$ $= \$12\,979.20$
End of the 3rd year	$I = Prn$ $= \$12\,979.20 \times 0.04 \times 1$ $= \$519.17$	$\$12\,979.20$ $+ \$519.17$ $= \$13\,498.37$

b $1498.37

GLOSSARY AND INDEX

12-hour time Time of day written in the usual way using a.m. or p.m. and the hours 1 to 12; for example, 9.27 p.m. (p. 426) *See also* **24-hour time**.

24-hour time Time of day written using 4 digits (instead of a.m. or p.m.) and the hours 0 to 23. For example, 1745 is the 24-hour time for 5.45 p.m. (p. 426) *See also* **12-hour time**.

absolute error The maximum possible error for a measurement, equal to $\pm\frac{1}{2}$ of the smallest unit marked on the measurement scale. (p. 185) *See also* **limits of accuracy**.

allowable tax deduction A part of a person's yearly income that is not taxed, such as work-related expenses or donations to charities. All deductions are subtracted from yearly income to determine **taxable income**. (p. 126).

allowance (government) Money paid by the government to support individuals for specific purposes (for example, to support the aged, unemployed, disabled, students and parents). (p. 109)

allowance (worker's) Money paid to a worker for expenses incurred as part of his or her job (for example, for travel, for special clothing, or for working in isolated or dangerous areas). (p. 100)

annual leave loading Extra payment to a worker based on a percentage (usually 17.5%) of 4 weeks annual leave. (p. 103)

Australian Central Standard Time (ACST) Standard time zone (UTC + 9.5) for central Australia: the Northern Territory and South Australia. (p. 437)

Australian Eastern Standard Time (AEST) Standard time zone (UTC + 10) for eastern Australia: NSW, Queensland, Victoria, ACT and Tasmania. (p. 437)

Australian Western Standard Time (AWST) Standard time zone (UTC + 8) for Western Australia. (p. 437)

bar chart (or **column graph**) A graph consisting of vertical or horizontal bars. (p. 4)

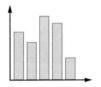

base (of a **prism**) One of the parallel end faces of a prism. (p. 274)

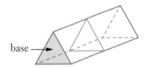

bias In statistics, an unwanted influence that stops a sample from being representative of a population. (p. 16)

bimodal distribution A statistical distribution with two peaks. (p. 415). *See also* **unimodal distribution**.

blood alcohol content (**BAC**) The concentration of alcohol in a person's blood, measured in g/100 mL. (p. 80)

bonus Extra pay for achieving high quality or a volume of work, such as meeting an important quota, goal or deadline. (p. 100)

box plot (or **box-and-whisker plot**) A diagram that displays the quartiles of a set of data as a box and the extremes as whiskers. (p. 404)

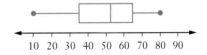

budget A plan for managing money. (p. 113)

calorie An older, non-metric unit of energy, equal to 4.2 kilojoules (kJ). (p. 67)

capacity Maximum volume of liquid that can be held by a container, usually measured in millilitres (mL), litres (L) or kilolitres (kL). (p. 276)

categorical data Information or data represented as a category rather than as a number (for example, the makes of cars, or the colours of eyes). Differs from **numerical data**. (p. 18)

census Collection of information about every member of a population. (p. 10)

circumference The perimeter of a circle. $C = \pi d$ or $C = 2\pi r$, where C is the circumference, π is pi (3.141 59 …), d is the **diameter** and r is the **radius** (see diagram at **diameter**). (p. 200)

class centre The centre of a class interval. For example, the class centre of the class interval 10–19 is 14.5. (p. 229)

class interval In statistics, when there are many data scores, they may be grouped into class intervals. For example, ages of people may be grouped into class intervals of 1–10, 11–20, 21–30, and so on. (p. 21)

column graph See **bar chart**.

commission The earnings of a salesperson or agent; usually a percentage of the value of items sold. (p. 105)

complementary event All the outcomes that are not the event. For example, the complementary event to rolling a 1 on a die is rolling a number that is not 1. (p. 155)

compound interest Interest paid on the principal invested as well as on any accumulated interest. Differs from **simple interest**. (p. 369)

comprehensive insurance Insurance that covers all damage to vehicles and property, including your own, in an accident in which you are at fault. (pp. 382, 388) See also **Compulsory Third Party insurance** or **Third Party Property insurance**.

Compulsory Third Party insurance or **CTP insurance** Insurance that covers personal injury or death to another person ('third party') in an accident in which you are at fault. Also called 'green slip' because the insurance certificate is green. (pp. 382, 388). See also **comprehensive insurance** or **Third Party Property insurance**.

constant A value that does not change. (p. 330) See also **variable**.

constant of variation (or **constant of proportionality**) The constant in a variation equation. For example, if y varies as x, the equation is $y = kx$ and the constant of variation is k. (p. 340)

continuous data Numerical data that can be measured on a smooth scale of values (without 'gaps'), such as heights of people. (p. 18)

conversion graph A graph that is used to convert between different units, such as between metric and imperial units of measurement, or between currencies in foreign currency exchange. (pp. 6, 345)

Coordinated Universal Time (**UTC**) See **UTC**.

cumulative frequency A running total of frequencies. (p. 228)

cumulative frequency histogram A histogram in which the height of each column represents the cumulative frequency of each score. (p. 228)

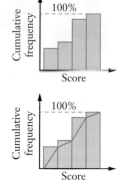

cumulative frequency polygon A line graph formed by joining the ends of the tops of the columns of the **cumulative frequency histogram**. (p. 228)

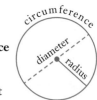

data Observations or facts which, when collected, organised and evaluated, become information. (p. 210)

daylight saving Scheme where clocks are turned forward an hour to take advantage of increased hours of daylight during the warmer half of the year. (p. 438)

deciles Values that divide a data set into 10 equal parts when the scores are arranged in order. (p. 218). See also **percentiles** and **quartiles**.

deduction (tax) See **allowable tax deduction**.

deductions (from pay) Amounts taken out of a person's gross pay (for example, union fees, superannuation, income tax). (p. 124)

depreciation Loss in value of an item or asset over time. (p. 366)

diameter The length of the interval passing through the centre of a circle and joining two points on the **circumference** of the circle. The diameter is double the **radius**. (p. 200)

direct linear variation (or **direct proportion**) The relationship between two variables (say x and y) by an equation of the form $y = kx$, where k is the **constant of variation**. (p. 340)

discrete data Numerical data that can be counted and whose values are separate and distinct, such as the numbers of pets owned, or the numbers of people in families. (p. 18)

ISBN 9780170413503

distribution The way the scores of a data set are arranged, especially when graphed. (p. 415). *See also* **shape of a distribution**.

dot plot A graph that uses dots to show frequencies of data scores (for example, the temperatures, in °C, of 10 hospital patients). Different from a **stem-and-leaf plot**. (p. 27)

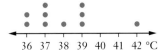

double time Wages paid at twice the normal rate (for example, for working on a Sunday or a public holiday). (p. 97)

equally likely Having an equal chance of occurring. (p. 148)

equation A mathematical statement that two quantities are equal. An equation contains an equals sign (for example, $4x - 5 = 11$). (p. 51)

event In probability, a result involving one or more outcomes. For example, when rolling a die, the event 'rolling an even number' contains the three outcomes $\{2, 4, 6\}$. (p. 148)

expense The cost of spending. (p. 113)

experimental probability *See* **relative frequency**.

five-number summary Lowest score, lower quartile, median, upper quartile and highest score of a data set. These five numbers are used to draw a **box plot**. (p. 404)

flat rate interest *See* **simple interest**.

formula A rule written as an algebraic equation, using variables. For example, the formula for the area of a triangle is $A = \frac{1}{2} bh$. (p. 46)

frequency The number of times a score or group of scores occurs in a data set. (p. 21)

frequency histogram A bar chart in which the height of each column represents the frequency of a single score or group of scores. (p. 25)

frequency polygon A line graph formed by joining the midpoints of the tops of the columns of a **frequency histogram**. (p. 25)

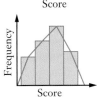

fuel consumption The rate at which fuel is used by a vehicle, usually measured in L/100 km. (p. 390)

goods and services tax (**GST**) The tax a consumer pays on any purchased item or service (for example, buying a car, or hiring a painter). (p. 133)

gradient (symbol m) Slope of a line.

$$\text{gradient} = \frac{\text{rise}}{\text{run}} = \frac{\text{change in } y}{\text{change in } x} \text{ (p. 329)}$$

gross pay (or gross wage) A person's pay before tax is deducted. Different from **net pay**. (p. 124)

histogram *See* **frequency histogram**.

hypotenuse The longest side of a right-angled triangle; the side opposite the right angle. (p. 195)

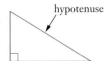

hypotenuse

income Money that is earned or gained (usually regularly). (p. 113)

income tax A tax on a person's income, paid to the government. (p. 126)

interest Money earned on an investment, or money paid to a financial institution for borrowing. (p. 360)

International Date Line The imaginary line that runs through the Pacific Ocean and is approximately the 180° meridian of longitude. A day is either gained or lost when this line is crossed. (p. 448)

interquartile range (**IQR**) The difference between the upper quartile and lower quartile of a data set $(Q_3 - Q_1)$. It is a measure of the spread of the data. (p. 214)

kilojoule (**kJ**) A unit of energy equal to 1000 joules. (p. 66)

kilowatt (**kW**) A unit of power equal to 1000 watts. (p. 304)

kilowatt hour (**kWh**) A unit of electrical energy equivalent to that used by one kilowatt of power in one hour. (p. 301)

latitude The angular distance north or south of the Equator of a point on the Earth's surface; parallel lines of latitude run across the Earth. (p. 441)

like terms Algebraic terms that have exactly the same pronumerals. For example, $5xy$ and $2xy$ are like terms, $3xy$ and $4x^2$ are not like terms. (p. 42)

likely Probably will happen; having a probability above $\frac{1}{2}$. (p. 148)

limits of accuracy The limits of a measured value; for example, the limits of accuracy of a measured height of 171 cm are 170.5 cm to 171.5 cm, meaning that the actual height lies within this range. The limits are found by adding to and subtracting from the measured value $\frac{1}{2}$ of the smallest unit marked on the measurement scale. (p. 187) *See also* **absolute error**.

linear function A function of the form $y = mx + c$, the graph of which is a straight line. (p. 328)

linear modelling Using a linear function to approximate a real-life situation. (p. 334)

longitude The angular distance east or west of the Greenwich meridian of a point on the Earth's surface; lines of longitude run from the North Pole to the South Pole, down the Earth. (p. 442)

market value Current sale value of a share or item. (p. 383)

mean The average of a set of scores.

$$\text{mean (or } \bar{x}) = \frac{\text{sum of scores}}{\text{number of scores}} = \frac{\Sigma x}{n} = \frac{\Sigma fx}{\Sigma f} \text{ (p. 210)}$$

measure of central tendency A statistical value, such as the mean, median or mode, that describes the centre or average of a set of data. (p. 210)

measure of spread A statistical value, such as the range, interquartile range or standard deviation, that describes the spread of a set of data. (p. 214)

median The middle score of a data set when scores are arranged in ascending order. If there are two middle scores, the median is the average of the two. (p. 210)

median class The class interval that contains the median score. (p. 233)

Medicare levy A tax to cover the costs of the public health system, calculated as a percentage (currently 2%) of a person's income. (p. 129)

modal class The class interval with the highest frequency. (p. 233)

mode The most common or frequent score(s) in a set of data. (p. 210)

negatively skewed *See* **skewed**.

net pay (or net wage) A person's pay after tax has been deducted. Differs from **gross pay**. (p. 124)

nominal data Categorical data that cannot be ordered; for example, colour of eyes. (p. 18)

numerical data Data that involve numbers, such as heights, or the number of children. (p. 18)

on-road costs After purchasing a motor vehicle, the additional costs required before it can be used on the road, such as registration, stamp duty and CTP insurance. (p. 382)

ordinal data Categorical data that can be ordered; for example, level of swimming class. (p. 18)

outlier An extreme (high or low) score in a data set that is much different than the other scores, either less than $Q_1 - 1.5 \times \text{IQR}$ or greater than $Q_3 + 1.5 \times \text{IQR}$. (p. 224)

overtime Time worked beyond usual working hours, usually paid at a higher rate. (p. 97)

Pareto chart A combined bar chart and line graph, where the columns are displayed in descending order and the line shows the cumulative frequency of the same data. Pareto charts are often used in quality control for businesses. (p. 28)

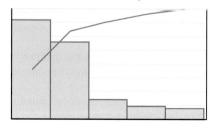

PAYG tax 'Pay As You Go' tax; income tax deducted from our pay in instalments each payday by our employer. (p. 124)

per annum (p.a.) (or annual) Per year. (p. 360)

percentage error The maximum possible error for a measurement, expressed as a percentage.

$$\text{percentage error} = \frac{\text{absolute error}}{\text{measurement}} \times 100\% \text{ (p. 185)}$$

percentiles Values that divide a data set into 100 equal parts when the scores are arranged in order. (p. 218). *See also* **deciles** and **quartiles**.

perimeter The distance around the outside of a shape, the sum of the lengths of its sides. (p. 198)

piecework Earnings based on the number of items processed, made or delivered, paid at a rate per item (rather than per hour). (p. 105)

positively skewed *See* **skewed**.

precision The smallest unit on the scale of a measuring device. For example, the precision of a ruler may be one millimetre. (p. 185).

prime meridian The 0° meridian of longitude. Also called the Greenwich meridian. (p. 446)

principal The original amount of money invested or borrowed. (p. 360)

prism A solid with flat faces and a uniform cross-section. (p. 264).

probability A measure of how likely an event E is to occur, written $P(E)$. Its value ranges from 0 to 1. (p. 148).

Pythagoras' theorem In a right-angled triangle, the square of the hypotenuse is equal to the sum of the squares of the two shorter sides. $c^2 = a^2 + b^2$ (p. 195)

quarterly Four times a year, every three months. (p. 307)

quartiles (**upper** and **lower**) The upper quartile is the 3rd **quartile** (Q_3) that cuts off the top 25% of scores in a data set, and the lower quartile is the 1st quartile (Q_1) that cuts off the bottom 25% of scores. (p. 218). *See also* **deciles** and **percentiles**.

radius (plural radii) The length of the interval joining the centre of a circle to the circumference (see diagram at **diameter**). The radius is half of the **diameter**. (p. 200)

random sample A sample for which every member of a population has an equal chance of selection. (p. 12)

range The difference between the highest score and the lowest score in a set of data. (p. 214)

reaction time The time between when a driver senses he needs to stop and when he applies the brakes. (p. 78)

registration *See* **vehicle registration**.

relative frequency (or **experimental probability**) The number of times an event or score occurs, written as a fraction of the total number of events or scores. (p. 151)

retainer A fixed amount paid to a salesperson before commission is added. (p. 106)

royalty Income earned by recording artists and authors, based on the number of copies of their work that are sold. (p. 105)

salary Fixed earnings quoted as a yearly amount, but paid weekly, fortnightly or monthly. (p. 92). *See also* **wage**.

sample A group of items selected from a population. (p. 10)

sample space A list of all the possible outcomes in a chance situation. (p. 157)

scientific notation A way of writing very large or very small numbers. For example, $98\,000\,000 = 9.8 \times 10^7$. (p. 192)

self-selected sample A sample in which people volunteer to be part of the sample, such as an SMS poll or a website survey, so it is not really random. (p. 12)

shape of a distribution The way the data in a frequency distribution is spread, can be symmetrical, positively skewed or negatively skewed. (p. 414)

significant figures Meaningful digits in a numeral that tell 'how many'. For example, $98\,000\,000$ has two significant figures: 9 and 8. (p. 191)

simple interest (or **flat rate interest**) Interest earned or charged only on the original amount of money (principal) invested or borrowed. Differs from **compound interest**. (p. 360)

skewed The shape of a statistical distribution when most of the data scores are either low (**positively skewed**) or high (**negatively skewed**). The tail indicates the direction of the skew. (p. 415)

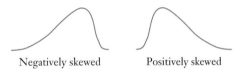

Negatively skewed   Positively skewed

speed A rate that compares distance travelled with time taken. Speed is often measured in kilometres per hour (km/h) or metres per second (m/s). (p. 74)

$$\text{average speed} = \frac{\text{distance travelled}}{\text{time taken}}$$

stamp duty A tax paid to the state government when buying a vehicle, calculated on the market value of the vehicle. (p. 382)

standard deviation (symbol σ) A statistical measure of the spread of a set of scores. (p. 237)

stem-and-leaf plot A 'number graph' that lists all the data scores, in groups. This stem-and-leaf plot shows 12 test scores, from 42 to 82. Different from a **dot plot**. (p. 27)

Stem	Leaf
4	2 5
5	0 2 8
6	6 7
7	3 5 7 7
8	2

stopping distance The distance travelled between when a driver senses he needs to stop and when the vehicle stops completely. (p. 78)

straight-line depreciation Method of depreciation in which an item's value decreases by the same amount each period. (p. 366)

stratified sample A sample consisting of a percentage of items from each 'strata' or 'layer' of a population. For example, a stratified sample from a population of 35% children and 65% adults should contain 35% children and 65% adults. (p. 12)

subject (of a formula): The variable on its own on the left side of the '=' sign of a formula, what the formula describes; for example, A (for area) in the formula $A = \frac{1}{2}bh$. (p. 58)

surface area The total area of all the faces of a solid shape. (p. 264)

symmetrical distribution A statistical distribution the graph of which is symmetrical. The left side and the right side are mirror images of each other. (p. 415)

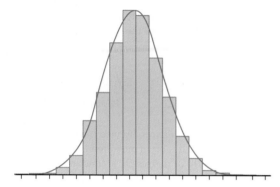

systematic sample A sample chosen by using a set pattern (for example, choosing every 10th number in a phone book). (p. 12)

tax debt The amount by which the amount of **PAYG tax** already paid is below the amount of tax due. This is owed by the taxpayer to the Australian Tax Office (ATO). (p. 126)

tax deduction *See* **allowable tax deduction**.

tax refund The amount by which the amount of **PAYG tax** already paid is above the amount of tax due. This is given back to the taxpayer by the Australian Tax Office. (p. 126)

tax return A form completed at the end of a financial year to account for income earned, allowable deductions and tax already paid. Used to calculate a **tax refund** or **tax debt**. (p. 126).

taxable income The part of a person's income that is taxed, equal to yearly income minus allowable deductions. (p. 126).

Third Party Property insurance Insurance that covers damage to another person's ('third party') vehicle or property in an accident in which you are at fault. It does not cover damage to your vehicle. (p. 388). *See also* **comprehensive insurance** or **Compulsory Third Party insurance**.

time-and-a-half Wages paid at 1.5 times the normal rate (for example, working on a Saturday). (p. 97)

time zone A zone of the world in which the time is the same for all places. (p. 444)

trapezoidal rule Formula for finding the approximate area of an irregularly shaped block using the area of a trapezium. $A \approx \frac{h}{2}\left(d_f + d_l\right)$. (p. 263)

tree diagram A diagram for listing all the possible outcomes of a multi-stage experiment such as tossing three coins together. (p. 159)

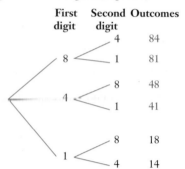

unimodal distribution A set of scores with only one peak (see diagram for **symmetrical distribution**, for example). (p. 415). *See also* **bimodal distribution**.

UTC (**Coordinated Universal Time**) (or Greenwich Mean Time, GMT) Local time at the prime meridian (0° longitude) time zone, from which other times in the world are measured. (p. 444)

variable A pronumeral that can take a range of values (that is, it is not a **constant**). (p. 334)

vehicle registration The yearly process of paying for and obtaining permission for using a vehicle on public roads. (p. 382)

vertical intercept *See* **y-intercept**.

volume The amount of space occupied by a solid, measured in cubic units. (p. 276)

wage The amount earned by an employee for a set number of working hours, usually paid weekly. (p. 92) *See also* **overtime** and **salary**.

watt (**W**) A unit of power equal to one joule of energy per second. (p. 304)

y-intercept (or **vertical intercept**) The value at which a straight line graph cuts the *y*-axis. For example, the *y*-intercept of this graph is 3. (p. 329)

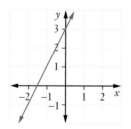